T0137439

Emergence, Complexity and Computation

Volume 36

The Emergence, Complexity and Computation (ECC) series publishes new developments, advancements and selected topics in the fields of complexity, computation and emergence. The series focuses on all aspects of reality-based computation approaches from an interdisciplinary point of view especially from applied sciences, biology, physics, or chemistry. It presents new ideas and interdisciplinary insight on the mutual intersection of subareas of computation, complexity and emergence and its impact and limits to any computing based on physical limits (thermodynamic and quantum limits, Bremermann's limit, Seth Lloyd limits...) as well as algorithmic limits (Gödel's proof and its impact on calculation, algorithmic complexity, the Chaitin's Omega number and Kolmogorov complexity, non-traditional calculations like Turing machine process and its consequences,...) and limitations arising in artificial intelligence. The topics are (but not limited to) membrane computing, DNA computing, immune computing, quantum computing, swarm computing, analogic computing, chaos computing and computing on the edge of chaos, computational aspects of dynamics of complex systems (systems with self-organization, multiagent systems, cellular automata, artificial life,...), emergence of complex systems and its computational aspects, and agent based computation. The main aim of this series is to discuss the above mentioned topics from an interdisciplinary point of view and present new ideas coming from mutual intersection of classical as well as modern methods of computation. Within the scope of the series are monographs, lecture notes, selected contributions from specialized conferences and workshops, special contribution from international experts.

More information about this series at http://www.springer.com/series/10624

Ramón Alonso-Sanz

Quantum Game Simulation

Springer

Ramón Alonso-Sanz
Technical Universty of Madrid
Madrid, Spain

ISSN 2194-7287   ISSN 2194-7295   (electronic)
Emergence, Complexity and Computation
ISBN 978-3-030-19636-3   ISBN 978-3-030-19634-9   (eBook)
https://doi.org/10.1007/978-3-030-19634-9

This Springer imprint is published by the registered company Springer Nature Switzerland AG
The registered company address is: Gewerbestrasse 11, 6330 Cham, Switzerland

Dedicado a Cris y Magui (correctoras) y a
Margarita, también *correctora*.

Preface

Quantum game theory (QGT) studies game theory with access to quantum information mathematical tools [1–4]. It arose at the very end of the past century [5, 6], ensuing since then a notable development [7–9]. Not exempt of criticism, as summarized in the review [10].

We adopt in this book one of the most celebrated quantum game implementations: the EWL protocol introduced in [5]. The EWL method entangles strategies *à la* quantum, framing the study in the so-called *correlated games* [11, 12], but allowing the players to decide *independently*.

Quantum mechanics and game theory have traditionally occupied utterly different realms. Quantum game theory was proposed in order to approach both scientific fields. Therefore, quantum games have allowed to connect quantum mechanics, which determines the behavior of systems at microscopic scales, with areas where game theory has historically attracted attention, such as mathematical economics and evolutionary biology. The application of QGT in mathematical economic areas in particular has supported the rise of an incipient quantum econophysics [13, 14].

The prevailing research trend in QGT is that of introducing features of quantum information into game theory, i.e., quantizing games, whereas the opposite perspective, that of introducing features of game theory into quantum mechanics, i.e., gaming the quantum, does not appear as prominent, albeit examples of this approach that have opened new routes for a better understanding of quantum information and quantum computation may be traced [15, 16].

It has been argued that classical games are being played at some microscopic level in the physical and biological worlds. Thus, the seminal paper [17] proposed that Nature plays classical dominant strategy games using clones of a virus that infects bacteria, whereas the reviews [18, 19] consider not only virus games but also catalytic RNA games, protein games, and, last but not least, gene games. To the best of our knowledge, quantum games have not properly been taken into account at this microscopic scale, the cellular and subcellular levels of life, e.g., they are not even merely mentioned in the challenges and future prospects section of [18, 19]. But it seems fairly conceivable the application of the conceptual toolkit of QGT to such "lower" levels of biology, i.e., to *quantum biology*, where *players* have not cognitive and rational capabilities. In fact, some authors [20, 21] speculate that it is not inconceivable that such [quantum] "games" are already being played at some microscopic [living] level in the real world. Nonetheless, notable difficulties and controversies have accompanied the attempts to apply quantum technology to the study of (potential) quantum effects in living organisms [22–24], far beyond the quantum original realm in *hard* physicochemical sciences.

So far, one can view quantum game theory as an exercise in pure mathematics. Nonetheless, quantum games are being studied using quantum hardware, such as a nuclear magnetic resonance quantum computer [25]. Thus, likely the emergence of landmark applications of QGT lies in the future (much as it is expected with quantum information and computation). In the meantime, tools are being provided that are expected to be useful to explain *how* quantum effects that take place in a microscopic world can give rise to the macroscopic world. This book aims to contribute to this goal, by simulating quantum iterative games where players interact in a local and synchronous manner, namely, according to the cellular automata paradigm.

The basic foundations of classical game theory are introduced in Chap. 1. In this chapter are also featured the two-person, 2×2 game-types that are later taken into consideration. The quantum approach to game theory according to the EWL method is described in Chap. 2. Chapter 3 deals with fair iterative collective games with the players arranged in a spatially structured two-dimensional lattice. Spatial collective unfair contests are simulated in Chap. 4. Games on networks are treated in Chap. 5. A probabilistic updating mechanism is implemented in the context of spatial simulations in Chap. 6. The disrupting effect of quantum noise on the dynamics of the spatial quantum formulation of quantum games is studied in Chap. 7. A spatial quantum relativistic game formulation is scrutinized in Chap. 8 and the effect of quantum memory in Chap. 9. Some generalizations of the EWL model are treated across the book, in particular in Chap. 10. Games with imperfect information and with imprecise payoffs are studied in Chap. 11. Last but not least, Chap. 12 studies correlated games constructed from independent players in a purely classical context.

Madrid, Spain   Ramón Alonso-sanz
February 2019

References

1. Flitney, A.P., Abbott, D.: An introduction to quantum game theory. Fluctuation Noise Lett. **2**(4), R175–R187 (2002)
2. Landsburg, S.E.: Quantum game theory. The Wiley Encyclopedia of Operations Research and Management Science. http://arxiv.org/pdf/1110.6237v1.pdf (2011)
3. Landsburg, S.E.: Quantum game theory. Not. AMS **51**(4), 394–399 (2004)
4. Piotrowski, E.W., Sladkowski, J.: An invitation to quantum game theory. Int. J. Theor. Phys. **42**(5), 1089–1099 (2003)
5. Eisert, J., Wilkens, M., Lewenstein, M.: Quantum games and quantum strategies. Phys. Rev. Lett. **83**(15), 3077–3080 (1999)
6. Meyer, D.A.: Quantum strategies. Phys. Rev. Lett. **82**(5), 1052 (1999)
7. Huang, D., Li, S.: A Survey of the Current status of research on quantum games. In: Proceedings of the 2018 4th International Conference on Information Management (ICIM2018), pp. 42–52. IEEE, The University of Oxford, U.K. (2018)
8. Liu, W., Cui, M., He, M.: An introductory review on quantum game theory. In: Fourth International Conference in Genetic and Evolutionary Computing (ICGEC), pp. 386–389. IEEE (2010)
9. Takahashi, T., Kim, S.J., Naruse, M.: A note on the roles of quantum and mechanical models in social biophysics. Prog. Biophys Mol. Biol. **130**, 103–105 (2017)
10. Khan, F.S., Solmeyer, N., Balu, R., Humble, T.: Quantum games: a review of the history, current state, and interpretation. Quantum Inf. Process. **17**, 309 (2018)
11. Osborne, M.J.: An Introduction to Game Theory. Oxford University Press, Oxford (2004)
12. Owen, G.: Game Theory. Academic Press, Cambridge (1995)
13. Guo, H., Zhang, J., Koehler, G.J.: A survey of quantum games. Decis. Support Syst. **46**(1), 318–332 (2008)
14. Schinckus, C.: A methodological call for a quantum econophysics. Quantum Interaction. *LNCS*, **8369**, 08–316 (2014)
15. Khan, F.S., Phoenix, S.J.D.: Gaming the quantum. Quantum Information & Computation **13**(3–4), 231–244 (2013)
16. Knebel, J., Weber, M.F., Kruger, T., Frey, E.: Evolutionary games of condensates in coupled birth-death processes. Nat. Commun. **6**, Article number: 6977 (2015)
17. Turner, P.E, Chao, L.: Prisoner's dilemma in an RNA virus. Nature **398**, 441–443 (1999)
18. Bohl, K., Hummert, S., Werner, S., Basanta, D., Deutsch, A., Schuster, S., Theissen, G., Schroeter, A.: Evolutionary game theory: Molecules as players. Mol. BioSyst. **10**, 3066–3074 (2014)
19. Hummert, S., Bohl, K., Basanta, D., Deutsch, A., Werner, S., Theissen, G., Schroeterc, A., Schuster, S.: Evolutionary game theory: cells as players. Mol. BioSyst. **10**, 3044–3065 (2014)
20. Kay, R., Johnson, N.F., Benjamin, S.C.: Evolutionary quantum game. J. Phys. A: Math.Gen. **34**(41), L54 (2001)
21. Koch, C., Hepp, K.: Quantum mechanics in the brain. Nature **440**, 611–612 (2006)
22. Alfonseca, M., Ortega, A., Cruz, M., Hameroff, S.R., Lahoz-Beltra, R.: A model of quantum-von Neumann hybrid cellular automata: principles and simulation of quantum coherent superposition and dehoerence in cytoskeletal microtubules. Quantum Inf.Comput. **15**(1–2), 22–36 (2015)
23. Ball, P.: Physics of life: The dawn of quantum biology. Nature **474**, 272–274 (2011)
24. Lambert, N., Chen, Y-N., Cheng, Y-C., Li, C-M., Chen, G-Y., Nori, F.: Quantum biology. Nat.Phys. **9**, 10–18 (2013)
25. Du, J.F., Li, H., Xu, X., Shi, M., Wi, J., Zhou, X., Han, R.: Experimental realization of quantum games on a quantum computer. Phys. Rev. Lett. **88**(13), 137902 (2002)

Contents

1 Classical Game Theory 1
   1.1   Two-Person Games 1
   1.2   Independent Players. Nash Equilibrium 5
   1.3   Correlated Games 7
   References 9

2 Quantum Approach to Game Theory 11
   2.1   The EWL Model 11
   2.2   Nash Equilibrium in the EWL-PD Model 16
   References 18

3 Spatial Quantum Game Simulation 21
   3.1   Spatial Quantum Games 21
   3.2   Two-Parameter Strategies 24
   3.3   Three-Parameter Strategies 38
   3.4   The Marinatto-Weber Models 45
   References 49

4 Unfair Contests 51
   4.1   Unfair Parameter Availability 51
   4.2   Unfair Strategy Updating 66
   4.3   Classical Memory 70
   References 72

5 Games on Networks 73
   5.1   Fair Contests on Random Networks 73
       5.1.1   Two-Parameter Strategies 73
       5.1.2   Three-Parameter Strategies 76
   5.2   Unfair Strategy Updating 78
   References 88

6 Probabilistic Updating................................... 91
  6.1 The QBOS with Probabilistic Updating 92
  6.2 The QPD with Probabilistic Updating 104
  References 114

7 Quantum Noise... 117
  7.1 The Density Matrix Formalism 117
  7.2 Amplitude-Damping Noise 118
  7.3 The QPD with Noise 123
  7.4 The QSD with Noise 132
  References 139

8 Quantum Relativistic Games............................. 141
  8.1 The Unruh Effect 141
  8.2 A Quantum Relativistic Prisoner's Dilemma.............. 143
  8.3 A Quantum Relativistic Battle of the Sexes 153
  8.4 Effect of Noise in a Quantum-Relativistic PD Game 161
  References 173

9 Quantum Memory 175
  9.1 Quantum Uncorrelated Noise.......................... 175
  9.2 Quantum Correlated Memory Noise..................... 177
  9.3 The Spatialized Quantum Prisoner's Dilemma
    with Memory 180
    9.3.1 Two Parameter Strategy Simulations 180
    9.3.2 Three-Parameter Strategy Simulation 187
  References 192

10 Games with Werner-Like States 193
  10.1 Werner-Like States in Quantum Games 193
  10.2 Spatial Games 194
    10.2.1 Either γ or δ Fixed........................ 195
    10.2.2 Both γ and δ Variable 198
  10.3 Games on Random Networks 201
  10.4 Three Quantum Parameter Strategies 203
  References 206

11 Imperfect Information and Imprecise Payoffs 209
  11.1 The Battle of the Sexes with Imperfect Information 209
    11.1.1 Imperfect Information in the BOS 210
    11.1.2 The Spatialized Imperfect QBOS................ 212
  11.2 Games with Imprecise Payoffs........................ 219
    11.2.1 Spatial Quantum Fuzzy Games 222
  References 229

12   Classical Correlated Games 231
   12.1   Classical Correlation from Independence 231
   12.2   Spatial Games 235
   12.3   Kgames on Random Networks 243
   12.4   Unfair Strategy Updating............................ 246
   References 250

Index 251

Chapter 1
Classical Game Theory

This chapter introduces the basic foundations of (classical) game theory. The games that are later taken into consideration to demonstrate how the implementation of quantum mechanic tools affects the field of game theory are featured in this chapter.

1.1 Two-Person Games

Game theory is concerned with rational choice in decisions involving two or more interdependent decision makers, called *players*. In the kind of static games to be considered in this book, the players must choose simultaneously a move of a set of possible moves. Associated with each combination of players' moves there is a collection of numerical payoffs, one to each player. In this book, only two players will be involved in the game (A and B), and only two moves are feasible. Thus, we will take into account two-person, 2×2 games. Four examples of such types of games are given in the tables of this section, where the payoff matrices of each game are given in their far left frames (the numbers under the diagonals are the payoffs of player A, and those over the diagonals are the payoffs to player B).

Each player wants to get as large a payoff as possible. He has a partial control on the outcome of the game, since his move choice will influence it. However, the outcome of the game is not determined by his choice alone, but also depends upon the choice of the other player. In such an interdependent scenario, it is not straightforward to define the concept of -solution- of a game. Namely, game theory does not provide a unique prescription for play in games. This is particularly so in games with players whose interests are not completely opposed, i.e., in non-zero sum games.

If the players act independently (non-cooperative games), the leading solution concept is that of Nash equilibrium that will be introduced in Sect. 1.2. But if the players are free to negotiate binding agreements (cooperative games), or even from the point of view of an external observer not involved in the game, the range of

© Springer Nature Switzerland AG 2019
R. Alonso-Sanz, *Quantum Game Simulation*, Emergence, Complexity
and Computation 36, https://doi.org/10.1007/978-3-030-19634-9_1

Table 1.1 The Prisoner's Dilemma (PD) game. Far left: Payoffs matrices. Center: Probabilities with independent players. Far right: Joint probabilities

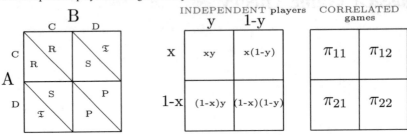

possible prospects expands which makes difficult to prescribe the best way to proceed. We will refer here to only two of such possible perspectives, Pareto-efficiency and social welfare. The outcome of a game is non-Pareto-efficient if there is another (better) outcome which would give both players higher payoffs, or would give one player the same payoff but a higher payoff to the other player. An outcome is Pareto-efficient (or Pareto-optimal) if there is no such other better outcome [1]. Social welfare (SW) functions may be envisaged as summarizing some particular conception of the *common good* [2]. In its simplest form, SW *solutions* maximize the sum of the payoffs of both players.

In the following description of some specific games, both the Pareto-efficiency and the social welfare solutions will be stated.

The Prisoner's Dilemma (PD) game

In the Prisoner's Dilemma (PD) game (Table 1.1), both players may choose either to cooperate (C) or to defect (D). Mutual cooperators each scoring the *reward R*, mutual defectors score the *punishment P* ; D scores the *temptation* $\mathfrak{T}$ against C, who scores S (*sucker*'s payoff) in such an encounter. In the PD it is: $\mathfrak{T} > R > P > S$. In this book the PD payoff values will be $\mathfrak{T} = 5$, $R = 3$, $P = 2$ and $S = 1$ except in Chap. 10, where $\mathfrak{T} = 5$, $R = 3$, $P = 1$ and $S = 0$.

The (C,C), (C,D), and (D,C) strategy pairs provide Pareto-efficient outcomes in the PD game. The (C,C) strategy pair is the only SW solution in the PD provided that $\mathfrak{T} + S < 2R$. This is the case of the PD(5,3,1,0) game. In the PD(5,3,2,1) game, the (C,D), and (D,C) strategy pairs are also SW solutions because $\mathfrak{T} + S = 6 = 2R$, albeit only (C,C) is balanced in the sense that both players get equal payoff.

The Hawk-Dove (HD) game

The Hawk-Dove (HD) game faces players that can choose to behave either aggressively (Hawk) or amicably (Dove). The structure of the payoffs matrices is similar to that in the PD, but in the HD it is $P < S$ instead of $P > S$ as is the PD. In Chap. 3 the HD will have the payoffs $\mathfrak{T} = 3$, $R = 2$, $S = 0$, $P = -1$, the latter payoff indicating that mutual aggressive choice has *negative* consequences for both players. In Chap. 10, the studied HD will have $\mathfrak{T} = 5$, $R = 3$, $P = 0$ and $S = 1$.

Table 1.2 The Hawk-Dove (HD) game. Far left: Payoffs matrices. Center: Response functions. Far right: Payoffs region

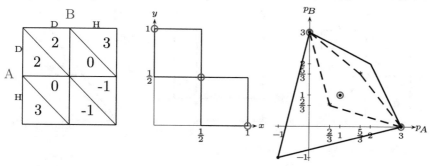

Table 1.3 The Samaritan's Dilemma (SD) game. Far left: Payoffs matrices. Center: Response functions. Far right: Payoffs region

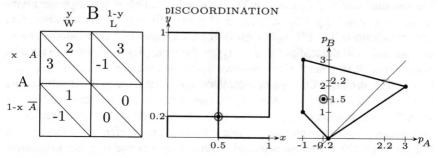

In the HD$(3,2,-1,0)$ game of Table 1.2, the (D,D), (D,H) and (H,D) strategy pairs provide Pareto-efficient outcomes, whereas the (D,D) strategy pair is the only SW solution (and balanced Pareto-efficient).

The Samaritan's Dilemma (SD) game

The Samaritan's Dilemma (SD) is a non-zero sum, asymmetric game played by two players: the charity (or Samaritan) player A and the beneficiary player B. Player A may choose Aid/No Aid, whereas player B may choose Work/Loaf. The Samaritan's dilemma arises in the act of charity. The Samaritan player wants to help (Aid) people in need. However, the beneficiary may simply rely on the handout (Loaf) rather than try to improve their situation (Work). This is not anticipated by the charity. Many people may have experienced this dilemma when confronted with people in need. Although there is a desire to help them, there is the recognition that a handout may be harmful to the long-run interests of the recipient [3–7]. Following the references [8–11], we adopt here the payoff matrices given in the far left panel of Table 1.3.

In the SD game of Table 1.3, the (A,W) and (A,L) strategy pairs provide Pareto-efficient outcomes, whereas the (A,W) strategy pair is the only SW solution.

Table 1.4 The Battle of the Sexes (BOS) game. Far left: Payoffs matrices. Center: Response functions. Far right: Payoffs region

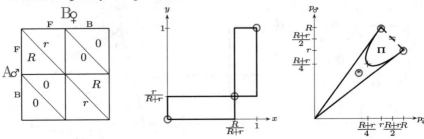

The Battle of the Sexes (BOS) game

The so-called *battle of the sexes* (BOS) is a simple example of a two-person asymmetric (or bi-matrix) game, i.e., a game whose payoff matrices are not coincident after transposition [12, 13]. In this game, the preferences of a conventional couple are assumed to fit the traditional stereotypes: the male player A prefers to attend a *F*ootball match, whereas the female player B prefers to attend a *B*allet performance.[1] Both players decide in the hope of getting together, so that their payoff matrices are given in the far left panel of Table 1.4, with rewards $R > r > 0$. There are both coordination and conflict elements in the BOS game. While both players want to go out together, the *conflict* element is present because their preferred activities differ, and the *coordination* element is present because they may end up going to different events. Thus, in the absence of preplay communication it is natural to expect that coordination failure (of ending up in one of the inefficient outcomes) will occur frequently [14, 15].

Both (F,F) and (B,B) strategy pairs are SW solutions and Pareto-efficient in the BOS game.

The Matching Pennies (MP) game

The zero-sum game defined by the payoff matrices given in the far left frame of Table 1.5 is often referred to as the matching pennies (MP) game: each player has a penny that can flip to either Head or Tail, and one unit is in play. Player A wins if both pennies match (either (H,H) or (T,T)), whereas player B wins if both players mismatch ((H,T) or (T,H)). Player A loses if both pennies mismatch, whereas player B loses if both pennies match [15–17].

In a zero-sum game, such as the MP, every strategy pair is Pareto-efficient, since every gain to one player means a loss to the other, and not any proper SW solution may be identified.

---

[1] A more politically correct description of the BOS game has been proposed in terms of attending a concert, either of Mozart or of Beethoven. Incidentally, player A and player B are often referred to as Alice and Bob in game theory and quantum information texts. In the same vein of correction, it has been proposed to name the players as Pat (Patrick, Pamela) and Chris (Christopher, Christine).

Table 1.5 The Matching Pennies (MP) game. Far left: Payoffs matrices. Center: Response functions. Far right: Payoffs segment

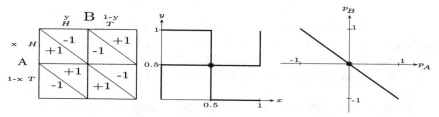

The PD and HD games are symmetric i.e., the payoff matrices of both players coincide after transposition, whereas the SD, MP and the BOS games are not symmetric. In symmetric games the roles of both players are somehow interchangeable, whereas in asymmetric games every player has to be studied separately.

1.2 Independent Players. Nash Equilibrium

In the somehow canonical approach to static game theory, both players choose independently from each other their probabilistic strategies $\mathbf{x} = (x, 1-x)'$ and $\mathbf{y} = (y, 1-y)'$, as sketched in the central frame of Table 1.1. Therefore, in a game with $\boldsymbol{\in}_A$ and $\boldsymbol{\in}_B$ payoff matrices, the expected payoffs (p) of both players are :

$$p_A(x, y) = \mathbf{x}'\boldsymbol{\in}_A\mathbf{y} \quad , \quad p_B(x, y) = \mathbf{x}'\boldsymbol{\in}_B\mathbf{y} \tag{1.1}$$

The pair of strategies $\big((x, 1-x), (y, 1-y)\big)$ referred here to as (x, y), are in Nash equilibrium if x is a best response to y and y is a best response to x. Formally: $p_A(x, y) \geq p_A(z, y)$, and $p_B(x, y) \geq p_B(x, z)$, $\forall z$. Nash equilibriums are shown with points where the two player's correspondences agree, i.e. cross, in the reaction response graphs.

In the PD game with (x, y) independent probabilistic strategies, mutual defection, i.e., $x^* = y^* = 0$, is the only pair in NE. This is so because Defection dominates Co-operation because $\mathfrak{T} > R$, $P > S$, and in consequence the reaction response graphs from the expected payoffs of both players given in Eq. (1.2) only cross at (0,0). The "dilemma" comes from the fact that each player tends to defect, though mutual co-operation provides substantially better results for the whole system (SW solution).

$$p_A(x, y) = Rxy + Sx(1-y) + \mathfrak{T}(1-x)y + P(1-x)(1-y) = \tag{1.2a}$$
$$(S - P + (R + P - (S + \mathfrak{T})y)x + P + (\mathfrak{T} - P)y$$
$$p_B(x, y) = Rxy + \mathfrak{T}x(1-y) + S(1-x)y + P(1-x)(1-y) = \tag{1.2b}$$

$$(S - P + (R + P - S + \mathfrak{T})x))y + P + (\mathfrak{T} - P)x$$

The formulas of both payoffs given in Eq. (1.2) have the same structure of the slope, either $S - P + (R + P - (S + \mathfrak{T})y$ or $S - P + (R + P - (S + \mathfrak{T})x$. Such a slope is negative in both cases and consequently x=0, y=0, i.e., (D,D), is the unique pair of best responses. Remarkably, the only combination of moves that is not Pareto-efficient.

The expected payoffs in the HD(3,2,0,-1) game, using uncorrelated probabilistic strategies (x, y) are given in Eq. (1.3). The HD game has three strategy pairs in NE, two of them with pure strategies: $(x^*=1, y^*=0)$ and $(x^*=0, y^*=1)$ (anti-coordination), and a third one with mixed strategies, which in the particular case of the HD(3,2,0,-1) of Table 1.2 turns out to be: $x^* = y^* = 1/2$, leading to $p_{AB} = 1$. In the HD(3,2,0,-1) game if $x = y$ it is $p_A = p_B$. Importantly, $(x=y=0)$, i.e., mutual Hawk, is not in NE in the HD game.

$$p_A = (x, 1-x) \begin{pmatrix} 2 & 0 \\ 3 & -1 \end{pmatrix} \begin{pmatrix} y \\ 1-y \end{pmatrix} = (-2y+1)x + 4y - 1 \qquad (1.3a)$$

$$p_B = (x, 1-x) \begin{pmatrix} 2 & 3 \\ 0 & -1 \end{pmatrix} \begin{pmatrix} y \\ 1-y \end{pmatrix} = (-2x+1)y + 4x - 1 \qquad (1.3b)$$

The expected payoffs in the SD game, using uncorrelated probabilistic strategies (x, y) are given in Eq. (1.4). Table 1.3 shows in its central frame the reaction functions whose intersection determines the NE: $(x = 0.5; y = 0.2)$, with associated payoffs $p_A = -0.2$, $p_B = 1.5$. Thus, the SD belongs to the class of the so called discoordination games, i.e., games with no pair of pure strategies in Nash equilibrium (NE). In the SD game, where the one player's incentive is to coordinate (charity (A, W)), while the other player tries to avoid this (beneficiary (A, L)). The far right frame of Table. 1.2 shows the payoffs region of the studied game, which turns out negative for player A in points such as the NE, whereas this does not happen for player B. Let us remark here then, that the payoffs of the SD are biased towards the beneficiary player B. In other words, the dilemma of the Samaritan game (sometimes referred to as the Welfare game) is somehow only that of the charity (or Samaritan) player A.

$$p_A = (x, 1-x) \begin{pmatrix} 3 & -1 \\ -1 & 0 \end{pmatrix} \begin{pmatrix} y \\ 1-y \end{pmatrix} = (5y-1)x - y \qquad (1.4a)$$

$$p_B = (x, 1-x) \begin{pmatrix} 2 & 3 \\ 1 & 0 \end{pmatrix} \begin{pmatrix} y \\ 1-y \end{pmatrix} = (-2x+1)y + 3x \qquad (1.4b)$$

The expected payoffs in the BOS(R, r) game, using uncorrelated probabilistic strategies $(x, 1 - x)$ and $(y, 1 - y)$ are: given in Eq. (1.5). Thus, according to the reaction correspondences given in Table 1.4, the BOS game has three strategy pairs in NE. Two of them with pure strategies: $(x^*=y^*=1)$ and $(x^*=y^*=0)$, and a third one with mixed strategies $(x^* = \dfrac{R}{R+r}, y^* = \dfrac{r}{R+r})$. In the latter NE it is $p_A(x^*, y^*) = p_B(x^*, y^*) = rR/(R+r) < r < R$, thus an egalitarian payoff but a payoff lower than that achievable with any of the NE equilibriums based on coordination. Both BOS players get the same payoff if $y = 1-x$, in which case, $p = (R+r)(1-x)x$. This egalitarian payoff is maximum when $x = y = 1/2$, with $p^+ = (R+r)/4$, the point marked + in the far right panel of Table 1.4. Thus, the set of payoffs which can be obtained by both players (or payoff region) is closed by the parabola passing by (R, r), (r, R), and (p^+, p^+), as shown in the payoff region panel of Table 1.4.

$$p_A(x; y) = (x, 1-x) \begin{pmatrix} R & 0 \\ 0 & r \end{pmatrix} \begin{pmatrix} y \\ 1-y \end{pmatrix} = ((R+r)y - r)x + r(1-y) \quad (1.5a)$$

$$p_B(y; x) = (x, 1-x) \begin{pmatrix} r & 0 \\ 0 & R \end{pmatrix} \begin{pmatrix} y \\ 1-y \end{pmatrix} = ((R+r)x - R)y + R(1-x) \quad (1.5b)$$

In the zero-sum MP game, the expected payoffs of both players verify $p_A + p_B = 0$, with,

$$p_A(x, y) = +xy - x(1-y) - (1-x)y + (1-x)(1-y) = 2(2y-1)x + 1 - 2y \quad (1.6a)$$

$$p_B(x, y) = -xy + x(1-y) + (1-x)y - (1-x)(1-y) = 2(1-2x)y - 1 + 2x \quad (1.6b)$$

According to (1.6), the MP is a discoordination game because its unique NE is based on mixed strategies: $x^* = y^* = 1/2$, leading to the player's payoffs $p_A^\bullet = p_B^\bullet = 0.0$ [17].

1.3 Correlated Games

In an alternative approach to that of the independent players considered in previous Sect. 1.2, an external probabilistic mechanism sends a signal to each player, so that, in principle, the players do not have any active role [15]. Thus, in the so called correlated games, an *joint* probability distribution $\Pi = \begin{pmatrix} \pi_{11} & \pi_{12} \\ \pi_{21} & \pi_{22} \end{pmatrix}$ assigns probability to every combination of player choices, as sketched in the far right frame of Table 1.1. This gives rise to the expected payoffs,

$$p_A(x, y) = \mathbf{1}' \mathbf{€}_A \circ \Pi \mathbf{1} , \quad p_B(x, y) = \mathbf{1}' \mathbf{€}_B \circ \Pi \mathbf{1} \quad (1.7)$$

where the symbol $\circ$ indicates the element-by-element matrix product, and $\mathbf{1}' = (1, 1)$. Incidentally, in the independent player's scenario it is $\Pi = \mathbf{xy}'$.

Thus, in the PD game it is $\begin{cases} p_A = R\pi_{11} + S\pi_{12} + \mathfrak{T}\pi_{21} + P\pi_{22} \\ p_B = R\pi_{11} + \mathfrak{T}\pi_{12} + S\pi_{21} + P\pi_{22} \end{cases}$; in the HD game

of Table 1.2 it is: $\begin{cases} p_A = \pi_{11}2 + \pi_{21}3 - \pi_{22} \\ p_B = \pi_{11}2 + \pi_{12}3 - \pi_{22} \end{cases}$, so that $p_A = p_B \leftrightarrow \pi_{12} = \pi_{21}$; in the

SD game of Table 1.3 it is: $\begin{cases} p_A = 3\pi_{11} - \pi_{12} - \pi_{21} \\ p_B = 2\pi_{11} + 3\pi_{12} + \pi_{21} \end{cases}$, whereas in the MP game it

is: $p_A = +\pi_{11} - \pi_{12} - \pi_{21} + \pi_{22}$, $p_B = -p_A$.

In the BOS(R, r) game it is: $\begin{cases} p_A = \pi_{11}R + \pi_{22}r \\ p_B = \pi_{11}r + \pi_{22}R \end{cases}$. In the particular case of

$\Pi = \begin{pmatrix} \pi & 0 \\ 0 & 1 - \pi \end{pmatrix}$, it is $\begin{cases} p_A = \pi R + (1 - \pi)r \\ p_B = \pi r + (1 - \pi)R \end{cases}$. As a result, the payoff region limited by the parabola and the segment that joins (R, r) and (r, R) in the payoff region
panel of Table 1.4 becomes accessible. In this scenario both players reach a maximum egalitarian payoff $p^= = (R + r)/2$ (the point marked = in the payoff region of
Table 1.4), with $\pi = 1/2$, i.e., fully discarding the mutually inconvenient FB and
BF combinations and adopting FF and BB with equal probability.

The joint probability distribution Π is in correlated equilibrium (CE) [18, 19] if
the players cannot gain by disobeying the signals given by this randomization device
Π.

In the BOS(R, r) game, $\Pi = \begin{pmatrix} \pi & 0 \\ 0 & 1 - \pi \end{pmatrix}$ is in correlated equilibrium. This is so
because if player A receives the signal F and obeys it, his expected payoff would be
$p_A^O/F = \pi_{11}R/(\pi_{11} + \pi_{12})$ whereas if he disobeys the expectation would be $p_A^D/F = \pi_{12}r/(\pi_{11} + \pi_{12})$. Equilibrium demands $p_A^O/F > p_A^D/F$, thus $\pi_{11}R \geq \pi_{12}r$. Analogously, $p_A^O/B = \pi_{22}r/(\pi_{21} + \pi_{22}) > p_A^D/B = \pi_{21}R/(\pi_{21} + \pi_{22}) \Rightarrow \pi_{22}r \geq \pi_{21}R$;
$p_B^O/F = \pi_{11}r/(\pi_{11} + \pi_{21}) > p_B^D/F = \pi_{21}R/(\pi_{11} + \pi_{21})\pi_{11}r \geq \pi_{21}R$;
$p_B^O/B = \pi_{22}R/(\pi_{12} + \pi_{22}) > p_B^D/B = \pi_{12}r/(\pi_{12} + \pi_{22}) \Rightarrow \pi_{22}R \geq \pi_{12}r$. The former inequalities are verified if $\pi_{12} = \pi_{21} = 0$, so that in the BOS with such as CE it
is: $p_A = \pi R + (1 - \pi)r$, $p_B = \pi r + (1 - \pi)R$.

Any convex combination of Nash equilibriums is in correlated equilibrium. In
the BOS(R, r) game the CE given by $\Pi = \begin{pmatrix} \pi & 0 \\ 0 & 1 - \pi \end{pmatrix}$ may be seen as a convex linear combination of the two Nash equilibriums with pure strategies, i.e.,
$\pi \begin{pmatrix} 1 & 0 \\ 0 & 0 \end{pmatrix} + (1 - \pi) \begin{pmatrix} 0 & 0 \\ 0 & 1 \end{pmatrix}$. From the NE with mixed strategies in the BOS(R, r)

game, i.e., $\Pi_m^* = \dfrac{1}{(R + r)^2} \begin{pmatrix} Rr & R^2 \\ r^2 & Rr \end{pmatrix}$, two alternative correlated equilibriums are:

$\Pi_I^* = \pi \Pi_m^* + (1 - \pi) \begin{pmatrix} 1 & 0 \\ 0 & 0 \end{pmatrix}$ and $\Pi_{II}^* = \pi \Pi_m^* + (1 - \pi) \begin{pmatrix} 0 & 0 \\ 0 & 1 \end{pmatrix}$. Under these cor-

related equilibriums it is $p_A^I = R(1 - \pi \dfrac{R}{R + r})$, $p_B^{II} = r(1 - \pi \dfrac{r}{R + r})$ and $p_A^{II} = p_B^I$, $p_B^{II} = p_A^I$.

In the HD game of Table 1.2, for correlated equilibrium is must be, $2\pi_{11}2 \geq 3\pi_{11} - \pi_{12}$, $3\pi_{21} - \pi_{22} \geq 2\pi_{21}$, $2\pi_{11} \geq 3\pi_{11} - \pi_{21}$, $3\pi_{12} - \pi_{22} \geq 2\pi_{12}$. If $\pi_{12} = \pi_{21} = \pi$ these inequalities reduce to $2\pi_{11} \geq 3\pi_{11} - \pi$, $3\pi - \pi_{22} \geq 2\pi$, thus $\pi \geq$

$\pi_{11}, \pi \geq \pi_{22}$. The latter inequalities are verified if $\pi_{11} = 0$, $\pi_{22} = \pi = 1/3$, in which case $p_A = p_B = 5/3$, or if $\pi_{22} = 0$, $\pi_{11} = \pi = 1/3$, in which case $p_A = p_B = 2/3$.

In the MP game, the only Π in correlated equilibrium is that one generated from independent players in NE, i.e., $\Pi = \begin{pmatrix} 1/4 & 1/4 \\ 1/4 & 1/4 \end{pmatrix}$.

References

1. Straffin, P.D.: Game Theory and Strategy. The Mathematical Association of America **36**, (1993)
2. Binmore, K.: Just Playing: Game Theory and the Social Contract II. MIT Press, (ISBN 0-262-02444-6). Russell Sage Foundation, N.Y.C (1998)
3. Buchanan, J.M.: The Samaritan's Dilemma. In: Phelps, E.S., Sage R. (eds.) Altruism, Morality, and Economic Theory, p. 71. Russell Sage Foundation, N.Y.C (1975)
4. Coate, S.: Altruism, the Samaritan's dilemma, and government transfer policy. Am. Econ. Rev. **85**(1), 46–57 (1995)
5. Lagerlof, J.: Efficiency-enhancing signalling in the Samaritan's dilemma. Econ. J. **114**(492), 55–69 (2004)
6. Raschky, P.A., Schwindt, M. Aid, Catastrophes and the Samaritan's Dilemma. Economica (2016). https://doi.org/10.1111/ecca.12194
7. Skarbek, E.C.: Aid, ethics, and the Samaritan's dilemma: strategic courage in constitutional entrepreneurship. J. Institutional Econ. **12**(2), 371–393 (2016)
8. Alonso-Sanz, R., Situ, H.: A quantum Samaritan's dilemma cellular automaton. R. Soc. Open Sci. **4**(6), 863–160669 (2017)
9. Huang, Z.M., Alonso-Sanz, R., Situ, H.: Quantum Samaritan's Dilemma under Decoherence. Int. J. Theor. Phys. **56**(3), 863–873 (2017)
10. Ozdemir, S.K., Shimamura, J., Morikoshi, F., Imoto, N.: Dynamics of a discoordination game with classical and quantum correlations. Phys. Lett. A **333**, 218–231 (2004)
11. Rasmussen, E.: Games and Information. An Introduction to Game Theory. Blackwell, Oxford (2001)
12. Hofbauer, J., Sigmund, K. (2003). *Evolutionary games and population dynamics*. (Cambridge University Press)
13. Maynard Smith, J.: Evolution and the Theory of Games. Cambridge University Press (1982)
14. Binmore, K.: Game Theory: A very short introduction. Oxford UP (2007)
15. Owen, G.: Game Theory. Academic Press (1995)
16. Gibbons, R.: Game Theory for Applied Economists, pp. 29–33. Princeton University Press (1992)
17. Iqbal, A., Abbott, D.: Quantum Matching Pennies Game. J. Phys. Soc. Jpn. **78**, 014803 (2009). https://arxiv.org/abs/0807.3599
18. Aumann, R.J.: Subjectivity and correlation in randomized strategies. J. Math. Econ. **1**, 67–96 (1974)
19. Aumann, R.J.: Correlated equilibrium as an expression of Bayesian rationality. Econometrica **55**(1), 1–18 (1987)

Chapter 2
Quantum Approach to Game Theory

The quantum approach to game theory described in this chapter combines both the independent players and the Π-correlated game models all at once.

2.1 The EWL Model

Within the EWL quantization scheme [1], the purely classical strategies $\hat{C}$ and $\hat{D}$, which will be hereafter respectively named $\hat{C}$ and $\hat{D}$, in order to explicitly manifest that we are playing a quantum game, act in a two level Hilbert space spanned by the two bases vectors $|0\rangle$ and $|1\rangle$.

The state of the game is a vector in the tensor product space spanned by the basis vectors $|00\rangle$, $|01\rangle$, $|10\rangle$, and $|11\rangle$. The EWL protocol starts with an initial entangled state $|\Psi_i\rangle = \hat{J}|00\rangle$, where the symmetric unitary operator equals $\hat{J}(\gamma) = \exp\left(i\frac{\gamma}{2}\hat{D}^{\otimes 2}\right) = \cos\frac{\gamma}{2}\hat{C} \otimes \hat{C} + i\sin\frac{\gamma}{2}\hat{D} \otimes \hat{D}$, being $\hat{D} = \begin{pmatrix} 0 & 1 \\ -1 & 0 \end{pmatrix}$, $\hat{C} = \begin{pmatrix} 1 & 0 \\ 0 & 0 \end{pmatrix}$, and $0 \leq \gamma \leq \pi/2$ the *entanglement factor*, which tunes the entanglement degree. The explicit forms of $\hat{J}$ and its transpose conjugate $\hat{J}^\dagger$ are given in Eq. (2.1), and the explicit form of $|\Psi_i\rangle$ in Eq. (2.2) below.

$$\hat{J} = \begin{pmatrix} \cos\frac{\gamma}{2} & 0 & 0 & i\sin\frac{\gamma}{2} \\ 0 & \cos\frac{\gamma}{2} & -i\sin\frac{\gamma}{2} & 0 \\ 0 & -i\sin\frac{\gamma}{2} & \cos\frac{\gamma}{2} & 0 \\ i\sin\frac{\gamma}{2} & 0 & 0 & \cos\frac{\gamma}{2} \end{pmatrix}, \quad \hat{J}^\dagger = \begin{pmatrix} \cos\frac{\gamma}{2} & 0 & 0 & -i\sin\frac{\gamma}{2} \\ 0 & \cos\frac{\gamma}{2} & i\sin\frac{\gamma}{2} & 0 \\ 0 & i\sin\frac{\gamma}{2} & \cos\frac{\gamma}{2} & 0 \\ -i\sin\frac{\gamma}{2} & 0 & 0 & \cos\frac{\gamma}{2} \end{pmatrix}$$

$$\tag{2.1}$$

$$|\Psi_i\rangle = \hat{J} \begin{pmatrix} 1 \\ 0 \\ 0 \\ 0 \end{pmatrix} = \begin{pmatrix} \cos\frac{\gamma}{2} \\ 0 \\ 0 \\ i\sin\frac{\gamma}{2} \end{pmatrix} = \cos\frac{\gamma}{2}|00\rangle + i\sin\frac{\gamma}{2}|11\rangle \tag{2.2}$$

© Springer Nature Switzerland AG 2019
R. Alonso-Sanz, *Quantum Game Simulation*, Emergence, Complexity and Computation 36, https://doi.org/10.1007/978-3-030-19634-9_2

The players perform independently their quantum strategies described by local Special Unitary operators $\hat{U}$ in the SU(2) group of matrices of order two with complex number elements. A 2×2 matrix of complex numbers would have eight free parameters. The special condition ($|\hat{U}| = 1$) subtracts a degree of freedom, and the unitary condition ($\hat{U}\hat{U}^\dagger = I$) subtracts four degrees of freedom as it implies four equations. Consequently the SU(2) group has only three free parameters. The explicit form of $\hat{U}$ is given in Eq. (2.3) below.

$$\hat{U}(\theta, \alpha, \beta) = \begin{pmatrix} e^{i\alpha}\cos\frac{\theta}{2} & e^{i\beta}\sin\frac{\theta}{2} \\ -e^{-i\beta}\sin\frac{\theta}{2} & e^{-i\alpha}\cos\frac{\theta}{2} \end{pmatrix}, \qquad \begin{array}{c} \theta \in [0, \pi] \\ \alpha \in [0, \pi/2], \ \beta \in [0, \pi/2] \end{array} \qquad (2.3)$$

Please, note that $\hat{C} = \hat{U}(0,0,0) = \begin{pmatrix} 1 & 0 \\ 0 & 1 \end{pmatrix} = I$, and $\hat{D} = \hat{U}(\pi, \alpha, 0) = \begin{pmatrix} 0 & 1 \\ -1 & 0 \end{pmatrix} = i\sigma_y$.

After the application of these strategies, the state of the game changes to $|\Psi\rangle = (\hat{U}_A \otimes \hat{U}_B)\hat{J}|00\rangle$. Prior to measurement, the $\hat{J}^\dagger$ gate is applied and then the state of the game becomes,

$$|\Psi_f\rangle = \begin{pmatrix} \Psi_1 \\ \Psi_2 \\ \Psi_3 \\ \Psi_4 \end{pmatrix} = \hat{J}^\dagger(\hat{U}_A \otimes \hat{U}_B)\hat{J}|00\rangle, \qquad (2.4)$$

This follows a pair of Stern-Gerlach type detector for measurement. Therefore, the complex probability amplitudes ($|\Psi_i|$, $i = 1\ldots4$) induce the conventional joint probabilities according to the Born rule. Thus, $\Pi = \begin{pmatrix} |\Psi_1|^2 & |\Psi_2|^2 \\ |\Psi_3|^2 & |\Psi_4|^2 \end{pmatrix}$.

If the α and β parameters of both players are zero, i.e., if there is no *imaginary* part in the quantum strategies $\hat{U}$, Π turns out factorizable (or separable) because it is, $\Pi = \begin{pmatrix} x \\ 1-x \end{pmatrix}(y \ 1-y)$, with $x = \cos^2\frac{\theta_A}{2}$, $y = \cos^2\frac{\theta_B}{2}$.[1] Therefore, θ may be termed the *classical* parameter. The same factorizable joint probability distribution

---

[1] $\hat{U}(\theta,0,0) = \begin{pmatrix} \cos\frac{\theta}{2} & \sin\frac{\theta}{2} \\ -\sin\frac{\theta}{2} & \cos\frac{\theta}{2} \end{pmatrix}$, $\hat{U}_A \otimes \hat{U}_B = \begin{pmatrix} \cos\frac{\theta_A}{2}\cos\frac{\theta_B}{2} & \cos\frac{\theta_A}{2}\sin\frac{\theta_B}{2} & \sin\frac{\theta_A}{2}\cos\frac{\theta_B}{2} & \sin\frac{\theta_A}{2}\sin\theta_B \\ -\cos\frac{\theta_A}{2}\sin\frac{\theta_B}{2} & \cos\frac{\theta_A}{2}\cos\frac{\theta_B}{2} & -\sin\frac{\theta_A}{2}\sin\frac{\theta_B}{2} & \sin\frac{\theta_A}{2}\cos\theta_B \\ -\sin\frac{\theta_A}{2}\cos\frac{\theta_B}{2} & -\sin\frac{\theta_A}{2}\sin\frac{\theta_B}{2} & \cos\frac{\theta_A}{2}\cos\frac{\theta_B}{2} & \cos\frac{\theta_A}{2}\sin\theta_B \\ \sin\frac{\theta_A}{2}\sin\frac{\theta_B}{2} & -\sin\frac{\theta_A}{2}\cos\frac{\theta_B}{2} & -\cos\frac{\theta_A}{2}\sin\frac{\theta_B}{2} & \cos\frac{\theta_A}{2}\cos\theta_B \end{pmatrix}$ $(\hat{U}_A\otimes$

$\hat{U}_B)\begin{pmatrix} \cos\frac{Y}{2} \\ 0 \\ 0 \\ i\sin\frac{Y}{2} \end{pmatrix} = \begin{pmatrix} \cos\frac{\theta_A}{2}\cos\frac{\theta_B}{2}\cos\frac{Y}{2} + i\sin\frac{\theta_A}{2}\sin\frac{\theta_B}{2}\sin\frac{Y}{2} \\ -\cos\frac{\theta_A}{2}\sin\frac{\theta_B}{2}\cos\frac{Y}{2} + i\sin\frac{\theta_A}{2}\cos\frac{\theta_B}{2}\sin\frac{Y}{2} \\ -\sin\frac{\theta_A}{2}\cos\frac{\theta_B}{2}\cos\frac{Y}{2} + i\cos\frac{\theta_A}{2}\sin\frac{\theta_B}{2}\sin\frac{Y}{2} \\ \sin\frac{\theta_A}{2}\sin\frac{\theta_B}{2}\cos\frac{Y}{2} + i\cos\frac{\theta_A}{2}\cos\frac{\theta_B}{2}\sin\frac{Y}{2} \end{pmatrix}$, $|\Psi_f\rangle = \begin{pmatrix} \cos\frac{Y}{2} & 0 & 0 & -i\sin\frac{Y}{2} \\ 0 & \cos\frac{Y}{2} & i\sin\frac{Y}{2} & 0 \\ 0 & i\sin\frac{Y}{2} & \cos\frac{Y}{2} & 0 \\ -i\sin\frac{Y}{2} & 0 & 0 & \cos\frac{Y}{2} \end{pmatrix}$ $(\hat{U}_A \otimes$

$\hat{U}_B)\hat{J}|00\rangle = \begin{pmatrix} \cos\frac{\theta_A}{2}\cos\frac{\theta_B}{2}\cos^2\frac{Y}{2} + i\sin\frac{\theta_A}{2}\sin\frac{\theta_B}{2}\sin\frac{Y}{2}\cos\frac{Y}{2} - i\sin\frac{\theta_A}{2}\sin\frac{\theta_B}{2}\cos\frac{Y}{2}\sin\frac{Y}{2} + \cos\frac{\theta_A}{2}\cos\frac{\theta_B}{2}\sin^2\frac{Y}{2} \\ -\cos\frac{\theta_A}{2}\sin\frac{\theta_B}{2}\cos^2\frac{Y}{2} + i\sin\frac{\theta_A}{2}\cos\frac{\theta_B}{2}\sin\frac{Y}{2}\cos\frac{Y}{2} - i\cos\frac{\theta_A}{2}\sin\frac{\theta_B}{2}\sin^2\frac{Y}{2} \\ -i\cos\frac{\theta_A}{2}\sin\frac{\theta_B}{2}\cos\frac{Y}{2}\sin\frac{Y}{2} - \sin\frac{\theta_A}{2}\cos\frac{\theta_B}{2}\sin^2\frac{Y}{2} - \sin\frac{\theta_A}{2}\cos\frac{\theta_B}{2}\cos^2\frac{Y}{2} + i\cos\frac{\theta_A}{2}\sin\frac{\theta_A}{2}\sin\frac{Y}{2}\cos\frac{Y}{2} \\ -i\cos\frac{\theta_A}{2}\cos\frac{\theta_B}{2}\cos\frac{Y}{2}\sin\frac{Y}{2} + \sin\frac{\theta_A}{2}\sin\frac{\theta_B}{2}\sin^2\frac{Y}{2} + \sin\frac{\theta_A}{2}\sin\frac{\theta_B}{2}\cos^2\frac{Y}{2} + i\cos\frac{\theta_A}{2}\cos\frac{\theta_B}{2}\sin\frac{Y}{2}\cos\frac{Y}{2} \end{pmatrix} =$

$= \begin{pmatrix} \cos\frac{\theta_A}{2}\cos\frac{\theta_B}{2} \\ -\cos\frac{\theta_A}{2}\sin\frac{\theta_B}{2} \\ -\sin\frac{\theta_A}{2}\cos\frac{\theta_B}{2} \\ \sin\frac{\theta_A}{2}\sin\frac{\theta_B}{2} \end{pmatrix}$.

is generated with no entanglement, i.e., with $\gamma = 0$.[2] Anyhow, without entanglement or without imaginary part in the quantum strategies, the game becomes *classical* with players adopting *independent* strategies, and we have not any added value from such a pretended quantum games.

Let us point out here that: (*i*) controversy over what exactly a quantum game really is may be traced in the specialized literature [2–9], and (*ii*) other alternative quantum game approaches have been also recently proposed, e.g., [10–12].

Note that this book only considers the EWL model in the context of discrete static games. Further work will address the simulation of the quantum approach to (*i*) games with continuous strategic space, a classical instance of which is the Cournot duopoly, that was pioneered by Li et al. [13] by means of a *minimal* quantization approach that is quite different from the EWL approach [14–16], and (*ii*) dynamic games such as the penny-flip game [17] or the Stackelberg duopoly [18].

Two-parameter strategies (2P)

General strategies in SU(2) with three parameters will be referred to as 3P, whereas the two-parameter subset of SU(2) with $\beta = 0$ will be referred to as 2P.

The 2P model may be criticized as being just a subset of the general SU(2) space of unitary strategies [19, 20]. Thus, the 2P-restriction in the original formulation of the EWL model [1] is a subject of continuing discussions. Anyhow, the 2P model shows interesting properties, such as allowing for proper Nash equilibrium strategies, as stressed in the reply to [21] given in [22], and has proved to be a good (and widely used) test-bed to show how the quantum approach in game theory may *solve* dilemmas, by allowing for Nash equilibrium strategies out of the scope of the classical approach [22, 23]. Last but not least, according to the study on purely classical correlated games [24, 25], the 2P-EWL (somehow *semi-quantum*), emerges as an excellent intermediate step from classical correlation to the full quantum approach (3P), and then it still deserves attention on its own.

If $\theta_A = \theta_B = \pi$, i.e., $U_A = U_B = \hat{U}(\pi, \alpha) = \hat{D} = \begin{pmatrix} 0 & 1 \\ -1 & 0 \end{pmatrix}$, it turns out that,

regardless of γ, $|\Psi\rangle = |11\rangle = (0\ 0\ 0\ 1)'$, and consequently $\Pi^{DD} = \begin{pmatrix} 0 & 0 \\ 0 & 1 \end{pmatrix}$.

If $\theta_A = \theta_B = 0$, both operator strategies are diagonal and also it is diagonal $\Pi = \begin{pmatrix} 1 - \pi_{22} & 0 \\ 0 & \pi_{22} = \sin^2(\alpha_A + \alpha_B)\sin^2\gamma \end{pmatrix}$.[3] Such a diagonal matrix is in general non-factorizable. The exception being the case of *degeneration* to $\pi_{11} = 1$ either if *i*)

---

[2] It is $J(0) = J(0)^\dagger = I$, and $|\Psi_i\rangle = (1\,0\,0\,0)'$. Therefore, $\psi_1 = e^{i(\alpha_A + \alpha_B)} \cos\frac{\theta_A}{2} \cos\frac{\theta_B}{2}$, $\psi_2 = -e^{i(\alpha_A - \beta_B)} \cos\frac{\theta_A}{2} \sin\frac{\theta_B}{2}$, $\psi_3 = -e^{i(\alpha_A - \beta_B)} \sin\frac{\theta_A}{2} \cos\frac{\theta_B}{2}$, $\psi_4 = e^{-i(\beta_A + \beta_B)} \sin\frac{\theta_A}{2} \sin\frac{\theta_B}{2}$. Please, recall that $|e^{ix}|^2 = 1^2 = 1$.

[3] $\hat{U}(0, \alpha) = \begin{pmatrix} e^{i\alpha} & 0 \\ 0 & e^{-i\alpha} \end{pmatrix}$, $\hat{U}_A \otimes \hat{U}_B = \begin{pmatrix} e^{i(\alpha_A + \alpha_B)} & 0 & 0 & 0 \\ 0 & e^{i(\alpha_A - \alpha_B)} & 0 & 0 \\ 0 & 0 & e^{i(-\alpha_A + \alpha_B)} & 0 \\ 0 & 0 & 0 & e^{-i(\alpha_A + \alpha_B)} \end{pmatrix}$,

$(\hat{U}_A \otimes \hat{U}_B) \begin{pmatrix} \cos(\gamma/2) \\ 0 \\ 0 \\ i\sin(\gamma/2) \end{pmatrix} = \begin{pmatrix} e^{i(\alpha_A + \alpha_B)} \cos(\gamma/2) \\ 0 \\ 0 \\ e^{-i(\alpha_A + \alpha_B)} i\sin(\gamma/2) \end{pmatrix}$, $|\psi_f\rangle = \begin{pmatrix} \cos(\gamma/2) & 0 & 0 & -i\sin(\gamma/2) \\ 0 & \cos(\gamma/2) & i\sin(\gamma/2) & 0 \\ 0 & i\sin(\gamma/2) & \cos(\gamma/2) & 0 \\ -i\sin(\gamma/2) & 0 & 0 & \cos(\gamma/2) \end{pmatrix}$

$\alpha_A = \alpha_B = 0$ or ii) $\alpha_A = \alpha_B = \pi/2$. In the scenario i), it is $\hat{U}_A = \hat{U}_B = \hat{U}(0,0) = I = \hat{C}$, in which case $|\Psi\rangle = \hat{J}^\dagger(\hat{I} \otimes \hat{I})\hat{J}|00\rangle\,|00\rangle = \hat{J}^\dagger\hat{J}|00\rangle = \hat{I}|00\rangle = |00\rangle$. In the

scenario ii), it is $\hat{U}_A = \hat{U}_B = \hat{U}(0, \dfrac{\pi}{2}) = \hat{Q} = \begin{pmatrix} i & 0 \\ 0 & -i \end{pmatrix} = i\sigma_z$. Therefore, $\Pi^{CC} =$

$\Pi^{QQ} = \begin{pmatrix} 1 & 0 \\ 0 & 0 \end{pmatrix}$.

The strategy with middle-level values of the quantum parameters will be referred to as $\overline{U} = \hat{U}(\pi/2, \pi/4)$. The joint probability matrices of the games facing cooperation and defection with $\overline{U}$ in the 2P model are given in Eqs. 5.1–5.2. With middle-level election of the quantum parameters for the two players ($\theta_A = \theta_B = \pi/2$, $\alpha_A = \alpha_B = \pi/4$), i.e., with $\hat{U}_A = \hat{U}_B\overline{U}$ it is[4]:

$$\Pi(\overline{U}, \overline{U}) = \frac{1}{4}\begin{pmatrix} 1 - \sin^2\gamma & 1 - \sin\gamma \\ 1 - \sin\gamma & (1 + \sin\gamma)^2 \end{pmatrix} \tag{2.5}$$

Thus, in the 2P-BOS game with middle-level election of the parameters it is:

$(\hat{U}_A \otimes \hat{U}_B)\hat{J}|00\rangle = \begin{pmatrix} e^{i(\alpha_A+\alpha_B)}\cos^2(\gamma/2) + e^{-i(\alpha_A+\alpha_B)}\sin^2(\gamma/2) \\ 0 \\ 0 \\ -ie^{i(\alpha_A+\alpha_B)}\cos(\gamma/2)\sin(\gamma/2) + ie^{-i(\alpha_A+\alpha_B)}\sin(\gamma/2)\cos(\gamma/2) \end{pmatrix}$

$= \begin{pmatrix} \cos(\alpha_A+\alpha_B) + i\sin(\alpha_A+\alpha_B)\cos\gamma \\ 0 \\ 0 \\ -i\sin(\alpha_A+\alpha_B)\sin\gamma \end{pmatrix}$

$[4]\,\hat{U}_A = \hat{U}_B = \begin{pmatrix} \frac{1}{\sqrt{2}}(1+i)\frac{1}{\sqrt{2}} & \frac{1}{\sqrt{2}} \\ -\frac{1}{\sqrt{2}} & \frac{1}{\sqrt{2}}(1-i)\frac{1}{\sqrt{2}} \end{pmatrix} = \frac{1}{2}\begin{pmatrix} 1+i & \sqrt{2} \\ -\sqrt{2} & 1-i \end{pmatrix},\quad (\hat{U}_A \otimes \hat{U}_B) =$

$\frac{1}{4}\begin{pmatrix} 2i & (1+i)\sqrt{2} & \sqrt{2}(1+i) & 2 \\ -\sqrt{2}(1+i) & 2 & -2 & (1-i)\sqrt{2} \\ -\sqrt{2}(1+i) & -2 & 2 & (1-i)\sqrt{2} \\ 2 & -\sqrt{2}(1-i) & -(1-i)\sqrt{2} & -2i \end{pmatrix},$

$(\hat{U}_A \otimes \hat{U}_B)\begin{pmatrix} \cos\frac{\gamma}{2} \\ 0 \\ 0 \\ i\sin\frac{\gamma}{2} \end{pmatrix} = \frac{1}{4}\begin{pmatrix} 2i\cos\frac{\gamma}{2} + 2i\sin\frac{\gamma}{2} \\ -(1+i)\sqrt{2}\cos\frac{\gamma}{2} + \sqrt{2}(1-i)i\sin\frac{\gamma}{2} \\ -(1+i)\sqrt{2}\cos\frac{\gamma}{2} + \sqrt{2}(1-i)i\sin\frac{\gamma}{2} \\ 2\cos\frac{\gamma}{2} + 2\sin\frac{\gamma}{2}) \end{pmatrix},$

$|\psi_f\rangle = \begin{pmatrix} \cos\frac{\gamma}{2} & 0 & 0 & -i\sin\frac{\gamma}{2} \\ 0 & \cos\frac{\gamma}{2} & i\sin\frac{\gamma}{2} & 0 \\ 0 & i\sin\frac{\gamma}{2} & \cos\frac{\gamma}{2} & 0 \\ -i\sin\frac{\gamma}{2} & 0 & 0 & \cos\frac{\gamma}{2} \end{pmatrix}(\hat{U}_A \otimes \hat{U}_B)\hat{J}|00\rangle = \frac{1}{4}$

$\begin{pmatrix} 2i\cos^2\frac{\gamma}{2} + 2i\sin\frac{\gamma}{2}\cos(\gamma/2 - 2i\cos\frac{\gamma}{2}\sin(\gamma/2 - 2i\sin^2\frac{\gamma}{2} \\ \sqrt{2}(-(1+i)\cos^2\frac{\gamma}{2} + (1-i)i\sin\frac{\gamma}{2}\cos\frac{\gamma}{2} - (1+i)\cos\frac{\gamma}{2}i\sin\frac{\gamma}{2} - (1-i)\sin^2\frac{\gamma}{2}) \\ \sqrt{2}(-(1+i)\cos\frac{\gamma}{2}i\sin\frac{\gamma}{2} - (1-i)\sin^2\frac{\gamma}{2} - (1+i)\cos^2\frac{\gamma}{2} + (1-i)i\sin\frac{\gamma}{2}\cos\frac{\gamma}{2}) \\ 2\cos\frac{\gamma}{2}\sin\frac{\gamma}{2} + 2\sin^2\frac{\gamma}{2} + 2\cos^2\frac{\gamma}{2} + 2\sin\frac{\gamma}{2}\cos\frac{\gamma}{2} \end{pmatrix} =$

$\frac{1}{4}\begin{pmatrix} 2i\cos\gamma \\ \sqrt{2}(-1 + \sin\gamma - i\cos\gamma) \\ \sqrt{2}(-1 + \sin\gamma - i\cos\gamma) \\ 2(1 + \sin\gamma) \end{pmatrix}.$

$$P \begin{Bmatrix} A \\ B \end{Bmatrix} = \begin{Bmatrix} R \\ r \end{Bmatrix} \frac{1}{4}(1 - \sin\gamma)^2 + \begin{Bmatrix} r \\ R \end{Bmatrix} \frac{1}{4}(1 + \sin\gamma)^2 \qquad (2.6)$$

As $\pi_{22} > \pi_{11}$ in Eq. (2.5) (recall that $\gamma \in [0, \pi/2]$), in Eq. (2.6) it is $p_B > p_A$. Thus, the 2P-model is somehow biased towards the player B in the BOS game [26]. This is not so in the 3P-model, in which case with middle-level values of the quantum parameters the probabilities in Π turns out equalized regardless of γ as shown later in this chapter.

In the 2P model, the (purely Quantum) strategy $\hat{Q} = \hat{U}(0, \pi/2) = \begin{pmatrix} i & 0 \\ 0 & -i \end{pmatrix}$ stands out, because the $\{\hat{Q}, \hat{Q}\}$ pair becomes in NE for high values of γ in the PD, HD and SD games yielding $\pi_{11}^{\hat{Q},\hat{Q}} = \pi_{11}^{\hat{C},\hat{C}} = 1$ regardless of γ (see Sect. 2.2). The Π matrices emerging when $\hat{Q}$ interacts with the pures $\hat{D}$ and $\hat{C}$ strategies are given in Eq. (2.7) and the graphs of the payoffs arising in the four particular game-types studied here are shown in Fig. 2.1. These matrices will be routinely referred when commenting the outputs arising in collective games simulations in this book.

$$\Pi^{\hat{Q},\hat{D}} = \begin{pmatrix} 0 & \cos^2\gamma \\ \sin^2\gamma & 0 \end{pmatrix}, \; \Pi^{\hat{D},\hat{Q}} = (\Pi^{\hat{Q},\hat{D}})', \; \Pi^{\hat{C},\hat{Q}} = \Pi^{\hat{Q},\hat{C}} = \begin{pmatrix} \cos^2\gamma & 0 \\ 0 & \sin^2\gamma \end{pmatrix}$$
$$(2.7)$$

Equation (2.8), were $\omega = \theta/2$, give the general formulas of the elements of Π for arbitrary quantum parameters and entanglement factor in the 2P model.

$$\pi_{11} = \cos^2\omega_A \cos^2\omega_B (\cos^2\gamma \sin^2(\alpha_A + \alpha_B) + \cos^2(\alpha_A + \alpha_B)) \qquad (2.8a)$$

$$\pi_{12} = (-\cos\omega_A \sin\omega_B \cos\alpha_A + \cos\omega_B \sin\omega_A \sin\gamma \sin\alpha_B)^2 + \cos^2\omega_A \sin^2\omega_B \sin^2\alpha_A \cos^2\gamma. \qquad (2.8b)$$

$$\pi_{21} = (-\cos\omega_B \sin\omega_A \cos\alpha_B + \cos\omega_A \sin\omega_B \sin\gamma \sin\alpha_A)^2 + \cos^2\omega_B \sin^2\omega_A \sin^2\alpha_B \cos^2\gamma. \qquad (2.8c)$$

$$\pi_{22} = (\sin\gamma \cos\omega_A \cos\omega_B \sin(\alpha_A + \alpha_B) + \sin\omega_A \sin\omega_B)^2 \qquad (2.8d)$$

Three-parameter strategies (3P)

Unlike in the 2P model, in the 3P-model, middle-level parameter values ($\theta_A = \theta_B = \pi/2$, $\alpha_A = \alpha_B = \beta_A = \beta_B = \pi/4$), lead to a uniform Π, i.e., $\Pi = \begin{pmatrix} 1/4 & 1/4 \\ 1/4 & 1/4 \end{pmatrix}$, regardless of γ.[5]

---

[5] $\hat{U}_A = \hat{U}_B = \hat{U}(\pi/2, \pi/4, \pi/4) = \frac{1}{\sqrt{2}} \begin{pmatrix} e^{i\pi/4} & e^{i\pi/4} \\ -e^{-i\pi/4} & e^{-i\pi/4} \end{pmatrix} = \frac{1}{\sqrt{2}} e^{i\pi/4} \begin{pmatrix} 1 & 1 \\ i & -i \end{pmatrix}$,

$$(\hat{U}_A \otimes \hat{U}_B)|\psi_i\rangle = \frac{1}{2} e^{i\pi/2} \begin{pmatrix} 1 & 1 & 1 & 1 \\ i & -i & i & -i \\ i & i & -i & -i \\ -1 & 1 & 1 & -1 \end{pmatrix} \begin{pmatrix} \cos(\gamma/2) \\ 0 \\ 0 \\ i\sin(\gamma/2) \end{pmatrix} = \frac{1}{2} i \begin{pmatrix} \cos(\gamma/2) + i\sin(\gamma/2) \\ i\cos(\gamma/2)\sin(\gamma/2) \\ i\cos(\gamma/2) + \sin(\gamma/2) \\ -\cos(\gamma/2) - i\sin(\gamma/2) \end{pmatrix}$$,

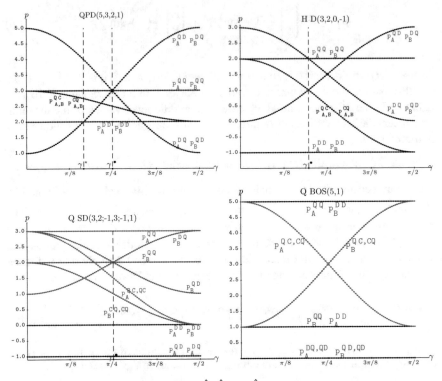

Fig. 2.1 Payoffs of the interactions of the $\hat{D}$, $\hat{C}$, and $\hat{Q}$ strategies. Top left: PD, Top right: SD, Bottom left: SD, Bottom right: BOS

2.2   Nash Equilibrium in the EWL-PD Model

Under the EWL model with two parameters, in the QPD game with low entanglement the unique strategy pair in NE remains to be mutual defection $\{\hat{D}, \hat{D}\}$, but with high entanglement the unique strategy pair in NE turns out to be $\{\hat{Q}, \hat{Q}\}$. One may say that the Q strategy resolves the PD dilemma in favor of the social welfare solution, though the skeptical reader might argue that as Q has no *real* component ($\theta = 0$), the dilemma is resolved in the *imaginary* world.

The γ-thresholds that mark off the intervals of the entanglement where mutual defection and mutual $\hat{Q}$ are in NE are given in Eq. (2.9). The thresholds given in

$$
|\psi_f\rangle = \tfrac{1}{2}i \begin{pmatrix} \cos^2(\gamma/2) + i\sin(\gamma/2)\cos(\gamma/2) + i\cos(\gamma/2)\sin(\gamma/2) - \sin^2(\gamma/2) \\ i\cos^2(\gamma/2) + \sin(\gamma/2)\cos(\gamma/2) - \cos(\gamma/2)\sin(\gamma/2) + i\sin^2(\gamma/2) \\ -\cos(\gamma/2)\sin(\gamma/2) + i\sin^2(\gamma/2) + i\cos^2(\gamma/2) + \sin(\gamma/2)\cos(\gamma/2) \\ -i\cos(\gamma/2)\sin(\gamma/2) + \sin^2(\gamma/2) - \cos^2(\gamma/2) - i\sin(\gamma/2)\cos(\gamma/2) \end{pmatrix}
$$

$$
= \tfrac{1}{2}\begin{pmatrix} i\cos\gamma - \sin\gamma \\ -1 \\ -1 \\ -1i\cos\gamma + \sin\gamma \end{pmatrix}.
$$

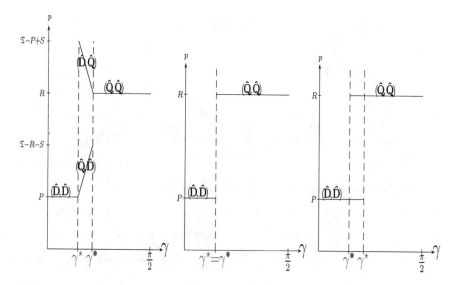

Fig. 2.2 Nash Equilibrium in the 2P-QPD($\mathfrak{T}$, R, P, S) game. Far left: $P - S < \mathfrak{T} - R$. Center: $P - S = \mathfrak{T} - R$. Far right: $P - S > \mathfrak{T} - R$

Eq. (2.9) share the same inner denominator, so that their ranking depends of that of the inner numerator. Thus, if, as in the particular case reported predominantly in this book, it is $P - S < \mathfrak{T} - R$ then $\gamma^* < \gamma^\bullet$. If $P - S = \mathfrak{T} - R$ it turns out $\gamma^* = \gamma^\bullet$, and if $P - S > \mathfrak{T} - R$ it turns out $\gamma^* > \gamma^\bullet$, If $P - S < \mathfrak{T} - R$ in the $(\gamma^*, \gamma^\bullet)$ interval both $\{\hat{Q}, \hat{D}\}$ and $\{\hat{D}, \hat{Q}\}$ are in NE. In the case of equalization of both thresholds, an abrupt $\{\hat{D}, \hat{D}\} \to \{\hat{Q}, \hat{Q}\}$ transition occurs at $\gamma^* = \gamma^\bullet$. If $P - S > \mathfrak{T} - R$ in the $(\gamma^\bullet, \gamma^*)$ interval both $\{\hat{D}, \hat{D}\}$ and $\{\hat{Q}, \hat{Q}\}$ are in NE [27, 28].

$$\gamma^* = \arcsin\sqrt{\frac{P - S}{\mathfrak{T} - S}} \quad , \quad \gamma^\bullet = \arcsin\sqrt{\frac{\mathfrak{T} - R}{\mathfrak{T} - S}} \tag{2.9}$$

The γ^* threshold is defined at the intersection of $p_B^{DD} = p_A^{DD} = P$ and $p_A^{QD} = p_B^{DQ} = S(1 - \sin^2 \gamma) + \mathfrak{T} \sin^2 \gamma$, whereas the $\gamma^\bullet$ threshold is defined at the intersection of $p_B^{QD} = p_A^{DQ} = \mathfrak{T}(1 - \sin^2 \gamma) + S \sin^2 \gamma$ and $p_B^{QQ} = p_A^{QQ} = R$ These NE ideas are sketched in Fig. 2.2 and proved via simulation in the particular case of the PD with (5,3,2,1) payoffs in Fig. 4.23.

Three particular examples, corresponding to the three scenarios in Fig. 2.2, are given in Fig. 2.3. The graphs of $p_B^{QD} = p_A^{DQ}$ and $p_B^{QD} = p_A^{DQ}$ intersect at $\gamma = \pi/4$ regardless of the payoff parameters of the PD,[6] rendering the common $(\mathfrak{T} + S)/2$ payoff, that equals 3.0, 2.5 and 2.0 in the three scenarios of Fig. 2.3. Incidentally,

---

[6] $S(1 - \sin^2 \gamma) + \mathfrak{T} \sin^2 \gamma = \mathfrak{T}(1 - \sin^2 \gamma) + S \sin^2 \gamma \Rightarrow (S - \mathfrak{T})(1 - \sin^2 \gamma) = (S - \mathfrak{T})$
$\sin^2 \gamma \Rightarrow 2 \sin^2 \gamma) \Rightarrow \gamma^= = \arcsin\sqrt{\frac{1}{2}} = \pi/4$.

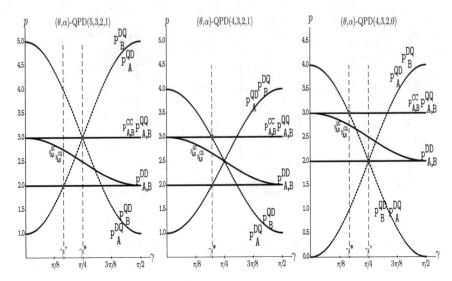

Fig. 2.3 Payoffs of the interactions of the $\hat{D}$, $\hat{C}$, and $\hat{Q}$ strategies in the QPD game. Far left: (5,3,2,1) payoffs. Center: (4,3,2,1) payoffs. Far right: (4,3,2,0) payoffs

with (5,3,2,1) payoffs, $(\mathfrak{T}+S)/2 = R = 3$, so that $p_A^{QD} = p_B^{DQ}$, $p_B^{DQ} = p_A^{DQ}$ and $p_{A,B}^{QQ}$ intersect in the same point; and with (4,3,2,0) payoffs, $(\mathfrak{T}+S)/2 = P = 2$, so that $p_A^{QD} = p_B^{DQ}$, $p_B^{DQ} = p_A^{DQ}$ and $p_{A,B}^{DD}$ intersect in the same point.

References

1. Eisert, J., Wilkens, M., Lewenstein, M.: Quantum games and quantum strategies. Phys. Rev. Lett. **83**(15), 3077–3080 (1999)
2. Bleiler,S.(2008). A formalism for quantum games and an application. arXiv preprint http://arxiv.org/abs/0808.1389
3. Branderburger, A.: The relationship between quantum and classical correlation games. Games Econ. Behav. **89**, 157–183 (2010)
4. Du, J., Ju, C., Li, H.: Quantum entanglement helps in improving economic efficiency. J. Phys. A Maths Gen. **38**, 1559–1565 (2005)
5. Frąiewicz, P.: The ultimate solution to the quantum battle of the sexes game. J. Phys. A Math. Theor. **42**(36), 365305 (2009)
6. Khan, F.S., Phoenix, S.J.D.: Mini-maximizing two qubit quantum computations. Quantum Inf. Process. **12**(12), 3807–3819 (2013)
7. Khan, F.S.: Dominant strategies in two-qubit quantum computations. Quantum Inf. Process. **14**(6), 1799–1808 (2015)
8. Levine, D.: Quantum games have no news for economists (2005). http://www.dklevine.com/papers/quantumnonews.pdf
9. Phoenix, S.J.D., Khan, F.S.: The role of correlations in classical and quantum games. Fluct. Noise Lett. **12**(3), 1350011 (2013)

10. Brunner, N., Linden, N.: Connection between Bell nonlocality and Bayesian game theory. Nat. Commumications **4**, 2057 (2013)
11. Pappa, A., Kumar, N., Lawson, T., Santha, M., Zhang, S., Diamanti, E., Kerenidis, I.: Nonlocality and conflicting interest games. Phys. Rev. Lett. **114**, 020401 (2015)
12. Zhang, S.: Quantum strategic game theory. In: Proceedings of the 3rd Innovations in Theoretical Computer Science Conference (ITCS '12) ISBN 978-1-4503-1115-1. ACM (New York, USA) (2012)
13. Li, H., Du, J., Massar, S.: Continuous-variable quantum games. Phys. Lett. A **306**, 73–78 (2002)
14. Quantum approach to Bertrand duopoly: Frąiewicz, P., Sladkowski. J. Quantum Inf. Process. **15**, 3636–3650 (2016)
15. Khan, S., Ramzam, M.R., Khan, M.K.: Quantum model of Bertrand duopoly. Chinese Phys. Lett. **27**, 080302 (2010)
16. Sekiguchi, Y., Sakahara, K., Sato, T.: Uniqueness of Nash equilibria in a quantum Cournot duopoly game. J. Phys. A Math. Theor. **43**(14), 145303 (2010)
17. Meyer, D.A.: Quantum strategies. Phys. Rev. Lett. **82**(5), 1052 (1999)
18. Lo, C.F., Kiang, D.: Quantum stackelberg duopoly. Phys. Lett. A **318**(4–5), 333–336 (2003)
19. Benjamin, S.C., Hayden, P.M.: Comment on "Quantum Games and Quantum Strategies". Phys. Rev. Lett. **87**, 069801 (2001)
20. Benjamin, S.C., Hayden, P.M.: Multiplayer quantum games. Phys. Rev. A **64**, 030301 (2001)
21. Benjamin, S.C., Hayden, P.M.: Comment on "Quantum games and quantum strategies". Phys. Rev. Lett. **87**(6), 069801 (2001)
22. Eisert, J., Wilkens, M., Lewenstein, M.: Comment on "Quantum games and quantum strategies"-Reply. Phys. Rev. Lett. **87**, 069802 (2001)
23. Eisert, J., Wilkens, M.: Quantum games. J. Mod. Opt. **47**(14–15), 2543–2556 (2000)
24. Alonso-Sanz, R.: Spatial correlated games. R. Soc. Open Sci. **4**(6), 171361 (2017)
25. Iqbal, A., Chappell, J.M., Abbott, D.: On the equivalence between non-factorizable mixed-strategy classical games and quantum games. R. Soc. Open Sci. **3**, 150477 (2016)
26. Flitney, A.P., Hollengerg, L.C.L.: Nash equilibria in quantum games with generalized two-parameter strategies. Phys. Lett. A **363**, 381–388 (2007)
27. Du, J.F., Xu, X.D., Li, H., Zhou, X., Han, R.: Entanglement playing a dominating role in quantum games. Phys. Lett. A **89**(1–2), 9–15 (2001)
28. Du, J.F., Li, H., Xu, X.D., Zhou, X., Han, R.: Phase-transition-like behaviour of quantum games. J. Phys. A Math. and Gen. **36**(23), 6551–6562 (2003)

Chapter 3
Spatial Quantum Game Simulation

This chapter deals with collective games in which the players are arranged in a spatially structured two-dimensional lattice as explained in Sect. 3.1. Sections 3.2 and 3.3 deal with two-parameter and three-parameter strategy simulations respectively. A variant of the canonical EWL model is considered in Sect. 3.4.

3.1 Spatial Quantum Games

In the spatial version of the collective quantum games we deal with in this book, each player occupies a site (i, j) in a two-dimensional $N \times N$ lattice. In order to compare different types of players, two types of players, termed A and B, are to be considered. A and B players alternate in the site occupation in a chessboard form, so that every player is surrounded by four partners (A–B, B–A), and four mates (A–A, B–B) as sketched in Table 3.1.

Cellular automata (CA) are spatially extended dynamical systems that are discrete in all their constitutional components: space, time and state-variable. Uniform, local and synchronous interactions, as assumed here, are landmark features of CA [1]. Spatialized quantum games are fairly unexplored, even in its simplest form, i.e., in the CA manner considered here. To the best of our knowledge, only the reference [2] may by properly cited in this respect, as the articles [3–6] are intended in *networks*, not in spatially structured *lattices*. The reader familiar with the CA literature should not confuse the quantum-game CA approach proposed here (updated in a deterministic way) with general quantum CA models [7]. Two main reasons support the CA approach: (*i*) Spatialization of games, both classical and quantum, seems to be a natural way of dealing with social interactions, (*ii*) Building the quantum devices needed to play quantum games may not be that difficult. Just to give an example of difficulty, implementing the Shor's quantum factoring algorithm would require hundreds or thousands of entangled particles. However, in simple games such as

© Springer Nature Switzerland AG 2019
R. Alonso-Sanz, *Quantum Game Simulation*, Emergence, Complexity and Computation 36, https://doi.org/10.1007/978-3-030-19634-9_3

Table 3.1 The chessboard players' lattice in the spatial simulations

A	B	A	B	A
B	A	B	A	B
A	B	A	B	A
B	A	B	A	B
A	B	A	B	A

the bipersonal ones studied here, only one particle per player is needed. In addition, quantum games do not require long sequences of coherent operations and hence are more likely to realize than large-scale quantum computations such as required to factor numbers large enough to be of cryptographic relevance. To summarize, we expect games to be much more widely played with quantum devices in the future, and that the CA approach will be relevant in this respect.

In the spatial simulations of this book, the game is played in the CA manner, i.e., with uniform, local and synchronous interactions [1]. In this way, at the generic time-step T, every (i, j) player plays with his four adjacent partners, and he will featured by the average payoff over these over these four interactions: $p_{i,j}^{(T)} = \frac{1}{4}(p_{(i,j),(i-1,j)}^{(T)} + p_{(i,j),(i+1,j)}^{(T)} + p_{(i,j),(i,j-1)}^{(T)} + p_{(i,j),(i,j+1)}^{(T)})$. The evolution is ruled by the (deterministic) imitation of the best paid neighbour, so that in the next generation, every generic player (i, j) will adopt the parameters of his mate player (himself included at this regard) with the highest payoff among their mate neighbors. The just described setup is reminiscent of that proposed in [8]. Please note that the transition rule is both memoryless and myopic, i.e., neither the past nor the future does matter, only the *present* time-step T determines $T + 1$.

Table 3.2 shows a simple example in the classical context ($\alpha = 0$) where $4p$ instead of p is shown. In this Table, initially every player cooperates ($\theta = 0 \equiv x = 1$), except for the player A located in the (3,4) cell which defects ($\theta = \pi \equiv x = 0$). Thus at $T = 1$ said player A gets the $4p = 20$ payoff instead of the common $4p = 12$ payoff. The imitation mechanism spreads the $\theta_A = \pi$ defection across the player A cells, whereas cooperation remains unaltered in player B cells because no player B defects.

As a rule, the simulations shown in this book are run up to $T = 200$ and the $[0, 2\pi]$ interval of variation of γ is sampled in one hundred equidistant points. Five different initial random assignment of the (θ, α, β) parameter values are implemented in every simulated game scenario, with the exception of simulations regarding the matching-pennies game, in which case only one initial random assignment of the quantum parameters suffices to represent the results under any other initial random parameter condition. The spatial simulations will be implemented in a 200×200 lattice with periodic boundary conditions.

The computations have been performed by a double precision Fortran code run on a mainframe. The *random_number* Fortran function has been used in randomization, making sure that the initial random assignation of the quantum parameters remains unaltered for every choice of the entanglement parameter by means of the *random_seed* Fortran function. The Fortran code has been worked out to facilitate the graphical output of the simulation results.

The results achieved at $T = 200$ in simulations with variable entanglement factor γ are shown in the figures that follow. They show the actual mean payoffs $(\overline{p})$, and the mean value of the quantum parameters $(\overline{\theta}, \overline{\alpha}, \overline{\beta})$. The standard deviations of these magnitudes are shown under the σ-label.

The figures also show the mean-field payoffs (p^*) achieved in a single hypothetical two-person game with players adopting the average parameter values appearing in the collective dynamic simulation. Namely, the $U_A^*(\overline{\theta}_A, \overline{\alpha}_A, \overline{\beta}_A)$ and $U_B^*(\overline{\theta}_B, \overline{\alpha}_B, \overline{\beta}_B)$ strategies of the form:

$$
U_A^* = \begin{pmatrix} e^{i\overline{\alpha}_A} \cos \frac{\overline{\theta}_A}{2} & e^{i\overline{\beta}_A} \sin \frac{\overline{\theta}_A}{2} \\ -e^{i\overline{\beta}_A} \sin \frac{\overline{\theta}_A}{2} & e^{-i\overline{\alpha}_A} \cos \frac{\overline{\theta}_A}{2} \end{pmatrix}, \quad U_B^* = \begin{pmatrix} e^{i\overline{\alpha}_B} \cos \frac{\overline{\theta}_B}{2} & e^{i\overline{\beta}_B} \sin \frac{\overline{\theta}_B}{2} \\ -e^{i\overline{\beta}_B} \sin \frac{\overline{\theta}_B}{2} & e^{-i\overline{\alpha}_B} \cos \frac{\overline{\theta}_B}{2} \end{pmatrix} \tag{3.1}
$$

As a rule, in the graphs the results regarding player A are shown in red colour, and those regarding player B are shown in blue colour. Also a rule for graphs, the mean-field payoff approaches are colored brown for player A and green for player B, somehow approaching the red and blue colors featuring the respective actual mean payoffs.

Table 3.2 The spatial QPD scenario. Far left: The (A,B) chessboard. Centre: A classical example where every player cooperates, except the defector player A located in the (3,4) cell. Far right: Parameters and payoffs at $T = 2$

							θ ($T=1$)						p ($T=1$)						θ ($T=2$)						p ($T=2$)					
A	B	A	B	A	B	0	0	0	0	0	0	12	12	12	12	12	12	0	0	0	0	0	0	12	10	12	10	12	12
B	A	B	A	B	A	0	0	0	0	0	0	12	12	10	12	12	12	0	π	0	π	0	0	10	20	6	20	10	12
A	B	A	B	A	B	0	0	π	0	0	0	12	10	20	10	12	12	0	0	π	0	0	0	12	6	20	6	12	12
B	A	B	A	B	A	0	0	0	0	0	0	12	12	10	12	12	12	0	π	0	π	0	0	10	20	6	20	10	12
A	B	A	B	A	B	0	0	0	0	0	0	12	12	12	12	12	12	0	0	0	0	0	0	12	10	12	10	12	12
B	A	B	A	B	A	0	0	0	0	0	0	12	12	12	12	12	12	0	0	0	0	0	0	12	12	12	12	12	12

3.2 Two-Parameter Strategies

Figure 3.1 deals with the results obtained at $T = 200$ in five QPD(5,3,2,1) spatial simulations with variable entanglement factor γ. Its left frame shows the mean payoffs of both the player A and B across the lattice, and its right frame the mean parameter values of both players across the lattice. The simulations of Fig. 3.1 with (5,3,2,1)-PD parameters, show how the CA imitation dynamics enables the emergence of NE in a manner qualitatively comparable to that described in Sect. 2.2, i.e., $\{\hat{D}, \hat{D}\} \rightarrow \left\{\{\hat{D}, \hat{Q}\}, \{\hat{Q}, \hat{D}\}\right\} \rightarrow \{\hat{Q}, \hat{Q}\}$ as γ increases, with thresholds given by Eqs. (2.9): $\gamma^* = \arcsin\left(\sqrt{\dfrac{2-1}{5-1}} = \dfrac{1}{4}\right) = \dfrac{\pi}{6} = 0.524$, and

$$\gamma^\bullet = \arcsin\left(\sqrt{\frac{3-1}{5-1}} = \frac{2}{4}\right) = \frac{\pi}{4} \simeq 0.785.$$

The left frame of Fig. 3.1 also shows the mean-field payoffs ($p^\star$) achieved in a single hypothetical two-person game with the virtual players adopting the strategies constructed according to Eqs. (3.1) from the mean quantum parameters given in the right panel of Fig. 3.1. The mean-field payoffs in Fig. 3.1 coincide with the actual mean payoffs with γ below $\gamma^\star$ and over $\gamma^\bullet$. But in the $(\gamma^\star, \gamma^\bullet)$ interval the mean-field payoff approaches do not coincide with the actual mean payoffs as a result of the spatial effects illustrated below. Both the mean-field estimation of player A (brown marked) and that of the player B (green marked), as a rule, underestimate the actual mean payoffs of player A (red marked) and of player B (blue marked) in the $(\gamma^\star, \gamma^\bullet)$ interval of Fig. 3.1. The standard deviation (σ) of both players rockets to circa 1.0 at $\gamma^\star$ and monotonically decreases down to zero at $\gamma^\bullet$.

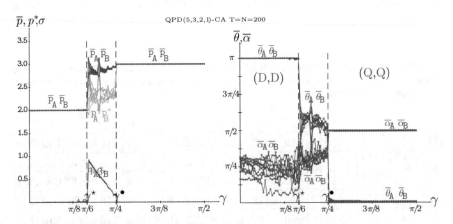

Fig. 3.1 The 2P-QPD(5,3,2,1) spatial game with variable entanglement factor γ. Five simulations at $T = 200$. Left: Mean and standard deviations of the actual payoffs ($\bar{p}$, σ), and mean-field payoffs ($p^\star$). Right: Mean quantum parameters

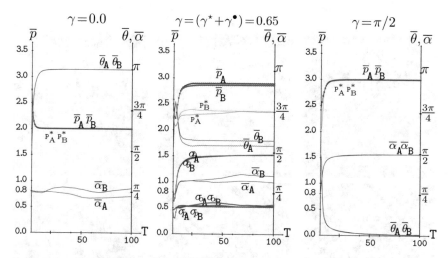

Fig. 3.2 Dynamics up to $T = 100$ in three simulations in the 2P-QPD(5,3,2,1)-CA scenario of Fig. 3.1. Left: $\gamma = 0$, Center: $\gamma = (\gamma^{\star} + \gamma^{\bullet})/2$, Right: $\gamma = \pi/2$

Figure 3.2 shows the dynamics up to $T = 100$ in simulations in the QPD(5,3,2,1)-CA scenario of Fig. 3.1 with $\gamma = 0$ (left), and $\gamma = (\gamma^{\star} + \gamma^{\bullet})/2$ (center), and $\gamma = \pi/2$ (right). As a result of the initial random assignment of the parameter values, it is initially in these frames: $\overline{\theta} \simeq \pi/2 = 1.570$, $\overline{\alpha} \simeq \pi/8 = 0.785$. A remarkable common feature in the three scenarios of Fig. 3.2 is that, despite the full range of parameters initially accessible in the CA local interactions, the parameters of the model are quickly selected so that, after a short initial transition period, the structure of the parameter patterns, and consequently those of the payoff patterns, promptly converge to a fairly stable structure. With $\gamma = 0$, both $\overline{\theta}$'s rocket to π, which makes $\overline{\alpha}$'s irrelevant as $\cos \pi/2 = 0$ annihilates the influence of α in (2.3). With $\gamma = \pi/2$, both $\overline{\alpha}$'s rocket towards $\pi/2$, both $\overline{\theta}$'s plummet to zero, i.e., the parameters of the Q strategy. In the central frame with γ in the middle of the $(\gamma^{\star}, \gamma^{\bullet})$ transition interval, the parameter stabilization is also quickly achieved (both $\overline{\theta}$'s close over $\pi/2$ and both $\overline{\alpha}$'s close to 1.0), but at variance with what happens in the $\gamma = 0$ and $\gamma = \pi/2$ scenarios, the mean-field payoff estimations do not coincide with the actual mean payoffs but underestimate them. Remarkably, the parameter tendencies heavily emerge, as a rule, from the very beginning, despite the full range of parameters initially accessible in the CA local interactions. This is so even in the case of the $\overline{p}$ vs. p^{*} divergence in the central frame of Fig. 3.2.

The parameter patterns at $T = 100$ with $\gamma = 0.654$ (the scenario of the central frame of Fig. 3.2) shown in Fig. 3.3 show some spatial structure, where a kind of *borders* separate fairly uniform clusters (in the case of the α pattern in a less crisp form).

The lower row of patterns of Fig. 3.3 zooms the 20×20 central part of the whole 200×200 lattice, so that the zoomed square has the A–B structure of the far left in lattice of Table 3.2, i.e., the two top-left cell is occupied by a player of A type. The

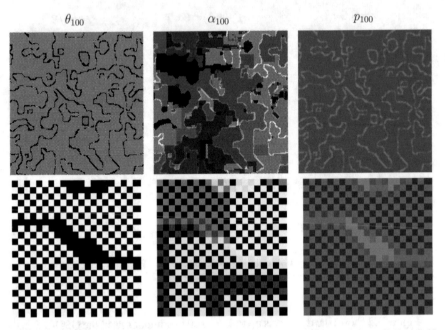

Fig. 3.3 Parameter and payoff patterns at $T = 100$ in a simulation with $\gamma = (\gamma^\star + \gamma^\bullet)/2 = 0.65$ in the 2P-QPD-CA scenario of Fig. 3.2. The upper row of patterns shows the whole 200×200 lattice. The lower row of patterns zooms its 20×20 central part. Increasing grey levels indicate increasing values

black spot in the central part of zoomed θ-pattern indicates that the players in the cells of this black spot defect ($\theta = \pi$), so that the players in the cells of the mentioned black spot (defectors surrounded by defectors) get a low 4×2 payoff, which is reflected in the zoomed payoff pattern in a clearer region. Above the central spot of the zoomed θ-pattern most of the A players defect whereas most of the B players have $\theta = 0$. The just mentioned B players have $\alpha = \pi/2$ (black cells) in the zoomed α-pattern. In consequence, (D,Q) is the strategy pair that predominates above the referred θ black spot. The values of the parameters in the cells below the central black are the reversed from those just described: A$\rightarrow$ $(0, \pi/2)$, B$\rightarrow$ $\theta = \pi$), so that (Q,D) is the strategy pair that predominates above the θ black spot. We may conclude that (Q,D) and (D,Q) (the two NE pairs in the γ transition interval) emerge in clusters separated by borders of mutual defection. In any case, the spatial structure of the patterns in the γ transition interval, far from the initial random configuration and of the fixed point reached with low or high γ, explains why the mean-field estimations of the payoffs (p^*) differ from the actual ones ($\overline{p}$) in the simulations with γ in the $(\gamma^\star, \gamma^\bullet)$ transition interval.

Figure 3.4 deals with spatial QPD simulations with payoffs alternative to the (5,3,2,1) set adopted canonically in this book, for which $\gamma^\star < \gamma^\bullet$. In the (4,3,2,1)

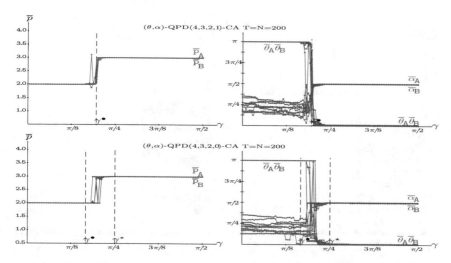

Fig. 3.4 The spatial 2P-QPD($\mathfrak{T}$,R,P,S) game with variable entanglement factor γ. Five simulations at T=200. Upper frames: (4,3,2,1) payoffs. Lower frames: (4,3,2,0) payoffs

simulations of the upper frames, it is $\gamma^* = \gamma^\bullet = \arcsin\left(\sqrt{\dfrac{2-1}{4-1}} = \dfrac{4-3}{4-1} = \dfrac{1}{3}\right)$.
An abrupt $\{\hat{D}, \hat{D}\} \rightarrow \{\hat{Q}, \hat{Q}\}$ transition occurs at $\gamma^* = \gamma^\bullet$. In the (4,3,2,0) simula-
tions of the lower frames it is $\gamma^* = \arcsin\left(\sqrt{\dfrac{2-0}{4-0}} = \dfrac{2}{4}\right) > \gamma^\bullet = \arcsin\left(\sqrt{\dfrac{4-3}{4-0}} = \dfrac{1}{4}\right)$.
In such a scenario, both $\{\hat{D}, \hat{D}\}$ and $\{\hat{Q}, \hat{Q}\}$ coexist in NE in the $(\gamma^\bullet, \gamma^*)$ interval as
sketched in Fig. 2.2, though the pair $\{\hat{D}, \hat{D}\}$ seems to predominate close to $\gamma^\bullet$ and
$\{\hat{Q}, \hat{Q}\}$ close to γ^*.

Figure 3.5 deals with the two-parameter, QHD(3,2,0,−1) spatial simulation
with variable entanglement factor γ. In parallel with what happens in the QPD,
mutual Q, rendering the outcome of mutual Dove, emerges after the threshold:
$\gamma^\bullet = \arcsin\left(\sqrt{\dfrac{\mathfrak{T}-R}{\mathfrak{T}-S}} = \dfrac{1}{3}\right) = 0.616$ (close to the center of the $(\pi/8, \pi/4)$ interval).
Spatial effects also arise in HD spatial simulations before $\gamma^\bullet$, so that the mean-field
estimates $(p^\star)$ underestimate the actual mean payoffs $(\overline{p})$. Much as happens with the
QPD in the transition interval of Fig. 3.1. The standard deviations of the payoffs of
both players are influenced by the increase of γ also as in the QPD, therefore they
decrease monotonically from a high value close to 1.5 with no entanglement down
to zero at $\gamma^\bullet$.

Figure 3.6 deals with the two-parameter, quantum SD(3,2,0,−1) spatial simulation
with variable entanglement factor γ. The beneficiary player B clearly overrates the
charity player A for low values of the entanglement, so that for γ up to just passed $\pi/8$,
$\overline{p}_A$ oscillates slightly below zero, whereas $\overline{p}_B$ oscillates slightly over around 1.5, thus
not far from $(-0.2, 1.5)$, the payoffs in classical NE. If $\gamma = 0$ (or if $\alpha_A = \alpha_B = 0$),
Π becomes factorizable as in the *classical* game with independent strategies (1.7),

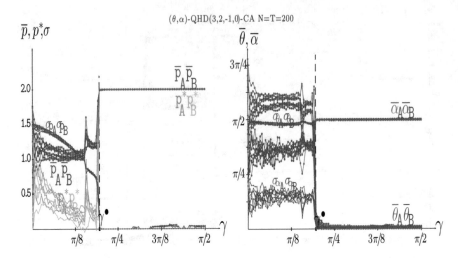

Fig. 3.5 The spatial 2P-QHD(3,2,0,−1) game with variable entanglement factor γ. Five simulations at $T = 200$. Left frame: Mean and standard deviations of the actual payoffs ($\overline{p}, \sigma$), and mean-field payoffs ($p^{\star}$). Right frame: Mean and standard deviation of the quantum parameters

i.e., $\Pi = \mathbf{xy}'$, with $x = \cos^2 \theta_A/2$, $y = \cos^2 \theta_B/2$, which makes the α parameters irrelevant. It is $\theta_A^{NE} = 2 \arccos(0.5) = \pi/2 = 1.570$, $\theta_B^{NE} = 2 \arccos(0.2) = 2.214$, not far from $3\pi/4 = 2.35$.[1] Over $\pi/8$ and before $\pi/4$, the right frame of Fig. 3.6 shows that $\overline{\theta}_A$ gradually tends to zero as γ increases, what is reflected in a stabilization of $\overline{p}_A$ in the left frame. From $\gamma = \pi/4$ and before $\pi/4$, $\overline{\theta}_A = \overline{\theta}_B = 0$, and $\overline{\alpha}_A = \overline{\alpha}_B = \pi/4$, i.e., both players adopt the $\hat{Q}$ strategy. Remarkably, the pair $\{\hat{Q}, \hat{Q}\}$ is the only pair in NE for $\gamma > \pi/4$. This is so because, (i) $p_B^{QQ} = 2$ and $p_B^{QD} = 3\cos^2 \gamma + \sin^2 \gamma = 3 - 2\sin^2 \gamma$, so that $p_B^{QQ} > p_B^{QD}$ for $\gamma > \gamma^{\star} = \pi/4$, and (ii) $p_A^{QQ} = 3$ and $p_A^{DQ} = -1\sin^2 \gamma - \cos^2 \gamma = -1$, so that $p_A^{QQ} > p_A^{DQ}$ $\forall \gamma$.

The mean-field payoff approaches do not coincide with the actual mean payoffs before $\gamma^{\bullet}$ in Fig. 3.6, albeit both actual and mean-field payoffs appear somehow superimposed. Anyhow, said two kinds of payoffs are more distant in the case of player B, which is reflected in higher values of σ_B. Thus, the main finding derived from Fig. 3.6 is that the imitation dynamics in CA spatial simulations enables the emergence of NE in the SD: The NE that there is in the classical game with low entanglement, and that of the pair (Q, Q) with high entanglement. The latter reverses the SD structural trend in favour of the beneficiary player, inducing the highest possible sum of payoffs (social welfare solution).

---

[1]In the five simulations of Fig. 3.6 at $\gamma = 0$ it is: $(\overline{\theta}_A, \overline{\theta}_B) = (1.568, 1.710), (0.741, 2.068), (1.239, 2.256), (1.2401.906), (1.606, 2.086)$, with associated mean-field payoffs: $(p_A^{\star}, p_B^{\star}) = (0.147, 1.503), (0.006, 2.414), (-0.238, 1.928), (0.113, 1.878), (-0.125, 1.456)$, very close to the actual payoffs: $(\overline{p}_A, \overline{p}_B) = (0.149, 1.502), (-0.054, 2.201), (-0.130, 1.813), (0.027, 1.798), (-0.087, 1.456)$, shown in the left frame of the figure.

$\overline{p}, p^*, \sigma$ (θ,α)-QSD(3,2;-1,3;-1,1)-CA   N=T=200

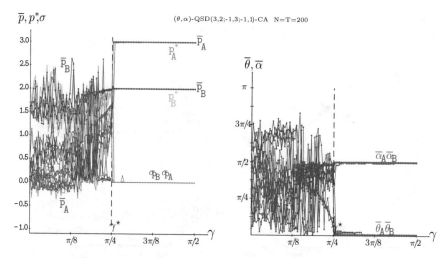

Fig. 3.6 The spatial 2P-QSD(3,2,0,−1) game with variable entanglement factor γ. Five simulations at T = 200. Left frame: Mean and standard deviations of the actual payoffs $(\overline{p}, \sigma)$, and mean-field payoffs (p^*). Right frame: Mean quantum parameters

The ♂, ♀ chessboard:

♂	♀	♂	♀	♂	♀
♀	♂	♀	♂	♀	♂
♂	♀	♂	♀	♂	♀
♀	♂	♀	♂	♀	♂
♂	♀	♂	♀	♂	♀
♀	♂	♀	♂	♀	♂

$\theta_♂, \theta_♀$ T = 1

0	π	0	π	0	π
π	0	π	0	π	0
0	π	π/2	π	0	π
π	0	π	0	π	0
0	π	0	π	0	π
π	0	π	0	π	0

$p_♂, p_♀$ T = 1

0	0	0	0	0	0
0	0	2.5	0	0	0
0	2.5	2.0	2.5	0	0
0	0	2.5	0	0	0
0	0	0	0	0	0
0	0	0	0	0	0

$\theta_♂, \theta_♀$ T = 2

0	π	0	π	0	π
π	π/2	π	π/2	π	0
0	π	π/2	π	0	π
π	π/2	π	π/2	π	0
0	π	0	π	0	π
π	0	π	0	π	0

$p_♂, p_♀$ T = 2

0	2.5	0	2.5	0	0
2.5	2.0	7.5	2.0	2.5	0
0	7.5	2.0	7.5	0	0
2.5	2.0	7.5	2.0	2.5	0
0	2.5	0	2.5	0	0
0	0	0	0	0	0

Fig. 3.7 The spatial BOS game scenario. Far left: The ♂, ♀ chessboard. Centre: A classical example where every player play its preferred choice, except the male player located in the (3,4) cell. Far right: Parameters and payoffs at T = 2

Much as before commented regarding the QPD game, the results shown in Figs. 3.5 and 3.6 regarding the QHD and QSD games with the particular HD and SD payoff sets here adopted are generalizable to any set of HD and SD payoffs. The only variation being the value of the γ^* threshold. So for example, in the CA simulations with (5,3,1,0) HD payoffs in Chap. 10 it is $\gamma^* = \arcsin\left(\sqrt{\frac{5-3}{5-1}} = \frac{1}{2}\right) = \pi/4$.

Figure 3.7 shows a simple spatial example in the classical BOS context where initially every player play its preferred choice ($\theta_♂ = 0 \equiv x = 1, \theta_♀ = \pi \equiv y = 0$), except the male player located in the (3,4) cell which plays $\theta_♂ = \pi/2 \equiv x = 1/2$. Thus at T = 1 only the players adjacent to the male player located in (3,4) get non-zero payoffs. The imitation mechanism spreads $\theta_♂ = \pi/2$ across the male cells, so that both male and female players get non-zero payoffs, with the females getting higher payoffs as her $\theta_♀ = \pi$, i.e., B, choice remains unaltered.

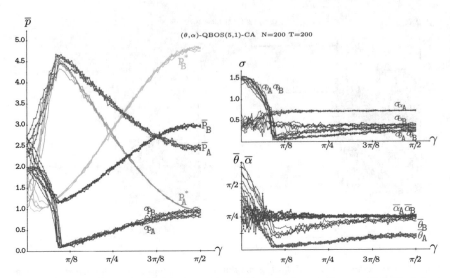

Fig. 3.8 The spatial 2P-QBOS(5,1) game with variable entanglement factor γ. Five simulations at $T = 200$. Left frame: Mean and standard deviations of the actual payoffs $(\overline{p}, \sigma)$, and mean-field payoffs $(p^{\star})$. Right frames: quantum parameters. Standard deviations (upper), mean (lower)

Figure 3.8 deals with the two-parameter, quantum BOS(5,1) spatial simulation with variable entanglement factor γ. The initial increase of the entangling factor from the classical $\gamma = 0$ context, leads to a dramatic bifurcation of the mean payoffs in favor to the player A, reaching a peak not far from (5,1) when γ approaches $\pi/8$. Before, but close, this value of γ, both mean payoffs commence a smooth approach as γ increases, reaching fairly equal values by $\gamma \simeq 3\pi/8$. The bias towards the player B in the model with maximal entangling (pointed out in Sect. 2.1) becomes apparent in Fig. 3.8 for high γ values, specifically for $\gamma > 3\pi/8$. With maximal entangling it is $\overline{p}_A = 2.419$, $\overline{p}_B = 2.949$, which is no a Pareto efficient solution. The bottom right frame of Fig. 3.8 shows that, (i) both $\overline{\alpha}_A$ and $\overline{\alpha}_B$ oscillate nearly $\pi/4$ after a noisy regime for low values of γ, (ii) both $\overline{\theta}_A$ and $\overline{\theta}_B$ initially decrease as γ increases, but close to $\pi/8$ both $\overline{\theta}'s$ commence to increase: $\overline{\theta}_B$ grows faster, reaching $\overline{\theta}_B \simeq \pi/4$ at $\gamma = \pi/2$, whereas $\overline{\theta}_A \simeq \pi/16$ with full entangling.

The left frame and the upper right frame of Fig. 3.8 also show the variation with γ of the standard deviation (σ) of the payoffs and of quantum parameters respectively. The standard deviation (σ) of both payoffs initially plummets from circa 2.0 with no entanglement down to close to zero before $\gamma = \pi/8$, then they recover up to circa 0.85 with full entanglement. This scheme of variation of $\sigma(\gamma)$ applies also in what respect to both θ parameters. The standard deviations of both α's appear fairly stable, being that of player A greater than that of player B.

Figure 3.9 deals with a simulation in the 2P-QBOS scenario of Fig. 3.8 with $\gamma = \pi/2$. The left panel of the figure shows up the evolution to $T = 100$ of the mean values across the lattice of θ and α, and the actual mean payoffs and the mean-field payoff approaches. As a result of the random assignment of the parameter values

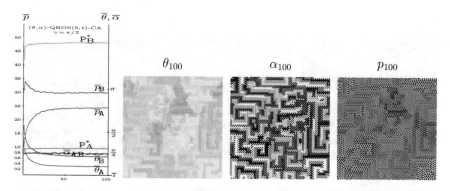

Fig. 3.9 The spatial 2P-QBOS(5,1) game of Fig. 3.8 with $\gamma = \pi/2$. Far left: Dynamics up to $T = 100$. Right: Quantum parameters and payoff patterns at $T = 100$

it is initially: $\overline{\theta}_A \simeq \overline{\theta}_B \simeq \pi/2 = 1.57$, and $\overline{\alpha}_A \simeq \overline{\alpha}_B \simeq \pi/4 = 0.78$. Consequently, the mean-field payoffs are initially: $p_A^*(0) \simeq 1.0$ and $p_B^*(0) \simeq 5.0$. The actual mean payoffs are initially biased as the theoretical payoffs are, but not in such dramatic extent: $\overline{p}_A(0) = 0.95$, $\overline{p}_B(0) = 3.03$. After the first round, both types of players tend to moderate this initial trend. Particularly in the case of the payoff of player A, which quickly grows, stabilizing around $p_A \simeq 2.56$; whereas the payoff of player B descends only to $p_B \simeq 3.06$ In any case, player B outperforms player A. In Fig. 3.9 the θ parameters of both players stabilize in low values, which constitutes an indicator of correlation (as indicated in Sect. 2.1), and, last but not least, the mean α parameter values of both kinds of players remain almost unaltered during the whole simulation, so that they remain notably coincident oscillating around the initial $\alpha \simeq \pi/4 = 0.78$. The maze-like structures emerging for both the parameter and the payoff patterns in the quantum simulation shown in Fig. 3.9 very much differ from the pattern structures that emerge in the purely classical context as commented regarding Fig. 3.13. Incidentally, such as maze-like patterns are much reminiscent of that found in patterns emerging in some reaction-diffusion phenomena [9].

Early patterns in a particular $\gamma = \pi/2$ simulation of Fig. 3.8 are shown in Fig. 3.10. The initial patterns at $T = 1$ shows the unstructured aspect that corresponds to the initial random parameter assignment. But as soon as at $T = 5$ the patterns how a sort of fine-grained *patchwork* aspect, whereas by $T=10$ a maze-like aspect commence to emerge, which is much more defined at $T = 40$, already advancing the general appearance achieved at $T = 100$ as shown in Fig. 3.9.

An example of the dynamics and long-term patterns in the 2P-QBOS scenario of Fig. 3.8 with the very low entanglement $\gamma = \pi/16$ is given in Fig. 3.11. The $\overline{\theta}$ values in the far left frame of this figure, evolve initially in opposite direction from their mean values, close to $\pi/2 = 1.57$, $\overline{\theta}_{\male}$ decreases up to 0.699, whereas $\overline{\theta}_{\female}$ grows up to 2.411. But the ulterior dynamics depletes this high $\overline{\theta}_{\female}$ value, so that it fairly stabilizes it in a low value, and $T = 200$ it is $\overline{\theta}_{\female} = 0.654$. The $\overline{\theta}_{\male}$ parameter decreases up to 0.364 at $T = 200$. The dynamics of both $\overline{\alpha}$ parameter values is smoother compared to that of $\overline{\theta}$, so from the initial $\overline{\alpha} \simeq \pi/4 = 0.78$, at $T = 200$

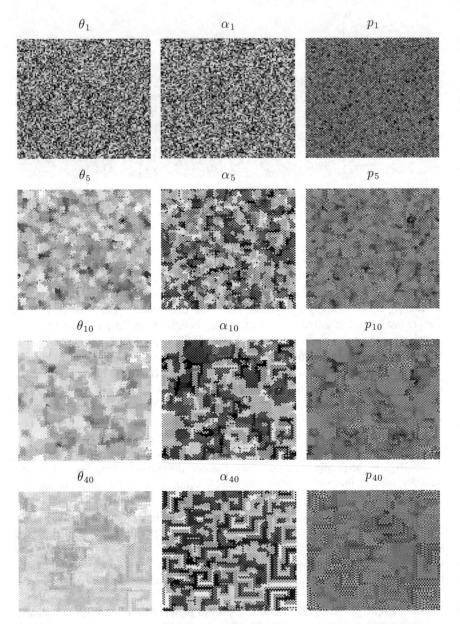

Fig. 3.10 Early patterns in a simulation in the scenario of Fig. 3.8. From top to bottom: $T = 1$, $T = 5$, $T = 10$, $T = 40$. Increasing grey levels indicate increasing values

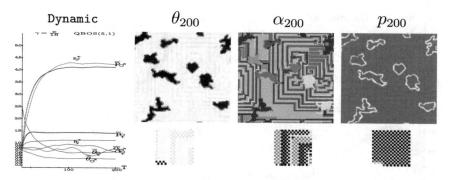

Fig. 3.11 The spatial 2P-QBOS(5,1) game of Fig. 3.8 with $\gamma = \pi/16$. Far left: Dynamics up to $T = 200$. Right: Quantum parameters and payoff patterns at $T = 200$

it is: $\overline{\alpha}_{\sigma} = 0.794$ and $\overline{\alpha}_{\varphi} = 0.661$. The parameter dynamics in Fig. 3.11 quickly drives the $\overline{p}$ mean payoffs to distant values, so that from approximately $T = 10$, the mean female payoff becomes stabilized around $\overline{p}_{\varphi} = 1.41$, and from approximately $T = 100$ the male payoff is stabilized around $\overline{p}_{\sigma} = 4.12$. Thus, the dynamics clearly favours the process to the male player, a kind of result contrasting with the bias towards the female player in the EWL model with full entangling. As a general rule, due to the spatial heterogeneity in the parameter values, the mean-field estimations of the mean payoffs are different to the actual ones. In the particular case of Fig. 3.11, the mean-field payoffs p^* estimate fairly well the trends of the actual mean payoffs $\overline{p}$, but p^*_{σ} over-estimates $\overline{p}_{\sigma}$ and p^*_{φ} under-estimates $\overline{p}_{\varphi}$. Figure 3.11 also shows the snapshots of the parameter and payoff patterns at $T = 200$, both for the full lattice and its zoomed 23×23 central part. Both maze-like structures and coordination clusters may be appreciated at some extent in the parameter patterns. In the θ parameter pattern, the maze-like structure is tenuously visible, whereas the coordination clusters stands up. In the α parameter pattern, the maze-like structure is predominant but also coordination clusters are appreciated. Coordination, favouring the male players, predominates in the payoff pattern, where the disagreement becomes apparent in the form of clear (meaning low payoff) closed narrow zones. For a given value of γ, e.g., $\gamma = \pi/16$ in Fig. 3.11, the variation of the initial parameter pattern configurations only alters minor details of the evolving dynamics, the main features of the dynamics and of the long-term patterns are preserved. As a general rule, as γ increases the maze-like appearance of the parameter patterns in the long-term becomes dominant, and even the payoffs patterns adopt this appearance. As the parameter patterns become fairly only maze-like (with no coordination clusters) as γ increases, the mean-field approaches p^* tend to poorly reflect the actual mean payoffs of both kinds of players. The spatial structure marks the difference, so that a mean-field approach tends to be less indicative of the actual spatial mean payoffs. A very good example of this may be that of $\gamma = 3\pi/8$, in which case both actual mean payoffs quickly converge to an equal value, whereas their corresponding p^* estimations stand far beyond this common actual mean payoff value.

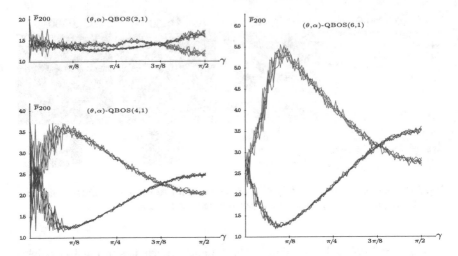

Fig. 3.12 Mean payoffs in the spatial 2P-QBOS(R, r) game. Variable entanglement factor γ. Five simulations at $T = 200$. Player A red-marked, player B blue-marked. Upper left frame: $R = 2, r = 1$, lower left frame: $R = 4, r = 1$, right frame: $R = 6, r = 1$

As before commented regarding the other game-types here studied, the results reported here regarding the QBOS game with the (5,1) particular parameters are generalizable to any set of BOS parameters. This is ascertained in the CA simulations with (2,1), (4,1) and (6,1) BOS parameters shown in Fig. 3.12. The features of these non-(5,1) QBOS simulations are commented regarding the simulations with probabilistic updating in Fig. 6.4.

Classical QBOS-CA Simulations

As stated in the introduction of the EWL model, if $\alpha = 0$ for both players the joint probabilities in the 2P model factorize so that the quantum component of the game vanishes and we are referred to the classical game employing independent strategies. Figure 3.13 deals with a (5,1)-BOS spatial simulation in such an $\alpha = 0$ classical scenario. The far left frame shows the evolution up to $T = 200$ of the mean values across the lattice of both θ parameters, as well as of the mean payoffs. As a result of the random assignment of θ-values it is initially $\overline{\theta}_A \simeq \overline{\theta}_B = \pi/2 \simeq 1.57$, and the mean payoffs commence at the arithmetic mean of the payoff values, i.e., $p^+ = (R+r)/4 = 1.5$. After the first round, both types of players drift to their preferred θ levels, i.e., nil for player A and π for the player B, and as consequence both payoffs plummet at $t = 2$, where $\overline{\theta}_A = 0.73, \overline{\theta}_B = 2.45, \overline{p}_A) = 0.78, \overline{p}_B = 0.84$. But immediately the θ-drift becomes moderated, and both $\overline{p}$ recover. After a fairly short transition period, the θ-values, and consequently the payoffs, stabilize. In the simulation of Fig. 3.13, the stable values of the mean payoffs are $\overline{p}_B = 2.96$ and $\overline{p}_A = 1.97$, a pair of payoff values which is not accessible in the uncorrelated formulation of the game, because the equation of the parabola closing the payoffs region in such scenario is: $3(p_A - p_B)^2 - 16(p_A + p_B) + 48 = 0$, so that the maximum feasible payoff of player B

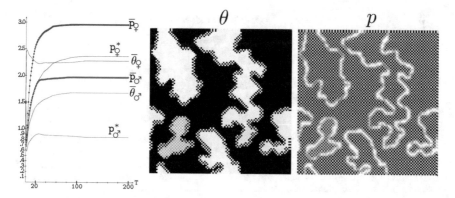

Fig. 3.13 A simulation of the purely classical (5,1)-BOS cellular automaton. Left: Dynamics up to $T = 200$ of the mean θ parameter values, of the mean payoffs $(\overline{p})$ and of the mean-field payoffs (p^*). Right: The patterns at $T = 200$ of θ and p. Increasing gray levels indicate increasing values

with fixed $p_A = 1.97$ is the value on the parabola, $p_B = 1.15$, which 2.96 notably exceeds. The actual mean payoffs of both kinds of players notably overestimate the mean-field payoffs in the left panel of Fig. 3.13, as an effect of the emergence of the *agreement* clusters, shown in the central panel of Fig. 3.13 as black (high θs, low x and y, so BB drift) and white (FF) regions, with interfaces of disagreement among the clusters translated into white borders in the payoffs snapshot in the far right panel of Fig. 3.13.

The study of spatial games was pioneered by Nowak and May [10, 11] with regards to the Prisoner's Dilemma (PD) [12]. They concluded in their original work that spatial structure (or territoriality) can facilitate the survival of cooperators. Thus, the spatialized PD has proved to be a promising tool to explain how cooperation can hold out against the ever-present threat of exploitation. The notable case of self-organization in the BOS just presented appears as a novel example of the boosting effect induced by the spatial ordered structure, which allows the access to payoffs which are feasible only with correlated strategies (Table 1.4 in Sect. 1.1).

In the 2P quantum matching pennies (MP) simulations of Fig. 3.14 the actual mean payoffs of both players are both close to zero regardless of γ, albeit that of player A is slightly positive and that of player B is slightly negative. The mean quantum parameters in the right frame of Fig. 3.14 are set very close to their middle-level values, i.e., $\overline{\theta}_A = \overline{\theta}_B = \pi/2$, $\overline{\alpha}_A = \overline{\alpha}_B = \pi/4$. From the join probability with middle-level parameters given in Eq. 2.5, the mean-field approach yields, $p_A^* = \sin \gamma$, $p_B^* = -\sin \gamma$, dramatically far from the actual mean payoffs as γ increases. Thus, the players *autoorganize* their strategies in such a way that the expectable zero actual mean payoffs are virtually achieved.

The standard deviations of the payoffs of both players in the simulations of Fig. 3.14 remain close over 0.50 regardless of γ. The standard deviation in a discrete uniform distribution with $\{a, b\}$ domain is: $\sigma\{a, b\} = (b - a)/2$. Thus, assuming independence, the standard deviation of the average of the four payoffs computed

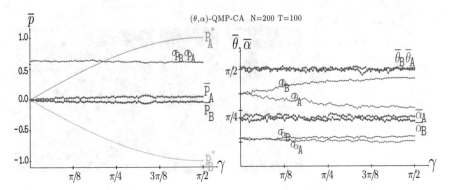

Fig. 3.14 The 2P-QMP game with variable γ in spatial simulations at $T = 100$. Left: Mean and standard deviations of the actual payoffs $(\overline{p}, \sigma)$, and mean-field payoffs $(p^\star)$. Right: Mean and standard deviations of the quantum parameters. In both frames the standard deviations (σ) of the plotted magnitudes are shown

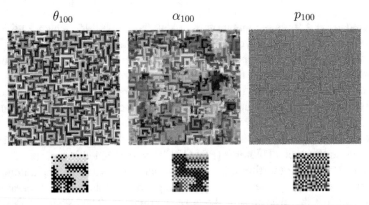

Fig. 3.15 The quantum parameter and payoff patterns in the 2P-QMP $\gamma = \pi/2$ scenario of Fig. 3.14. Upper: The whole 200×200 lattice. Lower: Zoom of the 21×21 central part. Increasing gray levels indicate increasing values of the magnitudes

for every player in the collective simulation will be $\sigma_p\{-1, 1\}/4 = 0.5$. The proximity of the actual standard deviations of the payoffs in Fig. 3.14 to said $\sigma_p\{-1, 1\} = 0.5$ indicates that the payoffs are distributed in a fairly discrete uniform manner. In other words, that the actual payoffs tend to be either $+1$ or -1.

The parameter patterns at $T = 100$ shown in Fig. 3.15 exhibit a characteristic maze-like aspect, particularly the θ-pattern. These spatial structures, far from the initial random assignments, somehow explain the divergence found between the mean-field and the actual mean payoffs in Fig. 3.14. The payoff pattern in Fig. 3.15 does not exhibit such apparent maze-like aspect, but a kind of fine-grain texture. Nevertheless, observed in detail, some fuzzy maze-like structures are revealed.

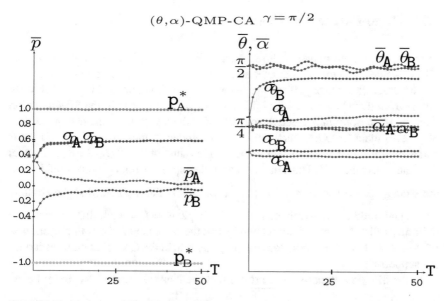

Fig. 3.16 Dynamics up to $T = 50$ of the mean payoffs (left) and mean quantum parameters (right) in the 2P-QMP $\gamma = \pi/2$ scenario of Fig. 3.14

The graphs in Fig. 3.16 show the dynamics up to $T = 50$ of the mean payoffs and mean quantum parameters in the 2P-QMP $\gamma = \pi/2$ scenario of Fig. 3.14. The left frame shows how, (i) both actual mean payoffs converge towards zero, from a initial notable predominance of p_A that only remains in a very low extent at the long-term, and (ii) the mean-field payoffs remain unaltered, with $p_A^\star = \sin \pi/2 = 1$, $p_B^\star = -\sin \pi/2 = -1$ in the full entanglement scenario of Fig. 3.16. The standard deviation in a continuous uniform distribution in the $[a, b]$ interval is: $\sigma[a, b] = (b - a)/2\sqrt{3}$, so that the standard deviation of the average of the four payoffs computed in the collective simulation will be $\sigma_p[-1, 1] = 2/4\sqrt{3} = 0.289$. The standard deviations of both payoffs start from approximately said $\sigma_p[-1, 1] = 0.289$ and quickly reach a $\sigma_p = 0.597$ value, virtually that in the left frame of Fig. 3.14 at $T = 100$, i.e., $\sigma_p(\gamma = \pi/2) = 601$. The right frame of Fig. 3.16 shows how the actual mean quantum parameters remain fairly unaltered from their initial mean values. Also the standard deviations of the $\alpha \in [0, \pi/2]$ parameters remain close to their initial value $\sigma_\alpha[0, \pi/2] = \pi/4\sqrt{3} = 0.453$. The standard deviations of the α parameters remain close to said $\sigma_\alpha = 0.453$ not only in Fig. 3.16, but for every value of γ as revealed in Fig. 3.14. Contrary to this, the standard deviations of the θ parameters are affected by the dynamics. Both start from the σ of a continuous uniform distribution in the $[0, \pi]$ interval: $\sigma_\theta = \pi/2\sqrt{3} = 0.907$, then σ_{θ_A} increases in a low extent, whereas σ_{θ_B} increases in a greater extent, towards $\pi/2$. This bifurcation in the values of both σ_{θ_A} and σ_{θ_B} reached at the long term is also appreciated in Fig. 3.14. The standard deviation in a discrete uniform distribution in the $\{0, \pi\}$ domain is $\sigma_\theta = \pi/2$. Thus, it seems that the θ_B parameter tends to its extreme values in the evolving dynamics.

3.3 Three-Parameter Strategies

In this section, both players follow general SU(2) strategies, i.e., three-parameter (3P) strategies with the β parameter active in the U structure given in Eq. (2.3).

The structure of the payoffs shown in the 3P-QPD simulations of Fig. 3.17 notably differs from that shown in its 2P-QPD counterpart in Fig. 3.1: In the 3P simulation there are no *discontinuities* and the payoffs of both players are coincident regardless of γ. Two main features characterize the parameter graphs in Fig. 3.17: (*i*) the $\bar{\theta}$ parameters drift to π which makes irrelevant the values of $\bar{\alpha}$, and (*ii*) the $\bar{\beta}$ parameters oscillate nearly the $\pi/4$. Thus, in the mean-field approach, both players would adopt the strategy $\hat{\mathfrak{x}} = U(\pi, \alpha, \pi/4) = \begin{pmatrix} 0 & e^{i\pi/4} \\ -e^{-i\pi/4} & 0 \end{pmatrix}$. Please, note that the strategy with equal-middle level parameters, i.e., $\theta = \pi/2$, $\alpha = \beta = \pi/4$, that generates an uniform Π distribution and consequently also the same payoff for both players $p = (\mathfrak{T} + R + P + S)/4 = 2.75$, does not play any role in the CA simulations as it does not support NE.

In the 3P-QPD scenario, it is proved that the pair $(\hat{\mathfrak{x}}, \hat{\mathfrak{x}})$ is in NE for γ below the threshold $\gamma^{\#} = \arcsin\left(\sqrt{\dfrac{P - S}{\mathfrak{T} + P - R - S}} = \dfrac{1}{3}\right) = 0.615$ [13].[2] In the 3P CA simulations of Fig. 3.17, $\gamma^{\#}$ appears to be a landmark for the appearance of instability of the $\bar{\theta}$ parameters, which is reflected in an erratic behaviour of the mean-field payoffs and in the rise of non-zero values of the standard deviation of both actual

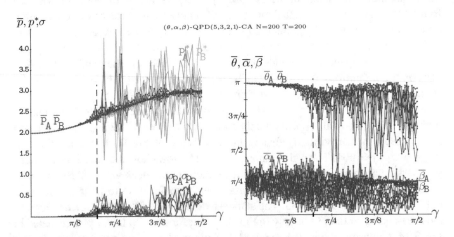

Fig. 3.17 The three-parameter QPD(5,3,2,1)-CA with variable entanglement factor γ. Five simulations at $T = 200$. Left: Mean and standard deviations of the actual payoffs $(\bar{p}, \sigma)$, and mean-field payoffs $(p^{\star})$. Right: Mean quantum parameters

---

[2]The reference [14] is also relevant at this respect, but the occasional reader should be warned about the variation of the α and β parameters in the $[-\pi, \pi]$ interval instead of in $[0, \pi/2]$, as proposed for α in the seminal EWL paper.

payoffs. But heavy spatial effects seem to be able to compensate these turbulences in order to make possible that the pair $(\hat{\mathfrak{X}}, \hat{\mathfrak{X}})$ roughly dominates the scene regardless of γ. Thus, the actual mean payoffs increase all along the γ variation, even for $\gamma > \gamma^{\#}$, according to the equation $p = P + (R - P) \sin^2 \gamma = 2 + \sin^2 \gamma$ induced by $\Pi^{\hat{\mathfrak{X}},\hat{\mathfrak{X}}} = \begin{pmatrix} \sin^2 \gamma & 0 \\ 0 & \cos^2 \gamma \end{pmatrix}$.[3] Figure 4.24 shows the emergence of the $\gamma^{\#}$ threshold in the 3P-QPD game when computing the best response to defection via CA simulation.

In the 3P-QHD(3,2,0,−1) simulations of Fig. 3.18, the payoff graphs show a very noisy appearance with strong pattern effects before the threshold $\gamma^{\star}$. In stark contrast with this, with entanglement over $\gamma^{\star}$ no pattern effects arise, so that the actual and mean-field payoffs coincide, and the payoffs of both players equalize and monotonically increase their common value in a crisp manner as γ increases. In parallel to this, the standard deviations (σ) of the actual payoffs decrease in a fairly smooth way as the entanglement increases from $\sigma_p \simeq 1.5$ with no entanglement down to nearly zero at $\gamma = \gamma^{\star}$. The $\bar{\alpha}$ and $\bar{\beta}$ mean parameters of both players oscillate close to $\pi/4$ all along the γ variation, at variance with this, both $\bar{\theta}$ behave erratically before $\gamma^{\star}$ and stabilize at π after $\gamma^{\star}$ in the right frame of Fig. 3.18. The standard deviations of the quantum parameters are not included in the right frame of Fig. 3.18, because they would be indistinguishable from the mean values. They oscillate before $\gamma^{\star}$ close below $\pi/2$ in the case of the θ parameters and around $\pi/8$ in the case of the α and β parameters.

Figure 3.19 deals with the results achieved in three-parameter strategies QSD-CA simulations. The actual mean payoffs of both players monotonically increase their values as the entanglement increases, in the case of player A from approximately zero up to approximately 1.5, in the case of player B from approximately 1.5 up to approximately 2.25. Thus, player B overrates player A all along the γ variation, albeit in a lower degree as γ grows. For low γ, the mean-field payoff estimations fit fairly well the actual mean payoffs, particularly regarding the player A, but as γ grows heavy spatial effects emerge, so that the actual mean payoffs $\bar{p}$ become fairly stabilized, in contrast with the variable behaviour of p^{*}, regarding the player A. No particular structure becomes apparent in the parameter patterns in the right frame of Fig. 3.19, which corresponds to a rather erratic behaviour of the mean-field estimations. The standard deviation of the payoffs of player A turns out rather

$$^3(\hat{U}_A \otimes \hat{U}_B)|\psi_i\rangle = \begin{pmatrix} 0 & 0 & 0 & e^{i\pi/2} = i \\ 0 & 0 & -1 & 0 \\ 0 & -1 & 0 & 0 \\ e^{-i\pi/2} = -i & 0 & 0 & 0 \end{pmatrix} \begin{pmatrix} \cos\frac{\gamma}{2} \\ 0 \\ 0 \\ i\sin\frac{\gamma}{2} \end{pmatrix} = \begin{pmatrix} -\sin\frac{\gamma}{2} \\ 0 \\ 0 \\ -i\cos\frac{\gamma}{2} \end{pmatrix},$$

$$|\psi_f\rangle = \begin{pmatrix} \cos\frac{\gamma}{2} & 0 & 0 & -i\sin\frac{\gamma}{2} \\ 0 & \cos\frac{\gamma}{2} & i\sin\frac{\gamma}{2} & 0 \\ 0 & i\sin\frac{\gamma}{2} & \cos\frac{\gamma}{2} & 0 \\ -i\sin\frac{\gamma}{2} & 0 & 0 & \cos\frac{\gamma}{2} \end{pmatrix} \begin{pmatrix} -\sin\frac{\gamma}{2} \\ 0 \\ 0 \\ -i\cos\frac{\gamma}{2} \end{pmatrix} = \begin{pmatrix} -\cos\frac{\gamma}{2}\sin\frac{\gamma}{2} - \sin\frac{\gamma}{2}\cos\frac{\gamma}{2} \\ 0 \\ 0 \\ i\sin^2\frac{\gamma}{2} - i\cos^2\frac{\gamma}{2} \end{pmatrix}$$

$$= \begin{pmatrix} -\sin\gamma \\ 0 \\ 0 \\ -i\cos\gamma \end{pmatrix}.$$

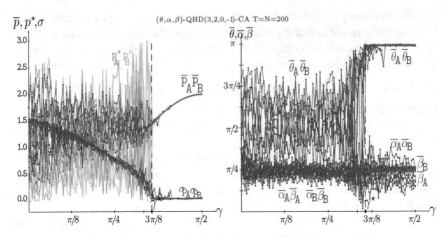

Fig. 3.18 The three-parameter $(3,2,0,-1)$-QHD-CA with variable entanglement factor γ. Five simulations at $T = 200$. Left: Mean and standard deviations of the actual payoffs $(\overline{p}, \sigma)$, and mean-field payoffs $(p^\star)$. Right: Mean quantum parameters

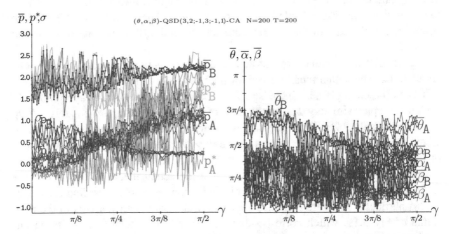

Fig. 3.19 The three-parameter QSD-CA with variable entanglement γ. Five simulations at $T = 200$. Left: Mean and standard deviations of the actual payoffs $(\overline{p}, \sigma)$, and mean-field payoffs $(p^\star)$. Right: Mean quantum parameters

indistinguishable from that of the actual mean payoff of player A, whereas that of player B becomes much more identifiable in the left frame of Fig. 3.19, decreasing for every simulation down to around 0.25.

Figure 3.20 shows the long-term mean payoffs and quantum parameters in the BOS(5,1) scenario of Fig. 3.8, but in the three-parameter strategies model. At variance with what happens in the two parameter scenario (Fig. 3.8), the mean payoffs are not dramatically altered by the variation of γ. The overall effect of the increase of γ being a moderation in the variation of the $\overline{p}$ values that oscillate nearly over 2.5. Please, note

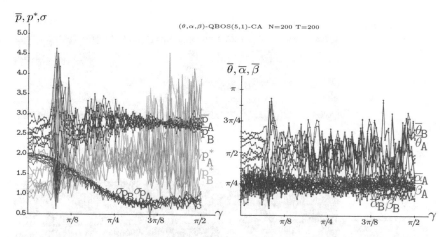

Fig. 3.20 The three-parameter QBOS(5,1)-CA with variable entanglement γ. Five simulations at $T = 200$. Left: Mean and standard deviations of the actual payoffs ($\bar{p}, \sigma$), and mean-field payoffs ($p^\star$). Right: Mean quantum parameters

that in the three-parameter strategies model, there is not any bias favoring the player B, so that there is not any tendency to any player over rating the other one. The spatial ordered structure induces a kind of self-organization effect, which allows to achieve fairly soon, approximately at $T = 20$, pairs of mean payoffs that are accessible only with correlated strategies in the two-person game. Recall that the maximum equalitarian payoff in the uncorrelated context is $p^+=(R+r)/4 = 1.5$, whereas the payoffs in the simulations of Fig. 3.20 with high entanglement factor reach values over 2.5, not far from the maximum feasible equalitarian payoff $p^= = (R + r)/2 = 3.0$.

Examples of the dynamics and long-term patterns in the scenario of Fig. 3.20 with $\gamma = \pi/2$ are shown in the four simulations of Fig. 3.21 (in [15] also the case of $\gamma = \pi/4$ is treated). Accordingly with the absence of bias favouring any of the types of players in the three-parameter simulations, both mean payoffs and parameter values evolve in a fairly parallel way in the referred dynamics, albeit the parameter patterns show a rich structure, enclosing both maze-like and nucleation regions, so that the mean-field estimations do not fit the actual payoffs. The mean α and β parameter values of both kinds of players remain fairly stable in the simulations in Fig. 3.21, not far from the initial $\pi/4 = 0.78$ mean value. In contrast, the $\bar{\theta}$ parameters exhibit a greater diversity in their behaviour. Thus, in the simulations -1- and -2- the $\bar{\theta}$ parameters tend to stabilization fairly soon, whereas in the simulations -3- and -4- appear increasing and decreasing tendencies. In the simulation -3- the $\bar{\theta}$ parameters still grow beyond $T = 200$, and nearly stabilize at $T = 900$, reaching at $T = 1000$: $\bar{\theta}_\female = 2.50, \bar{\theta}_\male = 2.39$. In the simulation -4-, the $\bar{\theta}$ parameters decrease very slowly, reaching at $T = 1000$: $\bar{\theta}_\female = 0.33, \bar{\theta}_\male = 0.36$.

Early patterns in a particular $\gamma = \pi/2$ simulation of Fig. 3.20 are shown in Fig. 3.22. The initial patterns at $T = 1$ shows the unstructured aspect that corresponds to the initial random parameter assignment. But as soon as at $T = 5$, the

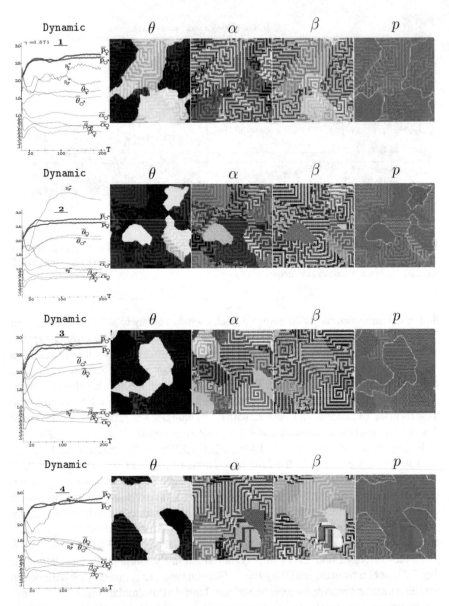

Fig. 3.21 Four simulations of a three-parameter QBOS(5,1)-CA game. Far Left: Evolving mean parameters, actual mean payoffs and mean-field payoffs. Center: Parameter patterns at $T = 200$. Far Right: Payoff patterns at $T = 200$. Increasing gray levels indicate increasing values

patterns show a sort of fine-grained *patchwork* aspect, and by $T = 40$ a maze aspect commence to emerge in the α and β patterns, whereas nucleation seems to predominate in what respect to the θ and payoff patterns. Please, note that at variance with this, in Fig. 3.10 no nucleation features are seemingly appreciable. These tendencies are reinforced when progressing in the dynamics, so that at $T = 200$ maze-like structures predominate in the α and β parameter patterns, whereas coordination clusters are predominant regarding the θ parameter, much like it happens in the classical spatialized BOS [16–18]. In any case, both maze-like structures and coordination clusters maybe appreciated at some extent in every parameter patterns and in the payoffs patterns, where the interfaces of disagreement in the parameters (leading to low $|\psi_1|^2$ and $|\psi_4|^2$) turn out tenuously visible.

The main features of the plots in the Fig. 3.20 have been checked to be preserved with the BOS parameters (2,1), (4,1) and (6,1). In these three cases, the general form of the $\overline{p}(\gamma)$ plots is that shown here for (5,1) in Fig. 3.20, reaching long-term $\overline{p}$ values not far from $(R + r)/2$. Thus, with $R=2$ around 1.5, with $R=4$ around 2.25, and with $R=6$ around 3.25.

In the 3P quantum matching pennies spatial simulations of Fig. 3.23, the actual mean payoffs of both players are both close to 0.5 regardless of γ, with $\overline{\theta}_A \simeq \overline{\theta}_B = \pi/2$, as in Fig. 3.14. But it turns out that $\overline{\alpha}_A \simeq \frac{\pi}{4} + \frac{\pi}{8} \sin \gamma, \overline{\beta}_A \simeq \frac{\pi}{4} - \frac{\pi}{8} \sin \gamma, \overline{\alpha}_B \simeq \overline{\beta}_A$, $\overline{\beta}_B \simeq \overline{\alpha}_A$. In the mean-field approach with $\hat{U}_A = U(\frac{\pi}{2}, \frac{\pi}{4} + \frac{\pi}{8} \sin \gamma, \frac{\pi}{4} - \frac{\pi}{8} \sin \gamma)$ and $\hat{U}_B = U(\frac{\pi}{2}, \frac{\pi}{4} - \frac{\pi}{8} \sin \gamma, \frac{\pi}{4} + \frac{\pi}{8} \sin \gamma)$ produces $|\psi_f\rangle = \frac{1}{2} \begin{pmatrix} -\sin \gamma + i \cos \gamma \\ -1 \\ -1 \\ \sin \gamma - i \cos \gamma \end{pmatrix}$,

so that $\Pi^\star = \frac{1}{4} \begin{pmatrix} 1 & 1 \\ 1 & 1 \end{pmatrix}$, leading to $p_A^* = p_B^* = 0$, mean-field payoffs fairly coincident with the actual mean payoffs ($\overline{p}$). Incidentally, middle level in all parameters also lead to uniform Π in the 3P-model, but unexpectedly the referred divergent $(\alpha(\gamma), \beta(\gamma))$-parameter structure emerges in the spatial dynamics.

In Fig. 3.23, the standard deviations (σ) of the actual payoffs and quantum parameters remain unaffected by γ close to the value of the deviation of a continuous uniform distribution in their domain of definition. Thus, the σ_α and σ_β remain close to $\sigma[0, \pi/2] = \pi/4\sqrt{3} = 0.453$, and, at variance with what happens in Fig. 3.14, in Fig. 3.23 the σ_θ remain at the $\sigma[0, \pi] = \pi/2\sqrt{3} = 0.907$ level. The standard deviations of the payoffs of both players in the simulations of Fig. 3.23 remain, as in Fig. 3.14, close over 0.5 regardless of γ.

The payoff and parameter patterns at $T = 100$ shown in Fig. 3.24 regarding the 3P-model resemble those shown in Fig. 3.15 regarding the 2P-model. Thus, the parameter patterns exhibit a characteristic maze-like aspect (particularly the θ-pattern) and the payoff pattern a fine-grain texture.

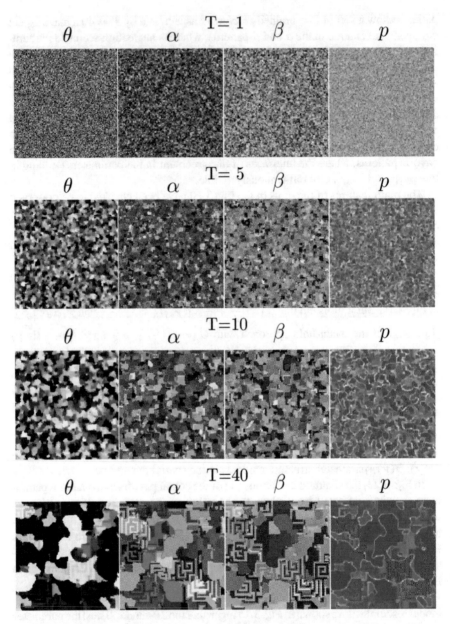

Fig. 3.22 Early patterns in a simulation in the scenario of Fig. 3.20. From top to bottom: $T = 1$, $T = 5$, $T = 10$, $T = 40$. Increasing grey levels indicate increasing values

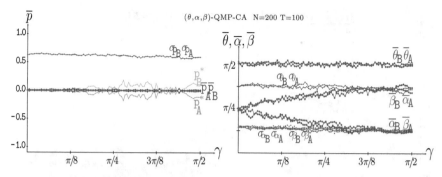

Fig. 3.23 The 3P-QMP game with variable γ in spatial simulations at $T = 100$. Left frame: Mean actual payoffs ($\overline{p}$), and mean-field payoffs ($p^\star$). Right frame: Mean quantum parameters. In both frames the standard deviations (σ) of the plotted magnitudes are shown

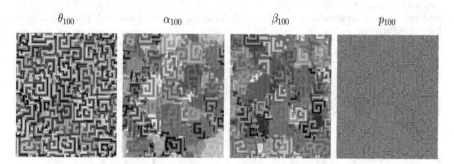

Fig. 3.24 The quantum parameter and payoff patterns in the 3P-QMP $\gamma = \pi/2$ scenario of Fig. 3.23

3.4 The Marinatto-Weber Models

Some variations of the EWL model were studied soon after its introduction, in particular by Marinatto and Weber [19]. Just to cite two of such a MW-variations, (i) the operator $J^\dagger$ may not be applied, and (ii) $|\Psi_0\rangle = |11\rangle = (0\ 0\ 0\ 1)'$ may be stated initially instead of $|\Psi_0\rangle = |11\rangle = (1\ 0\ 0\ 0)'$ as done in the EWL model. With $|\Psi_0\rangle = |11\rangle$ it is, $|\Psi_i\rangle = \hat{J}|11\rangle = \left(\cos\frac{\gamma}{2}\ 0\ 0\ i\sin\frac{\gamma}{2}'\right)$. Thus, the values of the no null elements of Eq. 2.2 are interchanged.

Figure 3.25 deals with a spatial simulation of the 2P-QPD(5,3,2,1) game starting from $|\Psi_0\rangle = |11\rangle$. The structure of the $\overline{p}$ graphs in Fig. 3.25 does not resemble that in 2P-QPD simulations in Fig. 3.1. Quite unexpectedly, the form of $\overline{p}(\gamma)$ in Fig. 3.25 resembles that shown in the 3P-QPD spatial simulations in Fig. 3.17. Even beyond, the absence of pattern effects in Fig. 3.25 (only minor perturbations appear in the $(\gamma^\star, \gamma^\bullet)$ interval) relates its $\overline{p}(\gamma)$ graph to that in the 3P-QPD network simulations shown later in Fig. 5.5.

In Fig. 3.25 the values of the mean α parameters tend to oscillate around $\pi/4$, whereas those of the θ parameters tend to the zero value. In a quantum game from

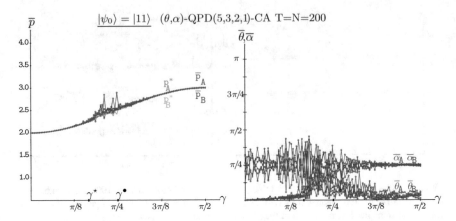

Fig. 3.25 The 2P-QPD(5,3,2,1) spatial game with variable entanglement factor γ starting from $|\Psi_0\rangle = |11\rangle$. Five simulations at $T = 200$. Left: Mean payoffs and mean-field approaches. Right: Mean quantum parameters

$|\Psi_0\rangle = |11\rangle$ with parameters $\theta_A = \theta_B = 0$, and $\alpha_A = \alpha_B = \pi/4$, it is, $\hat{U}_A = \hat{U}_B =$

$$\hat{U}(0, \pi/4) = \begin{pmatrix} e^{i\pi/4} & 0 \\ 0 & e^{-i\pi/4} \end{pmatrix} = e^{i\pi/4} \begin{pmatrix} 1 & 0 \\ 0 & -i \end{pmatrix}, \text{ so that } (\hat{U}_A \otimes \hat{U}_B)|\psi_i\rangle = e^{i\pi/2} \begin{pmatrix} 1 & 0 & 0 & 0 \\ 0 & -i & 0 & 0 \\ 0 & 0 & -i & 0 \\ 0 & 0 & 0 & -1 \end{pmatrix}$$

$$\begin{pmatrix} i\sin(\gamma/2) \\ 0 \\ 0 \\ \cos(\gamma/2) \end{pmatrix} = i \begin{pmatrix} i\sin(\gamma/2) \\ 0 \\ 0 \\ -\cos(\gamma/2) \end{pmatrix}. \text{ Consequently, } |\psi_f\rangle = \begin{pmatrix} -\sin\gamma \\ 0 \\ 0 \\ i\cos\gamma \end{pmatrix},^4 \text{ which induces}$$

$\Pi = \begin{pmatrix} \sin^2\gamma & 0 \\ 0 & \cos^2\gamma \end{pmatrix}$, the same joint probability distribution termed $\Pi^{\hat{x},\hat{x}}$ in the 3P-QPD spatial simulations in Fig. 3.17.

Figure 3.26 deals with a spatial simulation of the 2P-QHD(3,2,$-$1,0) game starting from $|\Psi_0\rangle = |11\rangle$. For high values of γ, approximately from $\gamma = 3\pi/8$, $\bar{\theta}_A = \bar{\theta}_B = 0$, and $\bar{\alpha}_A = \alpha_B = \pi/4$, as in Fig. 3.25. Moreover, after $\gamma = 3\pi/8$ the spatial effects vanish, so that the actual mean payoffs follow the equation: $p_A = p_B = 2\sin^2\gamma$.

Four pair of strategies in NE are reported in the QSD seminal reference [20] in the full entangled two-parameter model starting from $|11\rangle$. The four NE pairs induce the $(3, 2)$ payoffs and have in common that: $\alpha_A = 0$ and $\alpha_B = \pi/2$. Figure 3.27 shows that CA simulations tend to induce $\bar{\alpha}_A = 0$ and $\bar{\alpha}_B = \pi/2$ for not low γ and select the NE pair with $\theta_A = 3\pi/4$ and $\theta_B = \pi/4$ for high entanglement. The most distinctive differences of Fig. 3.27 compared to Fig. 3.6 are those of the fairly monotonic increase of $\bar{p}_A(\gamma)$ and the lower variation of the five values of $\bar{p}_A$ for low

---

$$^4|\psi_f\rangle = J^\dagger i \begin{pmatrix} i\sin(\gamma/2) \\ 0 \\ 0 \\ -\cos(\gamma/2) \end{pmatrix} = i \begin{pmatrix} i\cos(\gamma/2)\sin(\gamma/2) + i\sin(\gamma/2)\cos(\gamma/2) \\ 0 \\ 0 \\ \sin^2(\gamma/2) - \cos^2(\gamma/2) \end{pmatrix}.$$

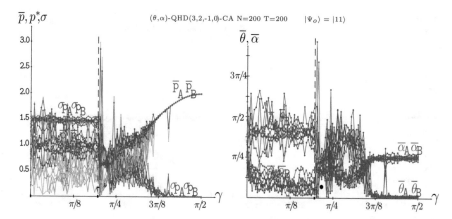

Fig. 3.26 The spatial 2P-QSD(3,2,−1,0) from $|\Psi_0\rangle = |11\rangle$. Variable entanglement factor γ Five simulations at $T = 200$. Left: Mean payoffs and mean-field payoffs. Right: Mean quantum parameters

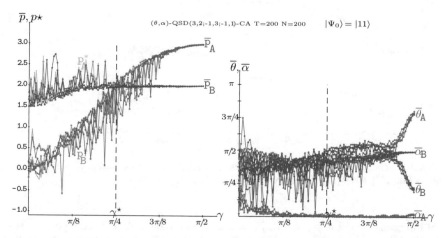

Fig. 3.27 The 2P-QSD-CA from $|\psi_o\rangle = |11\rangle$. Variable entanglement γ. Five simulations at $T = 200$. Left: Mean actual payoffs and mean-field payoff approaches. Right: Mean quantum parameters.

values of γ, together with a very good fit of the mean-field payoff approaches to the actual mean payoffs.

In the reference [20] two NE are reported in the 3P-QSD game with full entangling when considering the initial density matrices $\rho_i = (|00\rangle\langle00| + |11\rangle\langle11|)/2$ and $\rho_i = (|01\rangle\langle01| + |10\rangle\langle10|)/2$. The strategies in NE in these scenarios have in common the parameters $\theta_A = \theta_B = \pi/2$ and $\alpha_B + \beta_B = \pi/2$; with the first density matrix it is $\alpha_A + \beta_A = \pi$ and with the second one $\alpha_B + \beta_B = 0$. The results of their corresponding CA simulations are shown here in Fig. 3.28. These simulations detect the refereed 3P-NE in a straightforward manner, i.e., free of spatial effects, reporting in both cases non-negative payoffs to the Samaritan player. In the particular case

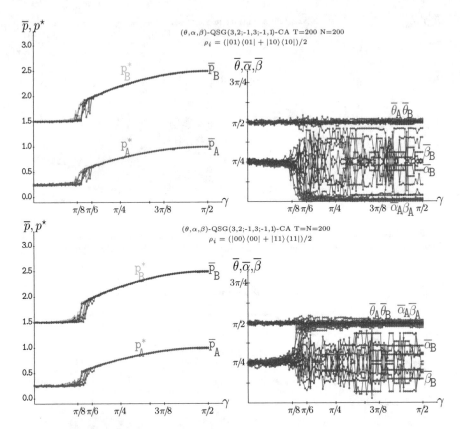

Fig. 3.28 The 3P-QSD-CA with variable entanglement γ. Five simulations at $T = 200$. Upper: $\rho_0 = (|00\rangle\langle 00| + |11\rangle\langle 11|)/2$. Lower: $\rho_0 = (|01\rangle\langle 01| + |10\rangle\langle 10|)/2$

of full entangling it is $\overline{p}_A = 1.0$, $\overline{p}_B = 2.5$ in both scenarios. Thus, the general 3P-SU(2) operators may become a powerful tool when the players share a classically correlated state.

The only effect of starting from $|\Psi_0\rangle = |11\rangle$ instead of from $|00\rangle$ in 2P-QBOS(5,1) simulations (not shown here) is that of reversing the roles of players A and B in Fig. 3.8.

References

1. Schiff, J.L.: Cellular Automata: A Discrete View of the World. Wiley (2008)
2. Miszczak, J.A., Pawela, L., Sladkowski, J.: General model for an entanglement-enhanced composed quantum game on a two-dimensional lattice. Fluct. Noise Lett. **13**(2), 1450012 (2014)
3. Li, Q., Iqbal, A., Perc, M., Chen, M., Abbott, D.: Coevolution of quantum and classical strategies on evolving random networks. PloS One **8**(7), e68423 (2013)
4. Li, Q., Iqbal, A., Chen, M., Abbott, D.: Evolution of quantum and classical strategies on networks by group interactions. New J. Phys. **14**(10), 103034 (2012)
5. Li, Q., Iqbal, A., Chen, M., Abbott, D.: Quantum strategies win in a defector-dominated population. Phys. A **391**, 3316–3322 (2012)
6. Li, A., Yong, X.: Entanglement guarantees emergence of cooperation in quantum prisoner's dilemma games on networks. Sci. Rep. **4**, 6286 (2014)
7. Wiesner, K.: Quantum cellular automata. In: Encyclopedia of Complexity and Systems Science, pp. 7154–7164 (2009). http://arxiv.org/0808.0679
8. Ellison, G.: Learning, local interaction, and coordination. Phys. Rev. Lett. **68–5**, 1047–1071 (1993)
9. Adamatzky, A., Martiez, G.J., Mora, J.C.S.T.: Phenomenology of reaction-diffusion binary-state cellular automata. Int. J. Bifurc. Chaos **16**(10), 2985–3005 (2006)
10. Nowak, M.A., May, R.A.: Evolutionary games and spatial chaos. Nature **359**, 826–829 (1992)
11. Nowak, M.A., May, R., M.: The spatial dilemmas of evolution. Int. J. Bifurc. Chaos **3**(11), 35–78 (1993)
12. Axelrod, R.: The Evolution of Cooperation, Revised edn. Basic Books (2008)
13. Du, J.F., Li, H., Xu, X.D., Zhou, X., Han, R.: Phase-transition-like behaviour of quantum games. J. Phys. A Math. Gen. **36**(23), 6551–6562 (2003)
14. Flitney, A.P., Hollengerg, L.C.L.: Nash equilibria in quantum games with generalized two-parameter strategies. Phys. Lett. A **363**, 381–388 (2007)
15. Alonso-Sanz, R.: On a three-parameter quantum battle of the sexes cellular automaton. Quantum Inf. Process. **12**(5), 1835–1850 (2013)
16. Alonso-Sanz, R.: The spatialized, continuous-valued battle of the sexes. Dyn. Games Appl. **2**(2), 177–194 (2012)
17. Alonso-Sanz, R.: Self-organization in the battle of the sexes. Int. J. Mod. Phys. C **22**(1), 1–11 (2011)
18. Alonso-Sanz, R.: Self-organization in the spatial battle of the sexes with probabilistic updating. Phys. A **390**, 2956–2967 (2011)
19. Marinatto, L., Weber, T.: A quantum approach to static games of complete information. Phys. Lett. A **272**, 291–303 (2000)
20. Ozdemir, S.K., Shimamura, J., Morikoshi, F., Imoto, N.: Dynamics of a discoordination game with classical and quantum correlations. Phys. Lett. A **333**, 218–231 (2004)

Chapter 4
Unfair Contests

This chapter deals with unfair contests in two scenarios. That of Sect. 4.1 where one of the players has not access to every quantum parameter available to the other player, and that of Sect. 4.2 where only one player type updates his strategies. This chapter will focus on the unfair parameter availability scenario [1, 2]. The case of unfair updating will be only presented in this chapter dealing with the Matching Pennies game, and the study of best response strategies in the Prisoner's Dilemma. Unfair updating will be treated more in depth in Chap. 5. The effect of classical memory in quantum games is scouted in an unfair QPD simulation in the last section of this chapter.

4.1 Unfair Parameter Availability

Figure 4.1 deals with a 2P-QPD(5,3,2,1)-CA contest involving a quantum (θ, α)-player A (red) versus a classical θ-player B (blue). Mutual defection remains in the left frame of this figure up to the same $\gamma^\star = \pi/6$ threshold of the fair scenario of Fig. 3.1. In the $(\gamma^\star, \gamma^\bullet)$ transition interval, the (Q,D) strategy clearly emerges isolated, because the (D,Q) pair, also in NE in the transition interval in fair contests, is now unfeasible as the player B may not resort to quantum strategies such as the Q one. The ordering of the payoffs of both players when the entanglement factor exceeds its middle value $\pi/4$ is somehow that expected in the studied unfair game: The quantum player out scores the classical player. But, what happens in the $(\gamma^*, \pi/4)$ interval is highly unexpected as the classical player out scores the quantum player, increasingly close to γ^*. This seemingly counter-intuitive advantage of the *handicapped* player (that will be also reported in other games) may help to explain some kind of odd "survival of the weakest" [3] phenomena.

At variance with what happens with the left $\gamma^\star$ threshold, the right threshold $\gamma^\bullet$ of the fair simulations in Fig. 3.1 does not keep unaltered in the unfair scenario of

© Springer Nature Switzerland AG 2019
R. Alonso-Sanz, *Quantum Game Simulation*, Emergence, Complexity
and Computation 36, https://doi.org/10.1007/978-3-030-19634-9_4

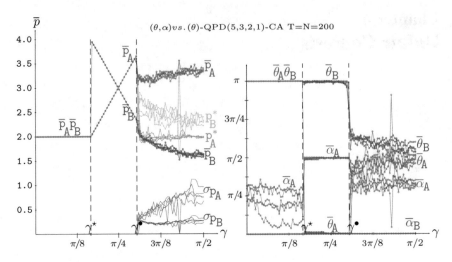

Fig. 4.1 The QPD(5,3,2,1)-CA unfair quantum (θ, α)-player A (red) versus classical θ-player B (blue). Variable entanglement γ. Five simulations at T=200. Left: Mean and standard deviations of the actual payoffs $(\bar{p}, \sigma)$, and mean-field payoffs $(p^\star)$. Right: Mean quantum parameters

Fig. 4.1. It is greater, and emerges in the intersection of $P_B^{Q,D} = \Im \cos^2 \gamma + S \sin^2 \gamma$ and $P_A^{Q,C} = R \cos^2 \gamma + P \sin^2 \gamma$, i.e., $\gamma^\bullet = \arcsin\left(\sqrt{\dfrac{\Im - R}{\Im - R + P - S}} = \dfrac{2}{3}\right) = 0.955$.

In the conventional (non CA) unfair QPD game with full entangling, the quantum player outperforms the classical player by means of the so called *miracle* strategy [4] defined as $\hat{M} = \hat{U}(\pi/2, \pi/2)$ in the way shown in the left frame of Fig. 4.2. This is so because $\Pi^{M, U(\theta_B)}(\gamma = \pi/2) = \dfrac{1}{2}\begin{pmatrix} 0 & 0 \\ 1 - \sin\theta_B & 1 + \sin\theta : B \end{pmatrix}$,[1] so that $p\begin{Bmatrix} M \\ U(\theta_B) \end{Bmatrix} =$

---

[1]
$$\hat{M} = \hat{U}\left(\frac{\pi}{2}, \frac{\pi}{2}\right) = \frac{1}{\sqrt{2}}\begin{pmatrix} i & 1 \\ -1 & -i \end{pmatrix}, \quad \hat{U}(\theta, 0) = \begin{pmatrix} \cos\frac{\theta}{2} & \sin\frac{\theta}{2} \\ -\sin\frac{\theta}{2} & \cos\frac{\theta}{2} \end{pmatrix}, \quad \hat{M} \otimes$$

$$\hat{U}(\theta, 0) = \frac{1}{\sqrt{2}}\begin{pmatrix} i\cos\frac{\theta}{2} & i\sin\frac{\theta}{2} & \cos\frac{\theta}{2} & \sin\frac{\theta}{2} \\ -i\sin\frac{\theta}{2} & i\cos\frac{\theta}{2} & -\sin\frac{\theta}{2} & \cos\frac{\theta}{2} \\ -\cos\frac{\theta}{2} & -\sin\frac{\theta}{2} & -i\cos\frac{\theta}{2} & -i\sin\frac{\theta}{2} \\ \sin\frac{\theta}{2} & -\cos\frac{\theta}{2} & i\sin\frac{\theta}{2} & -i\cos\frac{\theta}{2} \end{pmatrix}, \qquad (\hat{M} \otimes \hat{U}(\theta, 0))\hat{J}(\frac{\pi}{2})|00\rangle =$$

$$\frac{1}{2}\begin{pmatrix} i\cos\frac{\theta}{2} + i\sin\frac{\theta}{2} \\ -i\sin\frac{\theta}{2} + i\cos\frac{\theta}{2} \\ -\cos\frac{\theta}{2} + \sin\frac{\theta}{2} \\ \sin\frac{\theta}{2} + \cos\frac{\theta}{2} \end{pmatrix}, \qquad |\psi_f\rangle = \frac{1}{2\sqrt{2}}\begin{pmatrix} 1 & 0 & 0 & -i \\ 0 & 1 & i & 0 \\ 0 & i & 1 & 0 \\ -i & 0 & 0 & 1 \end{pmatrix} \quad (\hat{M} \otimes \hat{U})\hat{J}|00\rangle =$$

$$\frac{1}{2\sqrt{2}}\begin{pmatrix} i\cos\frac{\theta}{2} + i\sin\frac{\theta}{2} - i\sin\frac{\theta}{2} - i\cos\frac{\theta}{2} \\ -i\sin\frac{\theta}{2} + i\cos\frac{\theta}{2} - i\cos\frac{\theta}{2} + i\sin\frac{\theta}{2} \\ \sin\frac{\theta}{2} - \cos\frac{\theta}{2} - \cos\frac{\theta}{2} + \sin\frac{\theta}{2} \\ \cos\frac{\theta}{2} + \sin\frac{\theta}{2} + \sin\frac{\theta}{2} + \cos\frac{\theta}{2} \end{pmatrix} = \frac{1}{2\sqrt{2}}\begin{pmatrix} 0 \\ 0 \\ 2(\sin\frac{\theta}{2} - \cos\frac{\theta}{2}) \\ 2(\cos\frac{\theta}{2} + \sin\frac{\theta}{2}) \end{pmatrix}. \qquad \text{Note that}$$

$(\sin\frac{\theta}{2} \pm \cos\frac{\theta}{2})^2 = 1 \pm 2\sin\frac{\theta}{2}\cos\frac{\theta}{2} = 1 \pm \sin\theta.$

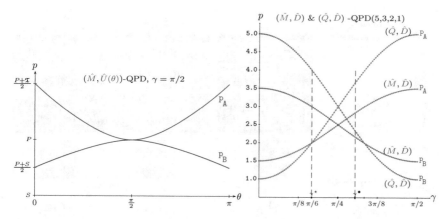

Fig. 4.2 Payoffs in unfair 2P-QPD contests. Left: $(\hat{M}, \hat{U}(\theta))$. Right: $(\hat{M}, \hat{D})$ and $(\hat{Q}, \hat{D})$ for variable γ

$$\frac{1}{2}\left(\begin{Bmatrix}\mathfrak{T}\\S\end{Bmatrix}(1 - \sin\theta_B) + P(1 + \sin\theta_B)\right),\text{ and consequently } p(\hat{M}) > p(\hat{U}(\theta_B)),\text{ except}$$

at $\theta_B = \pi/2$, where $p(\hat{M}) = p(\hat{U}(\pi/2)) = P$.

The miracle strategy seems to play some role with $\gamma > \gamma^{\bullet}$ in Fig. 4.1 as both θ_A and α_A are not far from $\pi/2$ in the right frame of Fig. 4.1. High spatial effects emerge with $\gamma > \gamma^{\bullet}$, so that the standard deviations of the payoffs are not null and the form of the graphs of the actual mean payoffs in the left frame of Fig. 4.1 reminds to those achieved with of the pair $(\hat{M}, \hat{D})$ shown in the right frame of Fig. 4.2, which in the particular case of full entangling provides the payoffs $\dfrac{\mathfrak{T} + P}{2} = 3.5$ and $\dfrac{S + P}{2} = 1.5$ to the quantum and classical player respectively, payoffs close to those arising in the unfair simulations of Fig. 4.1.

Figure 4.3 deals with two simulations in the unfair scenario of Fig. 4.1 with high γ factors above $\gamma = \pi/4$, the one above $\gamma = 3\pi/8$, the one below $\gamma = \pi/2$. As expected from the mean payoffs found in Fig. 4.1, in both cases the quantum player out scores the classical player. This occurs again from the very beginning, albeit the stabilization of the dynamics is not as straightforward as in the dynamics with low entangling shown in [5]. In any case, the parameter and p-values reached at $T = 100$ are very close to the final ones: Increasing the duration of the simulations does not change them significantly. Rich maze-like structures predominate for both θ and α spatial parameter patterns in Fig. 4.3. These structures are also shown in the payoff spatial pattern, though in a much fuzzier manner. The remarkable pattern structures emerging in the simulations of Fig. 4.3 induce the that the mean-field payoffs (p^*) are far distant to the corresponding actual mean payoffs (p), and even invert the ordering. Geometry plays a key role in this scenario, so that the mean-field approach does not properly operate.

Figure 4.4 deals with the case of an unfair 2P-QPD(5,3,2,1) game where the player B is allowed to resort only to the parameter α instead to θ as in the unfair

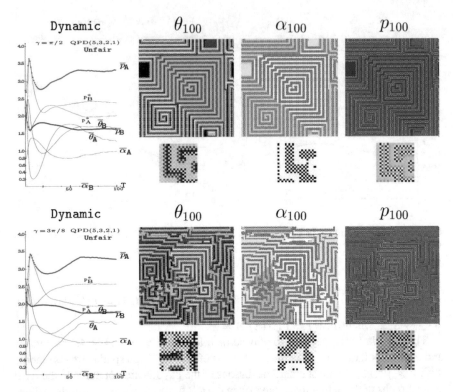

Fig. 4.3 Two simulations in the unfair scenario of Fig. 4.1. Upper: $\gamma = 3\pi/8$, Lower: $\gamma = \pi/2$. Far Left: Evolving mean parameters and payoffs. Center: Parameter patterns at $T = 100$. Far Right: Payoff patterns at $T = 100$. Increasing gray levels indicate increasing values

simulations just before considered. In such a scenario, the (Q,Q) pair emerges in NE beyond the threshold $\gamma^\star = \pi/4$. Before $\gamma^\star = \pi/4$, the full quantum player A defects ($\overline{\theta}_A = \pi$, irrelevant $\overline{\alpha}_A$), whereas the player B sets his free parameter to its maximum $\overline{\alpha}_B = \pi/2$. As a result, the (D,Q) pair emerges with associated $\Pi^{D,Q} = \begin{pmatrix} 0 & \sin^2\gamma \\ \cos^2\gamma & 0 \end{pmatrix}$, which induces a dramatic advantage of player A over player B, that can not resort to strategies such as D with $\theta > 0$. Remarkably, the quantum parameters perfectly converge to their steady states in the scenario of Fig. 4.4, so that no pattern effects emerge and consequently the standard deviations of the quantum parameters and so the payoffs are all null.

Figure 4.5 deals with five simulations of a quantum (θ, α)-player A (red) versus a classical θ-player B (blue) in a QSD-CA with variable entanglement factor γ. The structure of the asymptotic mean payoffs across the lattice $(\overline{p})$ of both players in this unfair scenario resembles that found in the fair scenario of Fig. 3.6 for not high γ, but extended now for all entanglement, without any discontinuity. Thus, rather unexpectedly, despite the fact that the beneficiary player B is restricted to classical

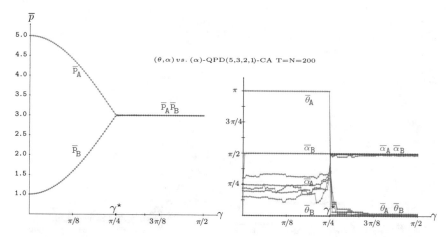

Fig. 4.4 The unfair QPD(5,3,2,1)-CA quantum (θ, α)-player A (red) versus (α)-player B (blue). Variable entanglement γ. Five simulations at T=200. Left: Actual mean payoffs $\overline{p}$. Right: Mean quantum parameters

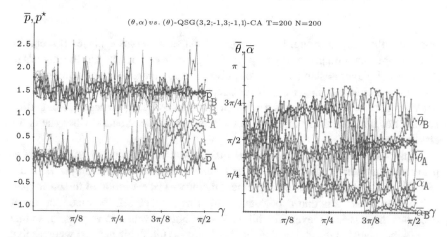

Fig. 4.5 The unfair QSD-CA quantum (θ, α)-player A (red) versus classical θ-player B (blue). Variable entanglement γ. Five simulations at T=200. Left: Mean payoffs $(\overline{p})$ and mean-field payoffs $(p^\star)$. Right: Mean quantum parameters

strategies, he overrates the charity player A regardless of γ as $\overline{p}_B$ oscillates around 1.5, whereas $\overline{p}_A$ oscillates close to zero in most cases. Nevertheless, two simulations with high γ show higher values of $\overline{p}_A$, that may grow up to 0.5, somehow resolving the dilemma of player A in the weak sense making $\theta_A = \pi/2$ in the conventional (non-CA) game [6], or $\overline{\theta}_A \simeq \pi/2$ in CA simulations, as shown and in the right frame of Fig. 4.5.

In simulations in the scenario of Fig. 4.5 but with reversed unfairness: θ-player A versus (θ, α)-player B, the structure of the mean payoffs versus γ (not shown here)

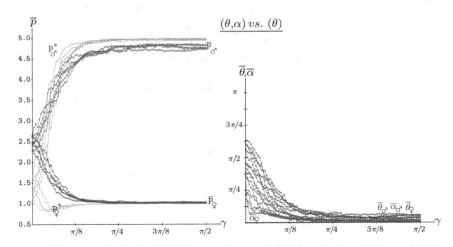

Fig. 4.6 The unfair QBOS(5,1)-CA quantum (θ, α)-player A versus classical θ-player B. Variable entanglement γ. Five simulations at T = 200. Left: Actual mean payoffs ($\bar{p}$) and mean-field payoffs (p^*). Right: Mean quantum parameters

resembles that shown in Fig. 4.5 in the (θ, α)-player A versus θ-player B scenario, although the quantum player B gets payoffs slightly over 1.5, whereas the payoffs of the classical player A stand slightly negative. The advantage of the quantum player (B) facing a classical player (A) is foreseeable, what may surprise here is that the quantum player B does not take a relevant advantage of his privileged role, getting payoffs not far from those achieved in the opposite unfair scenario where he is the classical one.

Figure 4.6 shows the asymptotic payoffs in five simulations of an unfair two-parameter quantum male player A (red) versus classical female player B (blue) (5,1)-BOS CA with variable entanglement factor γ. The evolution of the mean pay-offs of both players is that expected: As soon as γ takes off, the quantum player overrates the classical player. The bifurcation of both $\bar{p}$-values is very rapid, so that by $\pi/4$ the payoff of the classical player is lowered to the stable value 1, whereas that of the quantum player reaches an almost stable value close to 5. With the (4,1) BOS parameters, the plots in the scenario of Fig. 4.6 do not present a crisp bifurcation-like appearance: They are highly noisy for low γ values, and from nearby $\pi/8$ they change rather abruptly to p_A close to 4, and p_B close to 1. In the same vein, with the (2,1) BOS parameters, the plots are highly noisy, i.e., with no clear advantage for any of the players, up to γ close to $\pi/8$, only beyond this value of the entangling factor, the quantum player over rates the classical one, but in a notable way, e.g., with full entangling: $p_A \simeq 1.5$, $p_B \simeq 1.25$. By contrast, with $(R=6, r=1)$, the bi-furcation appearance is more defined than that shown here for (5,1), with the plots corresponding to the different initial quantum parameter randomizations very close all along their dynamics.

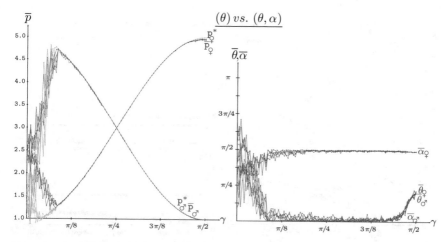

Fig. 4.7 The unfair QBOS(5,1)-CA classical player A versus (θ, α)-player B. Variable entanglement γ. Five simulations at T=200. Left: Actual mean payoffs $(\overline{p})$ and mean-field payoffs (p^*). Right: Mean quantum parameters

Figure 4.7 deals with the same unfair classical versus (θ, α) QBOS contest of Fig. 4.6 but with the role of both players reversed. Unlike in Fig. 4.6, where the quantum player A overrates the classical player B regardless of γ, in Fig. 4.7 the quantum player B overrates the classical player B only with $\gamma > \pi/4$. A rather unexpected finding. A common feature in both unfair QBOS scenarios of Fig. 4.6 and Fig. 4.7 is the absence of significant pattern effects, so that mean-field payoffs approach quite well the actual mean payoffs, particularly in Fig. 4.7.

In the unfair QBOS(5,1) scenario of Fig. 4.8 where the (θ, α)-player A plays with the α-player B, both players set their $\overline{\alpha}$ parameters to the middle value $\pi/4$ regardless of the value of γ (much as in the fair 2P-QBOS game in Fig. 3.8) and $\overline{\theta}_A$ monotonically increases its value from zero with no entanglement up to approximately $\pi/16$ with full entanglement. Thus, the initial bifurcation of the mean payoffs in favor to the player A from fairly similar payoffs for both players with no entanglement in the fair 2P-QBOS game in Fig. 3.8 as well as in the unfair scenarios of Figs. 4.6 and 4.7 is absent in Fig. 4.8 where the payoffs of both players evolve monotonically with γ. High spatial effects emerge in Fig. 4.8 as γ increases, quite comparable to those reported in the fair scenario of Fig. 3.8.

In the two types of two parameter quantum matching pennies (2P-QMP) unfair simulations of Fig. 4.9 where one of the players has not access to the α parameter, the actual mean payoffs of both players are fairly close to zero regardless of γ, with that of player A slightly positive and that of player B slightly negative $(\overline{p}_A(\gamma = \pi/2) = 0.041, \overline{p}_B(\gamma = \pi/2) = -0.041)$. It turns out that having access to the α parameter facing players that have not access to it does not imply a significant advantage in the MP game. This finding regarding the MP game dramatically contrasts with what happens in a fairly general manner in non-zero sum games such as the Prisoner's

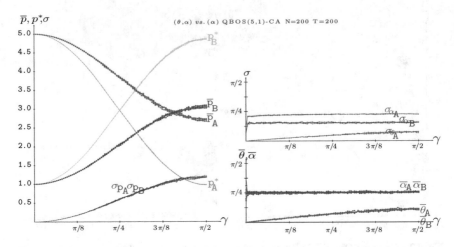

Fig. 4.8 The unfair QBOS(5,1)-CA (θ, α)-player A versus α-player B. Variable entanglement γ. Five simulations at T=200. Left frame: Mean and standard deviations of the actual payoffs $(\overline{p}, \sigma)$, and mean-field payoffs $(p^\star)$. Right frames: Quantum parameters. Standard deviations (upper), mean (lower)

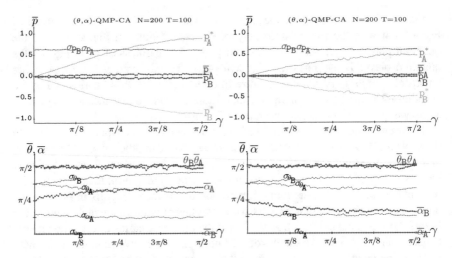

Fig. 4.9 The 2P-QMP game in unfair simulations at T=100. Variable entanglement γ. Left frame: $\alpha_B = 0.0$. Right frame: $\alpha_A = 0.0$

Dilemma or the Battle of the Sexes, where a (θ, α)-quantum player, as a rule, will overrate a (θ)-classical player.

In the $\theta_B = 0.0$ unfair 2P-QMP simulations of the left frame of Fig. 4.10 it is also $\overline{\theta}_A = 0.0$, and $\alpha_A \simeq \alpha_B \simeq \pi/4$. In games with $\theta_A = \theta_A = 0.0$ it is

$$\Pi = \begin{pmatrix} 1 - \pi_{22} & 0 \\ 0 & \sin^2 \gamma \sin^2(\alpha_A + \alpha_B) \end{pmatrix}, \text{ that if } \alpha_A + \alpha_B = \pi/2 \text{ becomes } \Pi =$$

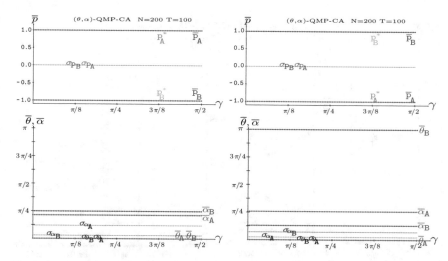

Fig. 4.10 The 2P-QMP game in unfair simulations at T = 100. Variable entanglement γ. Left frame: $\theta_B = 0.0$. Right frame: $\theta_A = 0.0$

$\begin{pmatrix} 1 - \pi_{22} & 0 \\ 0 & \sin^2 \gamma \end{pmatrix}$, leading to $p_A = 1$, $p_B = -1$ as in the upper left frame of Fig. 4.10.
In the $\theta_A = 0.0$ unfair 2P-QMP simulations of the right frame of Fig. 4.10 it is
$\theta_B = \pi$. In games with $(\theta_A = 0.0, \theta_B = \pi)$ it is $\Pi = \begin{pmatrix} 0 & 1 - \pi_{21} \\ \sin^2 \gamma \sin^2 \alpha_A & 0 \end{pmatrix}$,
that if $\alpha_A = 0.0$ becomes $\Pi = \begin{pmatrix} 0 & 1 \\ 0 & 0 \end{pmatrix}$, leading to $p_A = -1$, $p_B = 1$ as in upper
right frame of Fig. 4.10. Thus, contrary to what happens in the scenario of Fig. 4.9,
having access to the θ parameter facing players that have not access to it imply a full
advantage in the MP game. This advantage of the (θ, α)-quantum player facing an
(α)-imaginary player has been already reported regarding the PD and BoS games.

Three-Parameter Strategies

Figure 4.11 deals with the results achieved in 3P strategies QPD(5,3,2,1)-CA unfair
simulations, i.e., the 3P version of the 2P unfair simulations in Fig. 4.1. The structure
of the graphs of the actual mean payoffs (left frames) of Fig. 4.11 notably resembles
that of Fig. 4.1: equal payoffs for both players with not high γ, a transition γ-interval
inducing (Q,D), and prevalence of the actual mean payoff of the quantum player
over that of the classical player for high γ. In Fig. 4.11 pattern effects and non-null
standard deviation payoffs patently emerge from the $\gamma^\bullet$ entanglement landmark (in
parallel with happens in Fig. 4.1), and some disturbance is appreciated near $\gamma^\star$ (this
is not so in Fig. 4.1).

In the 3P-QPD unfair simulations in Fig. 4.11, below $\gamma^\star$ the player B adopts
the defection $\hat{D} = U(\pi, 0, 0) = \begin{pmatrix} 0 & 1 \\ -1 & 0 \end{pmatrix}$ strategy and the player A adopts the $\hat{\mathfrak{D}} =$

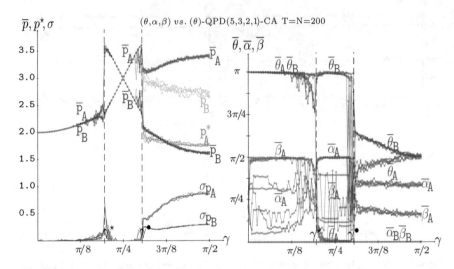

Fig. 4.11 The unfair QPD(5,3,2,1)-CA full quantum (θ, α, β)-player A (red) versus classical θ-player B (blue). Variable entanglement γ. Five simulations at T=200. Left: Mean and standard deviations of the actual payoffs $(\bar{p}, \sigma)$, and mean-field payoffs $(p^\star)$. Right: Mean quantum parameters

$U(\pi, \alpha, \pi/2) = \begin{pmatrix} 0 & i \\ i & 0 \end{pmatrix}$ strategy (recall that $\bar{\theta}_A = \pi$ makes irrelevant $\bar{\alpha}_A$). The pair $(\hat{\mathfrak{D}}, D)$ produces the same output (via the same Π) as the pair $(\hat{\mathfrak{x}}, \hat{\mathfrak{x}})$.[2] Therefore, the equations that define the common payoff for both players before γ^* in Fig. 4.11 are the same as in Fig. 7.10. In Fig. 4.11 it is $\gamma^* = \gamma^\# = 0.615$ and $\gamma^\bullet = 0.955$, i.e., the same $\gamma^\bullet$ of the 2P unfair simulations. Incidentally, mutual $\hat{\mathfrak{D}}$ produces the same output as mutual classical defection.

The form of the graphs in the unfair three-parameter QSD-CA simulations shown in Fig. 4.12 apparently differ from that of its two-parameter counterpart in Fig. 4.5. Thus, a soon as the entanglement takes off, the mean parameters, and the mean payoffs in consequence, become fairly stabilized: Both $\bar{\theta}$ close to $\pi/2$, and $\overline{\alpha_A}$ and

---

$$^2(\hat{\mathfrak{D}} \otimes \hat{D})|\psi_i\rangle = i \begin{pmatrix} 0 & 0 & 0 & 1 \\ 0 & 0 & -1 & 0 \\ 0 & 1 & 0 & 0 \\ -1 & 0 & 0 & 0 \end{pmatrix} \begin{pmatrix} \cos\frac{\gamma}{2} \\ 0 \\ 0 \\ i\sin\frac{\gamma}{2} \end{pmatrix} = \begin{pmatrix} i\sin\frac{\gamma}{2} \\ 0 \\ 0 \\ -\cos\frac{\gamma}{2} \end{pmatrix},$$

$$|\psi_f\rangle = \begin{pmatrix} \cos\frac{\gamma}{2} & 0 & 0 & -i\sin\frac{\gamma}{2} \\ 0 & \cos\frac{\gamma}{2} & i\sin\frac{\gamma}{2} & 0 \\ 0 & i\sin\frac{\gamma}{2} & \cos\frac{\gamma}{2} & 0 \\ -i\sin\frac{\gamma}{2} & 0 & 0 & \cos\frac{\gamma}{2} \end{pmatrix} \begin{pmatrix} i\sin\frac{\gamma}{2} \\ 0 \\ 0 \\ -\cos\frac{\gamma}{2} \end{pmatrix} = \begin{pmatrix} i\cos\frac{\gamma}{2}\sin\frac{\gamma}{2} + i\sin\frac{\gamma}{2}\cos\frac{\gamma}{2} \\ 0 \\ 0 \\ \sin^2\frac{\gamma}{2} - \cos^2\frac{\gamma}{2} \end{pmatrix} =$$

$$\begin{pmatrix} i\sin\gamma \\ 0 \\ 0 \\ \cos\gamma \end{pmatrix}.$$

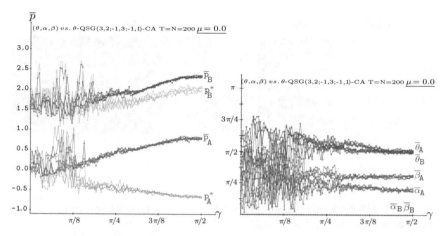

Fig. 4.12 The unfair QSD-CA three-parameter quantum player A versus classical player B. Variable entanglement γ. Five simulations at T = 200 in Left: Actual mean payoffs ($\bar{p}$) and mean-field payoffs (p^*). Right: Mean parameter values

$\bar{\beta}_A$ close to $\pi/4$. Besides, high spatial effects emerge in Fig. 4.12 in contrast with their absence in Fig. 4.5. Spatial effects are particularly relevant regarding the charity player A, as the increase of the entanglement supports the increasing of his actual mean payoffs up to over 0.5, whereas his mean-field estimations decrease below -0.5.

Figure 4.13 shows the dynamics up to $T = 100$, together with the parameter and payoff patterns at $T = 100$ in a simulation with $\gamma = \pi/2$ in the QSD-CA unfair scenario of Fig. 4.12. The far left frame of Fig. 4.13 indicates that both the quantum parameters and the payoffs quickly reach their fairly permanent values, without a relevant initial transition time. As a result, the actual (and mean-field) values shown at $T = 200$ in Fig. 4.12, do not significantly differ from those reached at $T = 100$ (and even before) in Fig. 4.13. The parameter patterns values (and the payoffs in consequence) in this figure show again, as in Fig. 7.15, a kind of *maze*-like aspect that is in the origin of the notable discrepancy between the actual and mean-field payoffs, particularly that of the charity player A.

In the simulations of Fig. 4.14 the unfairness of Fig. 4.12 is reversed, so that a θ-classical charity player A plays with a (θ, α, β)-player B. The classical player A gets negative payoffs regardless γ, but the payoffs of the beneficiary player B keep not far from 1.5 for all γ, thus, surprisingly, getting smaller payoffs than those achieved in the opposite unfair scenario of Fig. 4.12, where he is the classical one.

Figure 4.15 shows the long-term results in five BOS(5,1) CA simulations of an unfair three-parameter quantum male player A versus a classical female player B with variable entanglement factor γ. Initially, for low γ, the general appearance of the curves of the actual mean payoffs in Fig. 4.15 is the same as that of the Fig. 4.6. But before $\pi/8$, a kind of phase transition rockets the actual mean payoff of the player A and plummets that of the player B. Immediately after this episode, both types

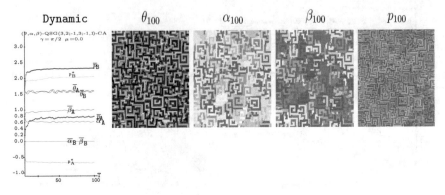

Fig. 4.13 The QSD-CA unfair scenario of Fig. 4.12 at $\gamma = \pi/2$. Far left: Dynamics up to T = 100. Right: Quantum parameters (θ, α, β) and payoff patterns (p) at $T = 100$

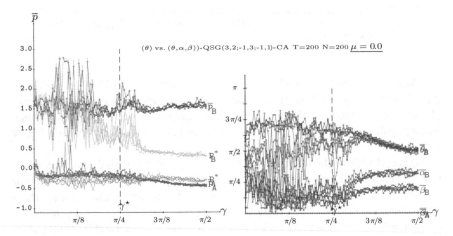

Fig. 4.14 The unfair QSD-CA classical player A versus three-parameter quantum player B. Variable entanglement γ. Five simulations at T = 200 in Left: Actual mean payoffs $(\overline{p})$ and mean-field payoffs (p^*). Right: Mean parameter values

of actual mean payoffs commence to approach, so that by $\gamma = \pi/4$ they equalize at the level $\overline{p} = 3.0$. After this middle-level of γ, the quantum player increasingly over rates the classical player. With very high levels of γ the monotone evolution of the $\overline{p}$-values appears disrupted, so that the (5,1) payoffs are not reached. Even, coordination fails close to maximal entangling so that the actual mean payoff of the classical player B goes slightly under $r = 1$ close to $\gamma = \pi/2$. The main features of the plots in Fig. 4.15 are preserved with the BOS parameters (4,1) and (6,1); even the small $\overline{p}_B < 1$-basin with very high γ. With $(R = 2, r = 1)$, (i) the classical player does not over rate the quantum one, (ii) both players equalize at $\overline{p} = 1.5$ by $\gamma = \pi/4$, (iii) the $\overline{p}_B < 1$ region enlarges compared to the scenarios with greater R, as it commences at $\gamma = 3\pi/8$.

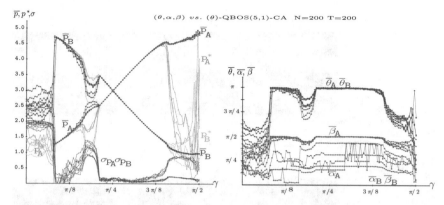

Fig. 4.15 The unfair QBOS(5,1)-CA three-parameter quantum player A versus classical player B. Variable entanglement factor γ. Five simulations at $T = 200$. Left: Mean and standard deviations of the actual payoffs ($\overline{p}$, σ), and mean-field payoffs ($p^\star$). Right: Mean quantum parameters

Examples of the dynamics and long-term patterns in the unfair scenario of Fig. 4.15 with maximal entangling are given in the four simulations of Fig. 4.16 [7]. As a result of the unfair scenario, the left panels of the figure shows how male players rapidly get mean payoffs near the maximum $R = 5$, whereas the female are induced to get mean values very close to $r = 1$. It is very remarkable the high coincidence of the evolving θ values of both kinds of players all along the evolution of the four simulations in Fig. 4.16. This corresponds with the very well defined areas of coordinated players (well defined black or blank regions) in the θ-patterns shown at $T = 200$ in the figure. Further dynamics is qualitatively different in the four simulations. In simulation -1- the parameter levels reached at $T = 200$ ($\overline{\theta}_{\sigma} = 1.93 \simeq \overline{\theta}_{\sigma} = 1.90, \overline{\alpha}_{\sigma} = 0.89, \overline{\beta}_{\sigma} = 1.20$) are almost stable, and consequently the patterns shown in the figure do not vary. In the simulation -2- the stabilization is reached soon after $T = 200$ ($\overline{\theta}_{\sigma} = \overline{\theta}_{\varphi} = 2.77, \overline{\alpha}_{\sigma} = 0.66, \overline{\beta}_{\sigma} = 1.41$), so that the parameter levels and patterns shown in the figure are nearly the steady-state ones. In simulation -3- the black rhomboid vanishes by $T = 220$, and only the small black θ-clusters shown in the figure at $T = 200$ survive in simulations up to $T = 400$ ($\overline{\theta}_{\sigma} = 0.047 \simeq \overline{\theta}_{\varphi} = 0.043, \overline{\alpha}_{\sigma} = 0.16, \overline{\beta}_{\sigma} = 0.07$). Somehow in the same vein, only the small blank areas in simulation -4- survive at $T = 400$ ($\overline{\theta}_{\sigma} = \overline{\theta}_{\varphi} = 3.10, \overline{\alpha}_{\sigma} = 0.66, \overline{\beta}_{\sigma} = 1.55$). Again, as in the fair context, the initial configuration does matter, and the details of the dynamics vary from one simulation to another. Albeit now the ordering $\overline{p}_{\sigma} > \overline{p}_{\varphi}$ keeps regardless the initial configuration, due to the bias inherent to the unfair scenario treated in Fig. 4.16.

Figures 4.17 and 4.18 consider the case of three-parameter quantum versus semi-quantum contests, semi-quantum referring to players that may not implement one of the *quantum* parameters, either α or β, but have access to the other one, β or α respectively (in both cases with the θ parameter operative). The changing payoffs predominance in Fig. 4.17 when the entanglement increases is highly surprising, and that of the (θ, α)-player in the highest range on γ fully unexpected. This is not the

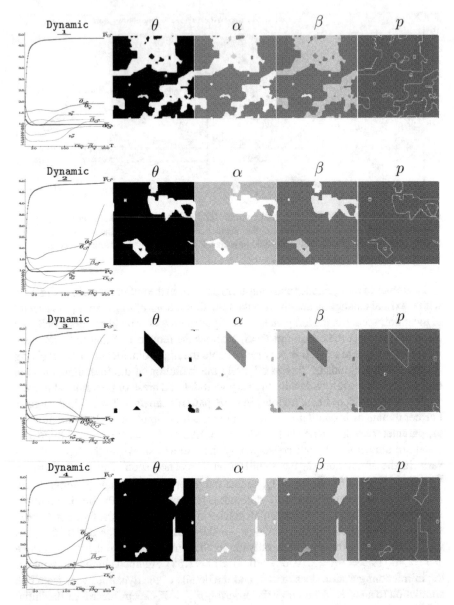

Fig. 4.16 Four simulations of the QBOS(5,1)-CA game in the unfair scenario of Fig. 4.15. Far Left: Evolving mean parameters, actual mean payoffs and mean-field payoffs. Center: Parameter patterns at $T = 200$. Far Right: Payoff patterns at $T = 200$. Increasing gray levels indicate increasing values

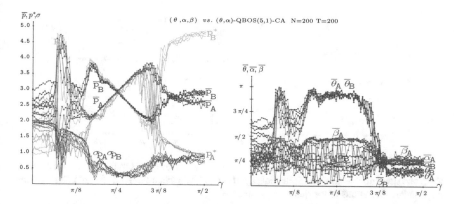

Fig. 4.17 The unfair QBOS(5,1)-CA three-parameter quantum player A versus (θ, α) player B. Variable entanglement factor γ. Five simulations at $T = 200$. Left: Mean and standard deviations of the actual payoffs $(\bar{p}, \sigma)$, and mean-field payoffs $(p^\star)$. Right: Mean quantum parameters

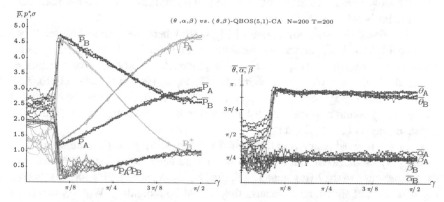

Fig. 4.18 The unfair QBOS(5,1)-CA three-parameter quantum player A versus (θ, β) player B. Variable entanglement factor γ. Five simulations at $T = 200$. Left: Mean and standard deviations of the actual payoffs $(\bar{p}, \sigma)$, and mean-field payoffs $(p^\star)$. Right: Mean quantum parameters

case in Fig. 4.18, in which case what is very notable is the wide range of γ values in which the (θ, β)-player over rates the quantum players, with no equalization at $\gamma = \pi/4$. Examples of the dynamics and long-term patterns in the unfair scenarios of Figs. 4.17 and 4.18 may be found in [7].

Again, it has been checked that the main features of the plots in the scenarios of Figs. 4.17 and 4.18 are preserved with the BOS payoffs (4,1) and (6,1). Nevertheless, the interval with *indefinite* plot appearance when γ is low varies at some extent: It is wider with the (4,1) payoffs, and narrower with the (6,1) payoffs. With $(R=2, r=1)$, the full quantum player over rates the (θ, α)-player from $\gamma = \pi/4$, whereas in the full quantum versus (θ, β) contest, the (θ, β)-player only slightly over-rates the full quantum player around the $[\pi/4, 3\pi/8]$ interval of γ.

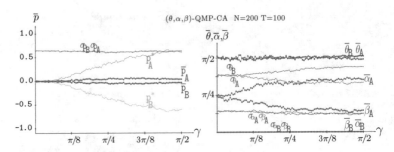

Fig. 4.19 The 3P-QMP game with variable γ in unfair $\alpha_B = \beta_B = 0.0$ simulations at T = 100. Left: Mean and standard deviations of the actual payoffs ($\overline{p}, \sigma$), and mean-field payoffs ($p^\star$). Right: Mean and standard deviations of the quantum parameters

Let us conclude this study on the unfair QBOS game by remarking that Figs. 4.17 and 4.18 make apparent that the role of the α and β parameters in the three-parameter model is fairly different.

In the 3P-QMP unfair simulation of Fig. 4.19, where the classical player B has not access to the (α, β) parameters, the actual mean payoffs of both players are fairly close to zero regardless of γ, with that of player A slightly positive and that of player B slightly negative ($\overline{p}_A(\gamma = \pi/2) = 0.036$, $\overline{p}_B(\gamma = \pi/2) = -0.036$). This result qualitatively agrees with that reported in Fig. 4.9, so that it may be concluded that (θ, α, β)-quantum players do not have a significant advantage when facing (θ)-classical players in the MP game.

Contrary to what happens in the scenario of Fig. 4.19, (θ, α, β)-quantum players facing players that have not access to the θ parameter imply a full advantage of the former ones in the MP game. The figures regarding such a 3P generalization of Fig. 4.10 are not shown here because they would be redundant. With $\theta_B = 0.0$ for example, the graph of payoffs would be exactly the same as in the upper-left frame of Fig. 4.10 (with as $p_A = -1$, $p_B = 1$), and the graph average parameters would be that in the lower-left frame of Fig. 4.10, with $\overline{\beta}_B = \pi/4$ added.

4.2 Unfair Strategy Updating

In the simulations of this section, only one player type updates his strategies whereas the other player type remains as with its initial spatial quantum parameter assignment. Unfair updating will be only presented in this section dealing with the matching pennies game, and the study of best response strategies in the prisoner's dilemma. Unfair updating will be treated more in depth in Chap. 5.

In the two types of 2P-QMP unfair strategy simulations of Fig. 4.20 it is, $\overline{\theta}_A = \overline{\theta}_B = \pi/2$, $\overline{\alpha}_A = \overline{\alpha}_B = \pi/4$, i.e., the mean quantum parameters are set to their middle-level values as in Fig. 3.14. Consequently, the mean-field approaches in Fig. 4.20 coincide with that in Fig. 3.14. But the actual mean payoffs do not. Thus, in

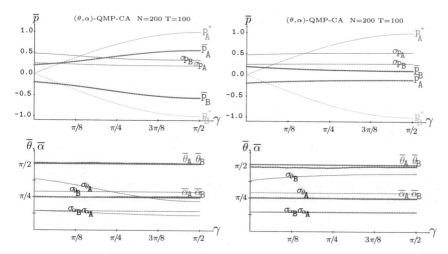

Fig. 4.20 The 2P-QMP game with variable γ in simulations where only one player type updates strategy in spatial simulations at T = 100. Left frames: Only player A updates. Right frames: Only player B updates. Upper: Mean and standard deviations of the actual payoffs ($\overline{p}$, σ), and mean-field payoffs (p^*). Bottom: Mean and standard deviations of the quantum parameters

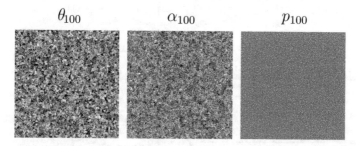

Fig. 4.21 The quantum parameter and payoffs patterns in the 2P-QMP $\gamma = \pi/2$ scenario of the left frames of Fig. 4.20

the left frame of Fig. 4.20, where only player A updates strategies, the active player overrates the passive player B. Whereas in the right frame of Fig. 4.20 in turn, where only player B updates strategies, the active player overrates the passive player A, though in a lower extent compared to the former active A scenario.

The payoff and parameter patterns at $T = 100$ in Fig. 4.21 show a kind of fine-grain texture aspect, maybe coarser in the case of the θ-pattern. Again, as in the fair contests, spatial local structures somehow explain the divergence found between the mean-field and actual mean payoffs.

In the 3P-MP spatial simulations of Fig. 4.22, where only player A type updates strategies, the actual mean payoff of player A is greater than that of player B, even at $\gamma = 0.0$, where $\overline{p}_A = 0.201$ and $\overline{p}_B = -0.201$. As γ increases, $\overline{p}_A$ increases and $\overline{p}_B$ decreases, so that at $\gamma = \pi/2$, it is $\overline{p}_A = 0.424$ and $\overline{p}_B = -0.424$. The graphs of the

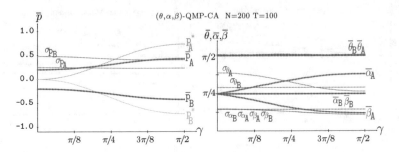

Fig. 4.22 The spatial 3P-QMP game with variable γ in simulations where only player A type updates strategy at $T = 100$. Left: Mean and standard deviations of the actual payoffs $(\bar{p}, \sigma)$, and mean-field payoffs $(p^\star)$. Right: Mean and standard deviations of the quantum parameters

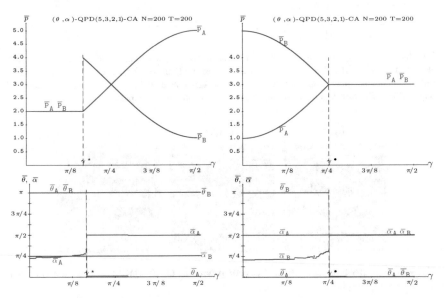

Fig. 4.23 Best response to pure strategies in the two-parameter QPD(5,3,2,1)-CA. Left: Player A updates his strategy, Player B is fixed to the D strategy. Right: Player B updates his strategy, Player A is fixed to the Q strategy

actual mean payoffs in the 3P-MP only-A updating contest of Fig. 4.22 are much alike to that in the 2P-MP only-A updating contest of the left frame of Fig. 4.20: In both figures the initial actual mean payoffs coincide at $\gamma = 0.0$, but at $\gamma = \pi/2$ in Fig. 4.20 it is $\bar{p}_A = 0.564 > 0.424$, $\bar{p}_B = -0.564 < -0.424$. In Fig. 4.22 it is $\bar{\theta}_A \simeq \pi/2$, and, as in Fig. 3.23, it turns out that $\bar{\alpha}_A \simeq \dfrac{\pi}{4} + \dfrac{\pi}{8}\sin\gamma$, and $\bar{\beta}_A \simeq \dfrac{\pi}{4} - \dfrac{\pi}{8}\sin\gamma$. The mean-field payoffs become close to $\pm\sin\dfrac{\gamma}{2}$,[3] so that p_A^* monotonically increases

---

[3] Their exact forms are $p_A^* = -p_B^* = \sin\gamma\,\sin(\frac{\pi}{4}\sin\gamma)$, because $\pi_{11}^* = \pi_{22}^* = \frac{1}{4}(1.0 + \sin\gamma\,\sin(\frac{\pi}{4}\sin\gamma))$, $\pi_{12}^* = \pi_{21}^* = \frac{1}{2}(1.0 - (\pi_{11}^* + \pi_{22}^*))$.

from zero up to $p_A^*(\pi/2) = 0.707^4 > \overline{p}_A(\pi/2) = 0.424$, and $p_B^* = -p_A^*$ decreases from zero down to $p_B^*(\pi/2) = -0.707 < \overline{p}_B(\pi/2) = -0.424$. Please, note that in the scenario of Fig. 4.22 the quantum parameters of player B are initially stated at random, so that their average values are their middle-values: $(\overline{\theta}_B \simeq \pi/2, \overline{\alpha}_B \simeq \overline{\beta}_B \simeq \pi/4)$. If the quantum parameters of every player B were fixed exactly to their middle values, i.e., $(\theta_B = \pi/2, \alpha_B = \beta_B = \pi/4)$, the best response of player A will be not that in Fig. 4.22, but $(\theta_A = \pi/2, \alpha_A = \pi/4, \beta_A = 0)$, that leads to $p_A = \sin \gamma$, $p_B = -\sin \gamma$.

Best Response

Figure 4.23 shows the best response to two fixed strategies using spatial simulations in the 2P-QPD(5,3,2,1) game in order to explain the emergence of the two γ thresholds introduced in Sect. 2.2. The left frames of the figure correspond to the scenario in which only player A updates his strategy whereas player B is fixed to the $\hat{D}$ strategy. In these frames it is shown how player A selects the $\hat{D}$ strategy to respond to player's B defection up to γ^*, but over this γ-threshold player A changes to the $\hat{Q}$ strategy. The right frames of the figure correspond to the scenario in which only player B updates his strategy whereas player A is fixed to the $\hat{Q}$ strategy. In these frames it is shown how player B selects the $\hat{\mathfrak{D}} = U(\pi, \alpha, \pi/2)$ strategy to respond to player's A quantum strategy up to $\gamma^\bullet$, but over this γ-threshold player B changes to the $\hat{Q}$ strategy.

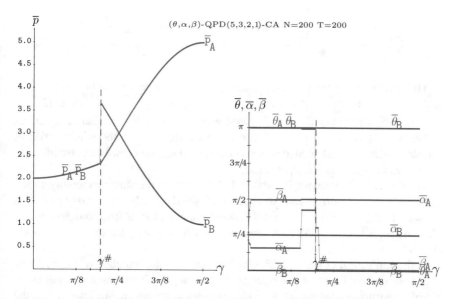

Fig. 4.24 Best response from player A to player's B defection in the three-parameter QPD(5,3,2,1)-CA. Left: Mean payoffs. Right: Mean quantum parameters

---

[4]It is $0.707 = 1/\sqrt{2}$. The article [8] finds the payoffs $\pm 1/\sqrt{2}$ as appearing with strategies in NE in the MP game in an EPR quantum setting that is not the adopted in this study.

Figure 4.24 shows the best response from player A to player's B defection using spatial simulations in the 3P-QPD(5,3,2,1) game in order show the emergence of the $\gamma^{\#}$ threshold introduced in Sect. 3.3. It is shown in the figure how player A responds to player's B defection with the $\hat{\mathfrak{D}} = U(\pi, \alpha, \pi/2)$ strategy up to $\gamma^{\#}$, but over this γ-threshold player A resorts to the $\hat{Q}$ strategy.

4.3 Classical Memory

As long as only the results from the last round are taken into account and the outcomes of previous rounds are neglected, the simulation model considered in this book may be termed *ahistoric*, although it is not fully memoryless as there is a chain (or Markovian) mechanism inherent in it, so that previous results affect further outcomes in an indirect way.

In the *historic* model introduced in this section, after the generic time-step T, and for every cell (i, j), both the payoffs (p) and the (θ, α) parameter values coming from the previous rounds are summarized by means of a geometric mechanism:

$$G_{i,j}^{(T)} = \frac{g_{i,j}^{(T)} = p_{i,j}^{(T)} + \sum_{t=1}^{T-1} \delta^{T-t} p_{i,j}^{(t)}}{\Omega^{(T)}} \quad (4.1a)$$

$$\Theta_{i,j}^{(T)} = \frac{\vartheta_{i,j}^{(T)} = \theta_{i,j}^{(T)} + \sum_{t=1}^{T-1} \delta^{T-t}\theta_{i,j}^{(t)}}{\Omega^{(T)}}, \quad \Lambda_{i,j}^{(T)} = \frac{\upsilon_{i,j}^{(T)} = \alpha_{i,j}^{(T)} + \sum_{t=1}^{T-1} \delta^{T-t}\alpha_{i,j}^{(t)}}{\Omega^{(T)}} \quad (4.1b)$$

The numerators of the weighted means in Eqs. 4.1 are sequentially generated as: $g_{i,j}^{(T)} = \delta g_{i,j}^{(T-1)} + p_{i,j}^{(T)}, \vartheta_{i,j}^{(T)} = \delta\vartheta_{i,j}^{(T-1)} + \theta_{i,j}^{(T)}, \upsilon_{i,j}^{(T)} = \delta\upsilon_{i,j}^{(T-1)} + \alpha_{i,j}^{(T)}$, with $\Omega^{(T)} = 1 + \delta\Omega^{(T-1)}$. Consequently, the proposed geometric memory mechanism is *accumulative* in its demand for knowledge of past history. Thus, the whole series of parameter values do not need to be kept in the computer memory. Just one additional real number per parameter: g, ϑ and υ.

The choice of the *memory factor* $0 \leq \delta \leq 1$ simulates the remnant memory effect: the limit case $\delta = 1$ corresponds to equally weighted records (*full* memory model), whereas $\delta \ll 1$ intensifies the contribution of the most recent iterations and diminishes the contribution of the past ones (*short-term* memory); the choice $\delta = 0$ leads to the ahistoric model.

In the simulation of games with classical memory, the imitation of the best rule will remain unaltered, but it will operate on the trait payoffs and parameters of every player, constructed from the previous rounds as above described. We have studied the effect of this kind of embedded memory in the classical spatial PD [9] and the Battle of the Sexes (BOS) [10, 11] games. As an overall rule, memory boosts cooperation in the PD and coordination in the BOS.

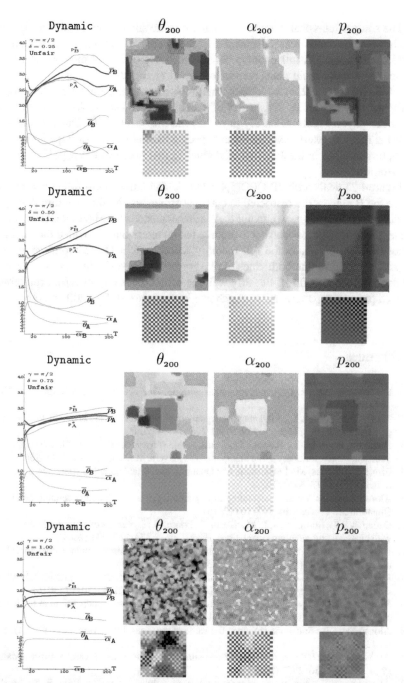

Fig. 4.25 The QPD(5,4,2,1)-CA with full entanglement, where the B players are restricted to classical strategies and the A players are endowed with memory of factor δ. From top to bottom: $\delta = 0.25$, $\delta = 0.50$, $\delta = 0.75$, $\delta = 1.00$

The study of classical memory in quantum game simulation remains to be done. Anyhow, some work was advanced in [12], where a simulation of a quantum PD(5,4,2,1)-CA, with full memory ($\delta = 1.0$) demonstrates that memory of past iterations slows the rapid convergence to mutual $\hat{Q}$ found in the ahistoric dynamic with full entanglement, in such a way that by $T = 200$, the parameter values appear fairly stabilized in levels distant from the extreme zero and π. As a result of this, the mean payoffs seem to hardly grow beyond $3.5 < 4.0 = R$. Thus, this kind of classical memory exerts a somehow unexpected inertial effect that impedes to reach the optimal output in the PD, that of mutual $\hat{Q}$ that provides the payoff on mutual cooperation.

Figure 4.25 deals with 2P-QPD(5,4,2,1)-CA simulations with full entanglement, where the B players are restricted to classical strategies and the quantum A players are endowed with memory of factor δ. As a consequence of the inertial retarder effect of memory, the classical players overcome the quantum ones provided that memory is not full, i.e., $\delta < 1$. In the full memory scenario ($\delta = 1$) the evolution appears *frozen* very soon so that both kinds of players get low payoffs. In these simulations, the actual mean payoffs $p^\star$ fit fairly well the actual mean payoffs ($\overline{p}$), even though the rich parameter pattern structures shown in the figure at $T = 200$ emerge.

References

1. Flitney, A.P., Abbott, D.: Quantum games with decoherence. J. Phys. A Math. Gen. **38**(2), 449 (2004)
2. Flitney, A.P., Abbott, D.: Advantage of a quantum player over a classical one in 2×2 quantum games. Proc. R. Soc. Lond. A **459**(2038), 2463–2474 (2003)
3. Hummert, S., Bohl, K., Basanta, D., Deutsch, A., Werner, S., Theissen, G., Schroeterc, A., Schuster, S.: Evolutionary game theory: cells as players. Mol. Biosyst. **10**, 3044–3065 (2014)
4. Eisert, J., Wilkens, M., Lewenstein, M.: Quantum games and quantum strategies. Phys. Rev. Lett. **83**(15), 3077–3080 (1999)
5. Alonso-Sanz, R.: Variable entangling in a quantum prisoner's dilemma cellular automaton. Quantum Inf. Process. **14**, 147–164 (2015)
6. Ozdemir, S.K., Shimamura, J., Morikoshi, F., Imoto, N.: Dynamics of a discoordination game with classical and quantum correlations. Phys. Lett. A **333**, 218–231 (2004)
7. Alonso-Sanz, R.: On a three-parameter quantum battle of the sexes cellular automaton. Quantum Inf. Process. **12**(5), 1835–1850 (2013)
8. Iqbal, A., Chappell, J.M., Abbott, D.: On the equivalence between non-factorizable mixed-strategy classical games and quantum games. R. Soc. Open Sci. **3**, 150477 (2016)
9. Alonso-Sanz, R.: Spatial order prevails over memory in boosting cooperation in the iterated prisoner's dilemma. Chaos **19**(2), 023102 (2009)
10. Alonso-Sanz, R.: Self-organization in the battle of the sexes. Int. J. Mod. Phys. C **22**(1), 1–11 (2011)
11. Alonso-Sanz, R.: Self-organization in the spatial battle of the sexes with probabilistic updating. Phys. A **390**, 2956–2967 (2011)
12. Alonso-Sanz, R.: A quantum prisoner's dilemma cellular automaton. Proc. R. Soc. A **470**, 20130793 (2014)

Chapter 5
Games on Networks

This chapter focuses on simulations where players are connected at random, instead of in the spatially structured manner as considered so far. Section 5.1 deals with simulations were both player-types update his strategies, whereas in Sect. 5.2 only one of the player-types updates his strategies. The simulations in networks are compared with those in spatial lattices in previous Chaps. 3 and 4. Both shared and distinctive features of the simulations in both types of layouts are scrutinized.

5.1 Fair Contests on Random Networks

In the simulations on networks (NW) studied in this chapter, each player is connected at random with four mates and four partners. Interactions remain uniform and synchronous as in the games played in the cellular automata (CA) manner studied so far. Thus, in each round (T) every player (i, j) plays with his four adjacent partners and he is featured by the average payoff over these four games, $p_{i,j}^{(T)}$. Every player adopts the quantum parameters of the adjacent mate (including himself) with the highest $p^{(T)}$. To allow for the spatial-network comparison, 200×200 players will be involved in the NW simulations here, half of each type. As usual across the simulations in this book, also in this chapter five initial random assignments of the quantum parameters are systematically implemented for every given random network [1].

5.1.1 Two-Parameter Strategies

The overall features of the network QPD(5,3,2,1) simulations of Fig. 5.1 are similar to those of spatial QPD(5,3,2,1) simulations of Fig. 3.1: Mutual defection arises below the lower $\gamma^* = \frac{\pi}{6} = 0.524$ threshold and mutual $\hat{Q}$ beyond the higher

© Springer Nature Switzerland AG 2019
R. Alonso-Sanz, *Quantum Game Simulation*, Emergence, Complexity
and Computation 36, https://doi.org/10.1007/978-3-030-19634-9_5

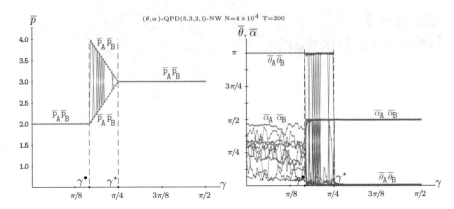

Fig. 5.1 The 2P-QPD(5,3,2,1) game on random networks. Variable entanglement γ. Five simulations at T=200. Left frame: Mean payoffs. Right frame: Mean quantum parameters

$\gamma^{\bullet} = \frac{\pi}{4} \simeq 0.785$ threshold. In the $(\gamma^{*}, \gamma^{\bullet})$ transition interval, where both $(\hat{Q}, \hat{D})$ and $(\hat{D}, \hat{Q})$ are in NE, both payoffs converge to $R = 3$ at $\gamma^{\bullet}$. Network and spatial QPD(5,3,2,1) simulations only differ in the behaviour of the system in the transition interval were the network simulations of Fig. 5.1 are free of the pattern effects of the spatial simulations reported in Fig. 3.1. In other words, unlike in the spatial simulations where both $\{\hat{D}, \hat{Q}\}$ and $\{\hat{Q}, \hat{D}\}$ coexist, in network simulations either $\{\hat{D}, \hat{Q}\}$ or $\{\hat{Q}, \hat{D}\}$ is selected as unique NE in the transition interval. The initial random parameter assignment induces which one of these two pairs is to be chosen.

The QHD(3,2,0,−1) network simulations of Fig. 5.2 behave qualitatively as the QPD ones in Fig. 5.1. In particular, in both scenarios a given $\gamma^{\star} = 0.616$ threshold actuates as a phase transition landmark: Before the $\gamma^{\star}$ threshold, both $(\hat{Q}, \hat{D})$ and $(\hat{D}, \hat{Q})$ are in NE, and after $\gamma^{\star}$ only $(\hat{Q}, \hat{Q})$ is in NE, yielding $p_A = p_B = 2.0$. Mutual defection is never in NE in the HD, so that in the simulations of Fig. 5.2 the graph payoffs of $(\hat{Q}, \hat{D})$ and $(\hat{D}, \hat{Q})$ emerge from $\gamma = 0.0$. The graphs in the networks simulations of Fig. 5.2 notably differ before $\gamma^{\star}$ from those reported in the QHD spatial simulations in Fig. 3.5, due, much as in the QPD, to that spatial effects are absent in the NW simulations here.

In the network QSD(3,2,1,−1) simulations of Fig. 5.3, before $\gamma^{\star}$, the beneficiary player B clearly overrates the charity player A, whose mean payoff seems fairly stable and close to its value in NE in the classical context, i.e., $\overline{p}_A = -0.2$, whereas the mean payoff of player B appears to be under its value in said classical NE, because $\overline{p}_B \simeq 1.0 < 1.5$. After the $\gamma^{\star}$ threshold $(\hat{Q}, \hat{Q})$ emerges in NE, yielding $p_A = 3.0$, $p_B = 2.0$. The $\gamma^{\star}$ threshold appears in Fig. 5.3 soon after the intersection of $p_B^{QQ} = 2$ and $p_B^{QD} = 3\cos^2 \gamma + \sin^2 \gamma = 3 - 2\sin^2 \gamma$ which happens at $\gamma = \pi/4$. Recall that in the spatial QSD(3,2,1,−1) simulations treated in Fig. 3.6, it is exactly $\gamma^{\star} = \pi/4$, with a much *noisier* appearance of the graphs before the $\gamma^{\star}$ threshold.

In the NW-QBOS(5,1) simulations of Fig. 5.4, the structure of both graphs marks the difference induced by the fact that the BOS game does not have a single Social

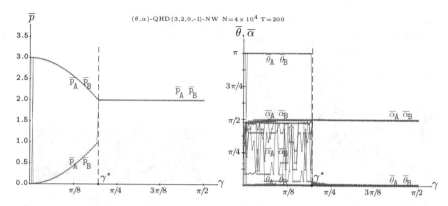

Fig. 5.2 The 2P-QHD(3,2,0,−1) game on random networks. Variable entanglement γ. Five simulations at T = 200. Left frame: Mean payoffs. Right frame: Mean quantum parameters

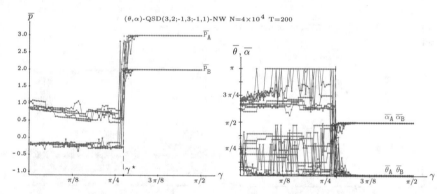

Fig. 5.3 The 2P-QSD(3,2,1,−1) game on random networks. Variable entanglement γ. Five simulations at T = 200. Left frame: Mean payoffs. Right frame: Mean quantum parameters

Welfare solution, i.e., not $(\hat{Q}, \hat{Q})$ emerging in NE. Although both types of graphs appear fairly *noisy*, with notable up-down oscillations, for player A the $\hat{Q}$ strategy seems to predominate, i.e., $\overline{\theta}_A$ drifts to zero and $\overline{\alpha}_A$ to $\pi/2$, and for player B the quantum parameters are fairly low. As a result, the overall broad structure of the payoffs is reminiscent of the crisp graphs associated to the pair $(\hat{Q}, \hat{C})$ in the low-right frame of Fig. 2.1, i.e., $p_A^{QC} = 5\cos^2\gamma + \sin^2\gamma = 5 - 4\sin^2\gamma$, $p_B^{QC} = \cos^2\gamma + 5\sin^2\gamma = 1 + 4\sin^2\gamma$. The noisy aspect of the graphs in the network QBOS simulations shown here in Fig. 5.4 very much contrasts with the crisp one in the spatial QBOS simulations shown in Fig. 3.8.

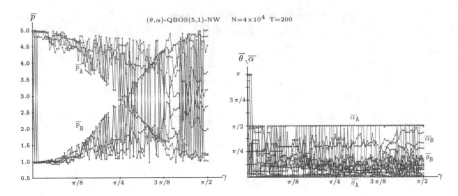

Fig. 5.4 The 2P-QBOS(5,1) game on random networks. Variable entanglement γ. Five simulations at T = 200. Left frame: Mean payoffs. Right frame: Mean quantum parameters

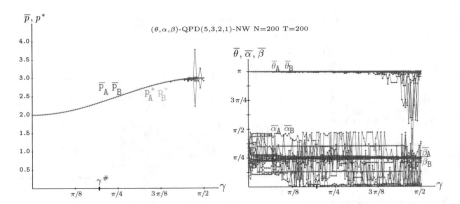

Fig. 5.5 The 3P-QPD(5,3,3,1) game on random networks. Variable entanglement γ. Five simulations at T = 200. Left frame: Mean payoffs. Right frame: Mean quantum parameters

5.1.2 Three-Parameter Strategies

In the simulations shown in this section, general the SU(2) strategies as given in Eq. 2.3 are considered.

The 3P-QPD(5,3,2,1) network simulations of Fig. 5.5 are to be compared to the spatial ones shown in Fig. 3.17. In comparison of said two figures it is remarkable that in the NW simulations of Fig. 5.5 the absence of spatial effects implies that for any given entanglement degree and initial quantum parameters assignment, the two player's type (either A or B) converge in the long-term dynamics to the same quantum parameters values and consequently get the same payoff. For this reason, unlike in the CA simulations of Fig. 3.17, the mean-field payoff approaches fully coincide with the actual mean payoffs and no standard deviations are to be examined in the NW simulations of Fig. 5.5.

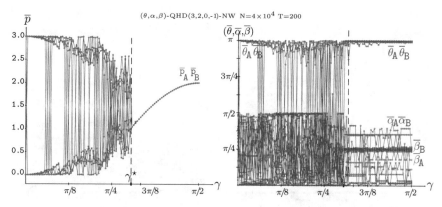

Fig. 5.6 The 3P-QHD(3,2,0,−1) game on random networks. Variable entanglement γ. Five simulations at T = 200. Left frame: Mean payoffs. Right frame: Mean quantum parameter values

The 3P-QHD(3,2,0,−1) network simulations of Fig. 5.6 are to be compared to the spatial ones shown in Fig. 3.18. In this comparison stands out that the $\gamma^\star$ threshold becomes lower and that the $\bar{\theta}$ parameter of both players oscillates close to π not only after $\gamma^\star$, but also with low entanglement. In the right frame of both Figs. 3.18 and 5.6, after $\gamma^\star$ it is $\bar{\theta}_A = \bar{\theta}_B = \pi/2, \bar{\beta}_A = \bar{\beta}_B = \pi/2$, which leads to $\pi_{11} = \sin^2 \gamma$, $\pi_{22} = 1 - \pi_{11}$ (please, recall that $\theta = \pi$ annihilates any influence of α in Eq. 2.3). In a two-person QHD(3,2,0,−1) game with these parameters it is $p_A = p_B = 3 \sin^2 \gamma - 1$, rendering $p_A = p_B = 2$ at maximal $\gamma = \pi/2$, as shown in the left frames of Figs. 3.18–5.6. In Fig. 3.18 it is $p_A = p_B = 1.43$ at $\gamma^\star = 1.12$, and in Fig. 5.6 it is $p_A = p_B = 1.0$ at $\gamma^\star = 0.96 \simeq (\pi/4 + 3\pi/8)/2 = 0.98$. Much as in the 3P-QPD simulations commented before, in the NW simulations of Fig. 5.6 the absence of spatial effects implies that for any given entanglement degree and initial quantum parameters assignment, all the players-type (either A or B) converge in the long term dynamics to the same quantum parameters and consequently get the same payoff. Therefore, no standard deviations are to be examined in the NW simulations of Fig. 5.6.

The structure of the graphs in the 3P-QBOS network simulations of Fig. 5.7 notably differs from those of its 3P spatial counterpart in Fig. 3.20. At variance with this, the structure of the payoff graphs in such 3P-QBOS network simulations of Fig. 5.7 is somehow reminiscent at a first glance to that of the 2P-QBOS network simulations of Fig. 5.4. Anyhow, in contrast to what happens in Fig. 5.4, in Fig. 5.7 no advantage on player A is significantly appreciated and the payoffs of both players approach in a fairly monotonic way towards $(R + r)/2 = (5 + 1) = 3$ as the entanglement increases.

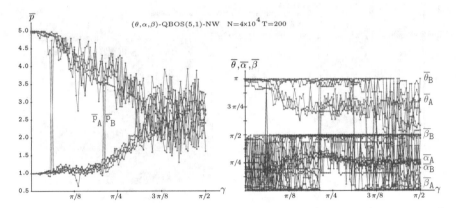

Fig. 5.7 The 3P-QHD(3,2,0,−1) game on random networks. Variable entanglement γ. Five simulations at T = 200. Left frame: Mean payoffs. Right frame: Mean quantum parameters. The 3P-QHD(3,2,0,−1) game on random networks. Variable entanglement γ. Five simulations at T = 200. Left frame: Mean payoffs. Right frame: Mean quantum parameters

5.2 Unfair Strategy Updating

This section studies an unfair scenario where only one player type updates his strategies in the manner indicated in Sect. 5.1. The two quantum parameter model is studied first, then the case of three quantum parameters is considered.

Two Quantum Parameters

Thus, in Figs. 5.8 and 5.9 only player A updates strategies in the PD and HD symmetric games. The asymmetric games SD and BOS are studied in Figs. 5.10 and 5.11 and Figs. 5.12, 5.13 and 5.14 respectively, where both players are treated separately.

In the figures of this section, the far left and central frames deal with spatial simulations and games on networks respectively, with initial random assignment of the quantum parameter values. In the far right frame, the quantum parameters of the player that does not update his strategies are fixed to their middle-level values instead of being assigned at random as is done with the player that updates strategy parameters. Thus, the far right frame with the passive player fixed to the middle-level strategy, referred here to as $\hat{M}$, provides a kind of theoretical reference of what is to be expected in the collective behaviour in the partial updating scenario studied in this section. The Defection and Quantum strategies facing the $\hat{M} = U(\pi/2, \pi/4)$ strategy in the 2P model produce the joint probabilities given below in Eqs. 5.1 and 5.2. Additionally, $\Pi^{M,Q} = (\Pi^{Q,M})'$.

$$\Pi^{D,M} = \frac{1}{2} \begin{pmatrix} 0 & \frac{1}{2}\sin^2\gamma \\ 1 - \frac{1}{2}\sin^2\gamma & 1 \end{pmatrix} \tag{5.1}$$

$$\Pi^{Q,M} = \frac{1}{2}\begin{pmatrix} 1 - \frac{1}{2}\sin^2\gamma & 1 - \sin^2\gamma \\ \sin^2\gamma & \frac{1}{2}\sin^2\gamma \end{pmatrix} \tag{5.2}$$

The lack of coincidence of both the mean-field and actual mean payoffs reflects the emergence of[1] quantum parameter patterns that impede an approach based on the mean values of said parameters (an example is given in Fig. 5.13). These patterns are fairly expected in spatial simulations, but unexpectedly also appear in NW simulations in the unfair scenarios treated in this section.[2]

In the QPD(5,3,2,1) simulations of Fig. 5.8, where only the player A type updates strategies, player A overrates player B regardless of γ. Much as expected when confronting active to passive players. According to the $\Pi^{D,M}$ and $\Pi^{Q,M}$ joint probability matrices given in Eqs. (5.1) and (5.2), in the QPD context of Fig. 5.8, it is: $p_A^{D,M} = \frac{1}{2}(\mathfrak{T} + P - \frac{1}{2}(\mathfrak{T} - S)\sin^2\gamma)$, $p_B^{D,M} = \frac{1}{2}(P + S + \frac{1}{2}(\mathfrak{T} - S)\sin^2\gamma)$; $p_A^{Q,M} = \frac{1}{2}(R + S + (\mathfrak{T} - R\frac{1}{2} + P\frac{1}{2} - S)\sin^2\gamma)$, $p_B^{Q,M} = \frac{1}{2}(\mathfrak{T} + R + (-\mathfrak{T} - R\frac{1}{2} + P\frac{1}{2} + S)\sin^2\gamma)$. The $p_A^{Q,M}$ and $p_A^{D,M}$ payoffs equalize at $\gamma^* = \arcsin\sqrt{\frac{2(\mathfrak{T} - R + P - S)}{3\mathfrak{T} - R + P - 3S}}$. In the particular case of the QPD(5,3,2,1) game it is: $p_A^{D,M} = 3.5 - \sin^2\gamma$, $p_B^{D,M} = 1.5 + \sin^2\gamma$; $p_A^{Q,M} = 2 + \frac{7}{4}\sin^2\gamma$, $p_B^{Q,M} = 4 - \frac{7}{4}\sin^2\gamma$; $\gamma^* = \arcsin\sqrt{6/11} = 0.831$; $p_A^{D,M}(\gamma^*) = p_A^{Q,M}(\gamma^*) = \frac{65}{22} \simeq 2.955$. In the far right frame of Fig. 5.8, the threshold $\gamma^* = 0.831$ indicates fairly well the emergence of the $\hat{Q}$ strategy adopted by players A, albeit before it, $\overline{\theta}_A$ does not keep at the π level featuring defection but slowly decays down to approximately $7\pi/8$ when close to γ^*. Notably, this $\overline{\theta}_A = 7\pi/8$ appears as the almost constant level of the $\overline{\theta}_A$ parameter before plummeting to low levels in both the spatial and networks simulations in Fig. 5.8. In parallel to the $\overline{\theta}_A$ decay when γ is high enough, $\overline{\alpha}_A$ increases up to close $\pi/2$ in these simulations, i.e., players A approach the $\hat{Q}$ strategy. Said approach

---

$${}^1\hat{U}_A \otimes \hat{U}_B = \begin{pmatrix} 0 & 1 \\ -1 & 0 \end{pmatrix} \otimes \begin{pmatrix} \frac{1}{2}(1+i) & \frac{1}{\sqrt{2}} \\ -\frac{1}{\sqrt{2}} & \frac{1}{2}(1-i) \end{pmatrix} = \begin{pmatrix} 0 & 0 & \frac{1}{2}(1+i) & \frac{1}{\sqrt{2}} \\ 0 & 0 & -\frac{1}{\sqrt{2}} & \frac{1}{2}(1-i) \\ -\frac{1}{2}(1+i) & -\frac{1}{\sqrt{2}} & 0 & 0 \\ \frac{1}{\sqrt{2}} & -\frac{1}{2}(1-i) & 0 & 0 \end{pmatrix}$$

$$|\psi_f\rangle = \begin{pmatrix} 0 \\ \cos(\gamma/2)\sin(\gamma/2) \\ -\frac{1}{2}(1+i\cos\gamma) \\ \frac{1}{\sqrt{2}} \end{pmatrix}, \text{ where } |-\frac{1}{2}(1+i\cos\gamma)|^2 = \frac{1}{4}(1+\cos^2\gamma) = \frac{1}{2}(1 - \frac{1}{4}\sin^2\gamma).$$

$${}^2\hat{U}_A \otimes \hat{U}_B = i\begin{pmatrix} 1 & 0 \\ 0 & -1 \end{pmatrix} \otimes \begin{pmatrix} \frac{1}{2}(1+i) & \frac{1}{\sqrt{2}} \\ -\frac{1}{\sqrt{2}} & \frac{1}{2}(1-i) \end{pmatrix} = i\begin{pmatrix} \frac{1}{2}(1+i) & \frac{1}{\sqrt{2}} & 0 & 0 \\ -\frac{1}{\sqrt{2}} & \frac{1}{2}(1-i) & 0 & 0 \\ 0 & 0 & -\frac{1}{2}(1+i) & -\frac{1}{\sqrt{2}} \\ 0 & 0 & \frac{1}{\sqrt{2}} & -\frac{1}{2}(1-i) \end{pmatrix}$$

$$|\psi_f\rangle = i\left(\frac{i}{2}(1+i\cos\gamma), -\frac{1}{\sqrt{2}}\cos\gamma, -i\frac{1}{\sqrt{2}}\sin\gamma, \frac{i}{2}\sin\gamma\right)'.$$

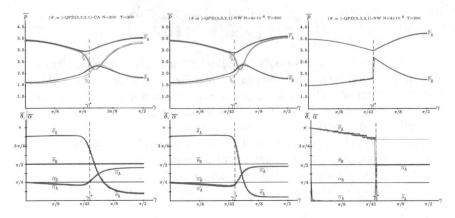

Fig. 5.8 The 2P-QPD(5,3,2,1) game in simulations where only players A update strategy. Far left frames: Spatial simulations. Central frames: Network simulations. Far right frames: $\theta_B = \pi/2$, $\alpha_B = \pi/4$. Variable entanglement γ. Five simulations at T=200

to the Q strategy with high entanglement applies better in the network simulations (central frame) than in the spatial simulations (left frame) where $\overline{\theta}_A$ and $\overline{\alpha}_A$ are more distant to zero and $\pi/2$ respectively.

In the QHD(3,2,0,−1) context of Fig. 5.9, the appearance of the graphs of the spatial and NW simulations (far left and central frames) qualitatively agree with that of the QPD in Fig. 5.9, so that, significantly, the active player A overrates the passive player B regardless of γ. But in its far right frame, players B adopt the $\hat{Q}$ strategy all across γ. Consequently, according to the $\Pi^{Q,M}$ given in Eq. (5.2) it is: $p_A^{Q,M} = 1.0 + 0.75 \sin^2 \gamma$, $p_B^{Q,M} = 2.5 - \dfrac{9}{4}\sin^2 \gamma$. These curves intersect at $\gamma^\star = \arcsin\sqrt{1/2} = \pi/4$, so that, unexpectedly it is $p_A^{Q,M} \leq p_B^{Q,M}$ for $\gamma < \gamma^\star$.

In the QSD(3,2,1,−1) context of Fig. 5.10 where only the player A type updates strategies, player B overrates player A regardless of γ due to the unfair payoffs structure of this game. Even so, player A gets positive payoffs as a reward to his active role. Incidentally, both players get smaller payoff as γ increases in the three panels of Fig. 5.10. The far right panel of the figure shows that $\overline{\alpha}_A$ is either zero or $\pi/2$, in both cases it is $\sin(\alpha_A + \pi/2) = \sin(\pi/2) = \cos(\pi/2) = \dfrac{1}{\sqrt{2}}$. In this context, the maximization of p_A induces a $\theta_A(\gamma)$ that smoothly increases from zero at $\gamma = 0$ to $\pi/2$ at $\gamma = \pi/2$.[3]

---

[3]With the notation, $\omega = \theta/2$, It is, $\pi_{11} = \frac{1}{2}\cos^2 \omega_A(-\sin^2 \gamma \frac{1}{2} + 1)$, $\pi_{12} = \frac{1}{2}(-\cos \omega_A + \sin \omega_A \sin \gamma \frac{1}{\sqrt{2}})^2$, $\pi_{21} = \frac{1}{2}(\sin^2 \omega_A \frac{1}{2} + \sin^2 \omega_A \cos^2 \gamma)$, so that, $2p_A = 3\cos^2 \omega_A(-\sin^2 \gamma \frac{1}{2} + 1) - (-\cos \omega_A + \sin \omega_A \sin \gamma \frac{1}{\sqrt{2}})^2 - \sin^2 \omega_A(\frac{1}{2} + \cos^2 \gamma)$. Consequently, $2p'_A = \sin \theta_A(\frac{4}{2}\sin^2 \gamma - \frac{7}{2}) + 2\cos \theta_A \sin \gamma \frac{\sqrt{2}}{2}$, so that $p'_A = 0$ leads to $\theta_A = \arctan\left(\dfrac{2\sqrt{2}\sin \gamma}{7 - 4\sin^2 \gamma}\right)$, with $\theta_A(0.0) = \arctan 0.0 = 0.0$ and $\theta_A(\pi/2) = \arctan \frac{2\sqrt{2}}{3} = 0.756 \simeq 0.785 = \pi/4$.

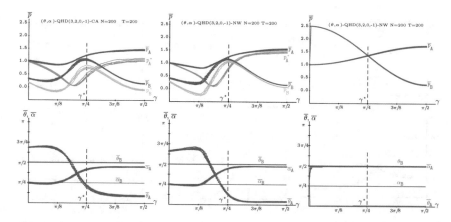

Fig. 5.9 The 2P-QHD(3,2,0,−1) game in simulations where only players A update strategy. Far left frames: Spatial simulations. Central frames: Network simulations. Far right frames: $\theta_B = \pi/2$, $\alpha_B = \pi/4$. Variable entanglement γ. Five simulations at T=200

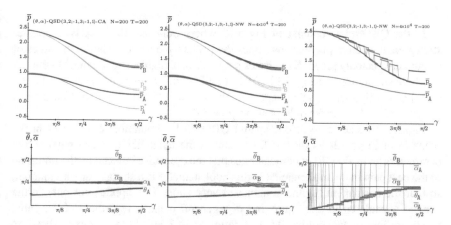

Fig. 5.10 The 2P-QSD(3,2,1,−1) game in simulations where only players A update strategy. Far left frames: Spatial simulations. Central frames: Network simulations. Far right frames: $\theta_B = \pi/2$, $\alpha_B = \pi/4$. Variable entanglement γ. Five simulations at T=200

In the QSD(3,2,1,−1) context of Fig. 5.11 where only the player B type updates strategies, player B overrates player A regardless of γ. In its far right frame, players B adopt the $\hat{Q}$ strategy, thus according to $\Pi^{M,Q}$ obtained as the transpose of the $\Pi^{M,Q}$ given in Eq. (5.2) it is: $P_A^{M,Q} = 1.0 - 0.75 \sin^2 \gamma$, with $P_A^{M,Q}(0) = 1.0$, $P_A^{M,Q}(\pi/2) = 0.25$, and $P_B^{M,Q} = 1.5 + 0.5 \sin^2 \gamma$, with $P_B^{M,Q}(0) = 1.5$, $P_A^{M,Q}(\pi/2) = 2.0$. In the far left and central frames of Fig. 5.11, the quantum parameters depend on γ much as in the QHD(3,2,0,−1) simulations of Fig. 5.9, with players B strategy approaching the $\hat{Q}$ one with high entanglement.

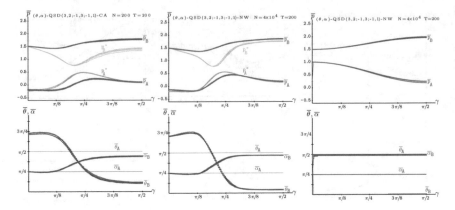

Fig. 5.11 The QSD(3,2,1,−1) game in simulations where only players B update strategy. Far left frames: Spatial simulations. Central frames: Network simulations. Far right frames: $\theta_A = \pi/2$, $\alpha_A = \pi/4$. Variable entanglement γ. Five simulations at T = 200

In the QBOS(5,1) context of Fig. 5.12 where only the player A type updates strategies, the active player A overrates the passive player B for not high values of γ, but unexpectedly player B overrates player A with high entanglement.[4] In the CA and NW simulations of Fig. 5.12 (far left and central frames) $\overline{\alpha}_A = \pi/4$, whereas $\theta_A(\gamma)$ smoothly increases from approximately $\gamma = \pi/8$ at $\gamma = 0$ to a value similar to that achieved in the far right panel at $\gamma = \pi/2$.

Figure 5.13 takes care of one spatial QBOS(5,1) simulation at $\gamma = \pi/4$ in the scenario of Fig. 5.12. Its far left frame shows the dynamics of mean payoffs and parameters up to $T = 20$. It demonstrates that the dynamics induced by the imitation of the best paid mate implemented in this book actuates in a straightforward manner, so that the permanent regime is achieved very soon. This applies not only in the context considered in this figure, but in a general manner, regardless of the game and conditions under scrutiny. Thus, iterating up to $T = 200$ may be excessive, less iterations would suffice in most cases in order to reach stable configurations. The initial lattice of α_A values shown in the center of Fig. 5.12 reflects the initial random assignment of values of this parameter, consequently no patterns are found on it. Contrary to this, the lattice of α_A values at $T = 200$ shown in the far right exhibits a very apparent patchy aspect. The darker cells indicate high α_A, close to the maximum $\pi/2$, whereas clearer cells indicates α_A close to zero. Their mean value is then $\pi/4$, as it is reflected in the dynamics graph (left frame). Here relies the origin of the

---

[4]It is, $\pi_{11} = \frac{1}{2}\cos^2\frac{\theta_A}{2}(-\sin^2\gamma\frac{1}{2}+1)$, $\pi_{22} = \frac{1}{2}(\sin\gamma\cos\frac{\theta_A}{2}\frac{1}{\sqrt{2}}+\sin\omega_A)^2$, so that $2p_A = 5\cos^2\omega_A(-\sin^2\gamma\frac{1}{2}+1)+(\sin\gamma\cos\omega_A\frac{1}{\sqrt{2}}+\sin\omega_A)^2$, $2p'_A = \sin\theta_A(2\sin^2\gamma-4)+\frac{2}{\sqrt{2}}\sin\gamma\cos\theta_A$. Thus, $p'_A = 0$ implies $\theta_A = \arctan\left(\frac{\sqrt{2}\sin\gamma}{2(2-\sin^2\gamma)}\right)$. It is, $\theta_A(\gamma = 0) = \arctan(0) = 0.0$, $\theta_A(\gamma = \pi/2) = \arctan\frac{\sqrt{2}}{2} = 0.615$.

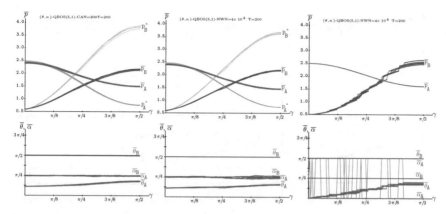

Fig. 5.12 The 2P-QBOS(5,1) game in simulations where only players A update strategy. Far left frames: Spatial simulations. Central frames: Network simulations. Far right frames: $\theta_B = \pi/2$, $\alpha_B = \pi/4$. Variable entanglement γ. Five simulations at T=200

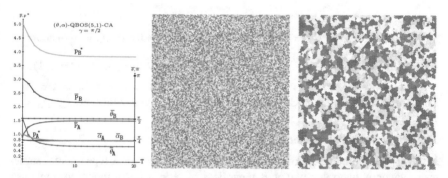

Fig. 5.13 A spatial 2P-QBOS(5,1) simulation at $\gamma = \pi/2$ in the scenario of Fig. 5.12. Far left frame: Dynamics up to $T = 20$. Central frame: Initial α_A lattice. Far right frame: α_A lattice at $T = 200$. Increasing grey levels indicate increasing α_A values in the lattices

failure of the mean-field payoff approach, that uses $\overline{\alpha}_A = \pi/4$, in the estimation of the actual payoffs, computed from the two-person games using the α_A's in the lattice.

In the QBOS(5,1) context of Fig. 5.14, where only the player B type updates strategies, this active player B overrates player A regardless of γ. Notable pattern effects emerge for player B in both the CA and NW simulations of Fig. 5.14, so that the actual mean payoff of player B are increasingly below the mean-field approach as the entanglement increases. In its far right frame it turns out that $\alpha_B = \pi/2$ and consequently it is $\pi_{11} = \frac{1}{2}\cos^2\frac{\theta_B}{2}(1 - \sin^2\gamma)$, $\pi_{22} = \frac{1}{2}(\sin\gamma\cos\frac{\theta_B}{2} + \sin\frac{\theta_B}{2})^2$. With these probabilities, the maximization of $p_B(\theta_B)$ leads to $\theta_A = \pi + \arctan\left(\dfrac{2.5\sin\gamma}{\sin^2\gamma - 1}\right)$. Remarkably this exact formula for θ_B is very well approached from the maximization of only

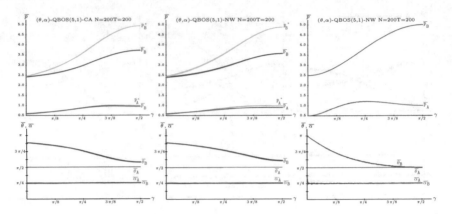

Fig. 5.14 The 2P-QBOS(5,1) game in simulations where only players B update strategy. Far left frames: Spatial simulations. Central frames: Network simulations. Far right frames: $\theta_A = \pi/2, \alpha_A = \pi/4$. Variable entanglement γ. Five simulations at T=200

$\pi_{22}(\theta_B)$, that leads to $\theta_B = 2\arctan(1/\sin\gamma)$,[5] and finally to $p_A = \frac{1}{2}(4\cos\theta_B + \sin\gamma\sin\theta_B - 4\sin^2\gamma + 5)$, $p_B = \frac{1}{2}(-4\cos\theta_B + 5\sin\gamma\sin\theta_B + 4\sin^2\gamma + 1)$. As a result, $p_B > p_A$ for all γ, with maximal difference at $\gamma = \pi/2$, where $\theta_B = 2\arctan(1/\sin(\pi/2)) = \pi/2$ which leads to $\pi_{22} = 1$ and consequently $p_A = 1$, $p_B = 5$.

Three Quantum Parameters

In the 3P-QPD simulations of Fig. 5.15, where only the player A type updates strategies, player A overrates player B regardless of γ in a fairly monotonic manner. Without any kind of *phase-transition* as happens in Fig. 5.8. In the CA and NW simulations (far left and central frames) both $\overline{\alpha}_A$ and particularly $\overline{\beta}_A$ are set close to the middle value $\pi/4$, but $\overline{\theta}_A \simeq 7\pi/8$, a high value of $\overline{\theta}_A$ in which relies the origin of the advantage of player A. In fact, in a game with strategies $U_A(\theta_A, \pi/4, \pi/4)$ and $U_B(\pi/2, \pi/4, \pi/4))$ it is,[6]

---

[5] $\pi'_{22} = (\sin\gamma\cos\frac{\theta_B}{2} + \sin\frac{\theta_B}{2})(-\sin\gamma\sin\frac{\theta_B}{2} + \cos\frac{\theta_B}{2})\frac{1}{2} = 0 \to \sin\gamma\sin\frac{\theta_B}{2} = \cos\frac{\theta_B}{2}$.

[6] It is $\hat{U}_A(\theta_A, \frac{\pi}{4}, \frac{\pi}{4}) = \begin{pmatrix} e^{i\pi/4}\cos(\theta_B/2) & \sin(\theta_A/2)e^{i\pi/4} \\ -e^{-i\pi/4}\sin(\theta_A/2) & e^{-i\pi/4}\cos(\theta_B/2) \end{pmatrix} = e^{i\pi/4}\begin{pmatrix} \cos(\theta_B/2) & \sin(\theta_A/2) \\ i\sin(\theta_A/2) & -i\cos(\theta_B/2) \end{pmatrix}$,

$\hat{U}_B(\pi/2, \frac{\pi}{4}, \frac{\pi}{4}) = \frac{1}{\sqrt{2}}\begin{pmatrix} e^{i\pi/4} & e^{i\pi/4} \\ -e^{-i\pi/4} & e^{-i\pi/4} \end{pmatrix} = \frac{1}{\sqrt{2}}e^{i\pi/4}\begin{pmatrix} 1 & 1 \\ i & -i \end{pmatrix}$,   leading   to $(\hat{U}_A \otimes \hat{U}_B) =$

$\frac{1}{\sqrt{2}}e^{i\pi/2}\begin{pmatrix} \cos(\theta_A/2) & \cos(\theta_A/2) & \sin(\theta_A/2) & \sin(\theta_A/2) \\ i\cos(\theta_A/2) & -i\cos(\theta_A/2) & i\sin(\theta_A/2) & -i\sin(\theta_A/2) \\ i\sin(\theta_A/2) & i\sin(\theta_A/2) & -i\cos(\theta_A/2) & -i\cos(\theta_A/2) \\ -\sin(\theta_A/2) & \sin(\theta_A/2) & \cos(\theta_A/2) & -\cos(\theta_A/2) \end{pmatrix}$.   Finally, $|\psi_f\rangle =$

$\frac{1}{\sqrt{2}}i\begin{pmatrix} \cos(\theta_A/2)\cos\gamma + i\sin(\theta_A/2)\sin\gamma \\ i\cos(\theta_A/2) \\ i\sin(\theta_A/2) \\ -\sin(\theta_A/2)\cos\gamma - i\cos(\theta_A/2)\sin\gamma \end{pmatrix}$.

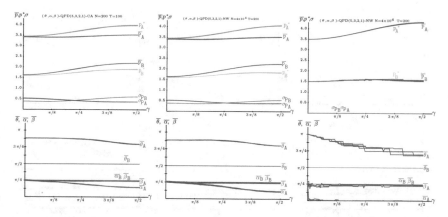

Fig. 5.15 The 3P-QPD(5,3,2,1) game in simulations where only players A update the strategy. Far left frames: Spatial simulations. Central frames: Network simulations. Far right frames: Network simulations with $\theta_B = \pi/2$, $\alpha_B = \pi/4 = \beta_B = \pi/4$. Variable entanglement γ. Five simulations at $T=200$

$$\Pi = \frac{1}{2} \begin{pmatrix} \cos^2(\theta_A/2) - \cos\theta\sin^2\gamma & 1 - \pi_{21} \\ \sin^2(\theta_A/2) & 1 - \pi_{11} \end{pmatrix} \qquad (5.3)$$

and consequently, in the PD($\mathfrak{T}$,R,P,S) it is: $p_A(\gamma) = \frac{1}{2}(C(\gamma) + S + (\mathfrak{T} - S)\sin^2 (\theta_A/2))$, $p_B(\gamma) = \frac{1}{2}(C(\gamma) + \mathfrak{T} - (\mathfrak{T} - S)\sin^2(\theta_A/2))$, with $C(\gamma) = C(0) - (R - P)\cos\theta\sin^2\gamma$, $C(0) = P + (R - P)\cos^2(\theta_A/2)$.

In the PD(5,3,1,0) with $\theta = 7\pi/8$ context of Fig. 5.15 it is $C(0) = 2 + (3 - 2)0.038 = 2.038$, $C(\pi/2) = 2.038 - 1(-0.924) = 2.982$, which renders $p_A(0) = \frac{1}{2}(2.038 + [1 + ((5 - 1)0.962 = 3.848) = 4.884] = 5.886) = 3.443$, $p_B(0) = \frac{1}{2}(2.038 + [5 - ((5 - 1)0.962 = 3.848) = 1.152] = 3.190) = 1.595$, whereas $p_A(\pi/2) = \frac{1}{2}(2.982 + 4.884) = 3.993$, $p_B(\pi/2) = \frac{1}{2}(2.982 + 1.152) = 2.067$. These payoffs for the extreme values of the entanglement factor agree fairly well with those achieved with the mean-field approaches of the CA and NW simulations. Thus, in the particular case of the NW simulations it is: $p_A^*(0) = 3.462$, $p_B^*(0) = 1.562$, $p_A^*(\pi/2) = 4.076$, $p_B^*(\pi/2) = 1.818$. The mean-field estimations fairly coincide with the actual mean-payoffs with no entanglement ($\overline{p}_A(0) = 3.431$, $\overline{p}_B(0) = 1.628$), but diverge from them as the entanglement increases, either over-estimating the actual payoff of player A ($\overline{p}_A(\pi/2) = 3.528$) or underestimating that of player B ($\overline{p}_B(\pi/2) = 2.231$).

As expected, the 3P-QHD simulations of Fig. 5.16 are qualitatively comparable to the 3P-QPD simulations of Fig. 5.15 Again in the CA and NW simulations of Fig. 5.16 (far left and central frames) both $\overline{\alpha}_A$ and particularly $\overline{\beta}_A$ are set close to

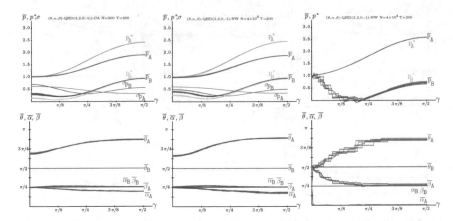

Fig. 5.16 The 3P-QHD(3,2,0,−1) game in simulations where only players A update the strategy. Far left frames: Spatial simulations. Central frames: Network simulations. Far right frames: Network simulations with $\theta_B = \pi/2$, $\alpha_B = \pi/4 = \beta_B = \pi/4$. Variable entanglement γ. Five simulations at T = 200

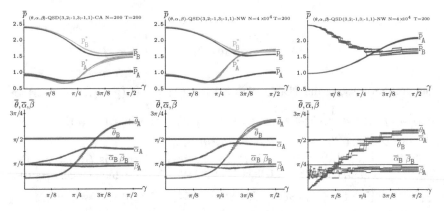

Fig. 5.17 The 3P-QSD(3,2,1,−1) game in simulations where only players A update the strategy. Far left frames: Spatial simulations. Central frames: Network simulations. Far right frames: Network simulations with $\theta_B = \pi/2$, $\alpha_B = \beta_B = \pi/4$. Variable entanglement γ. Five simulations at T = 200

the middle value $\pi/4$, so that the joint probabilities given in Eq. (2.3) may serve as a good tool to explain the payoffs found in the simulations of Fig. 5.16. In both contexts, $\overline{\theta}_A(\gamma)$ is kept at a high value, which slightly decreases with γ in Fig. 5.15 whereas increases in Fig. 5.16. In the NW simulations with the quantum parameters of player B fixed to the middle value (far right frame) in both Figs. 5.15 and 5.16, $\overline{\beta}_A$ remains close to the middle value $\pi/4$ and $\overline{\theta}_A$ achieves a high value (particularly with high entanglement), but $\overline{\alpha}_A$ is led to zero. As a result of this, the advantage of player A is greater in these simulations (particularly with high entanglement).

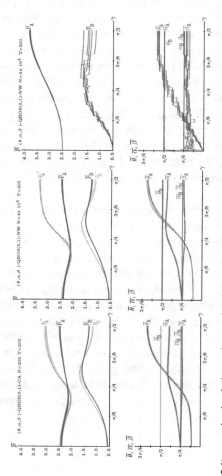

Fig. 5.18 The 3P-QBOS(5,1) game in simulations where only players A update the strategy. Far left frames: Spatial simulations. Central frames: Network simulations. Far right frames: Network simulations with $\theta_B = \pi/2$, $\alpha_B = \beta_B = \pi/4$. Variable entanglement γ. Five simulations at T=200

In the 3P-QSD(3,2,1,−1) context of Fig. 5.17 where only the player A type up-dates strategies, player B overrates player A regardless of γ in the spatial and NW simulations due to the unfair payoffs structure of this game. In the far right frame, player A manages to overcome the player B in simulations with high entanglement, but notable pattern effects emerge for player A in both the CA and NW simulations from $\gamma \simeq \pi/4$ preventing the growing of the actual mean payoff of player A as γ increases.

In the 3P-QBOS(5,1) context of Fig. 5.18 where only the player A type updates strategies, the active player A overrates player B regardless of γ. Thus, player B never overrates player A, not even with high entanglement as happens in the 2P model in Fig. 5.12. This happens because the EWL model with three parameters, in contrast with the basic 2P-EWL, is not biased towards any player B: If the quantum parameters of both players are set to their middle-values, Π is uniform. Similarly to what happens in Fig. 5.17, in Fig. 5.18 notable pattern effects emerge for player A in both the CA and NW simulations from $\gamma \simeq \pi/4$, whereas player B is almost unaffected at this respect. Said pattern effects impede the growing of the actual mean payoff of player A as γ increases, as does happen in the simulations in the far right frame. In the far right frame of both Figs. 5.17 and 5.18, $\overline{\alpha}_A = \pi/2 \; \forall \gamma$, it is $\overline{\beta}_A \simeq \pi/4 \; \forall \gamma$, whereas $\overline{\beta}_A$ grows from zero up to close over $\pi/2$ with high entanglement.

Let us conclude this chapter by stressing that only random networks are considered here. The extension to more realistic networks together with structurally dynamic networks, asynchronous updating, or spatial dismantling is due. In fact quantum games on networks is currently a fairly active area of research. Likely due not on only to its purely scientific interest but also due to the extraordinary importance that networks have nowadays: Networks are somehow ubiquitous nowadays. To the best of our knowledge, the articles produced so far in this area are mainly devoted to the PD with full entanglement (e.g., [2–4]). References considering full entanglement may also be found in the literature (e.g., [5, 6]), but in this case typically involving only the $\hat{C}$, $\hat{D}$ and $\hat{Q}$ strategies (much in the realm of [7]), not the whole set of available strategies. In our opinion, other games than the PD, variable entanglement and quantum general strategies also deserve to be considered in the study of quantum games on networks. A kind of general approach that we have addressed in part regarding classical games [8–12].

References

1. Alonso-Sanz, R.: On collective quantum games. Quantum Inf. Process. **18**, 64 (2019)
2. Li, Q., Chen, M., Perc, M., Iqbal, A., Abbott, D.: Effects of adaptive degrees of trust on coevolution of quantum strategies on scale-free networks. Sci. Rep. **3**, 2949 (2013)
3. Li, Q., Iqbal, A., Perc, M., Chen, M., Abbott, D.: Coevolution of quantum and classical strategies on evolving random networks. PloS one **8**(7), e68423 (2013)
4. Li, Q., Iqbal, A., Chen, M., Abbott, D.: Evolution of quantum strategies on a small-world network. Eur. Phys. J. B **85**(11), 376 (2012)

5. Li, A., Yong, X.: Emergence of super cooperation of prisoner's dilemma games on scale-free networks. PloS one **10**(2), e0116429 (2015)
6. Li, A., Yong, X.: Entanglement guarantees emergence of cooperation in quantum prisoner's dilemma games on networks. Sci. Rep. **4**, 6286 (2014)
7. Nowak, M.A., May, R.A.: Evolutionary games and spatial chaos. Nature **359**, 826–829 (1992)
8. Alonso-Sanz, R.: Self-organization in the battle of the sexes. Int. J. Mod. Phys. C **22**(1), 1–11 (2011)
9. Alonso-Sanz, R.: Self-organization in the spatial battle of the sexes with probabilistic updating. Phys. A **390**, 2956–2967 (2011)
10. Alonso-Sanz, R.: Memory versus spatial disorder in the support of cooperation. Biosytems **97**, 90–102 (2009)
11. Alonso-Sanz, R.: Spatial order prevails over memory in boosting cooperation in the iterated prisoner's dilemma. Chaos **19**(2), 023102 (2009)
12. Alonso-Sanz, R.: Memory boosts cooperation in the structurally dynamic prisoner's dilemma. Int. J. Bifurc. Chaos **19**(9), 2899–2926 (2009)

Chapter 6
Probabilistic Updating

The deterministic imitation of the best rule is adopted across this book, with the only exception of the present chapter, where a probabilistic updating mechanism is implemented in the context of spatial simulations (PCA). The battle of the sexes (BOS) and the prisoner's dilemma (PD) games are analyzed in this chapter with probabilistic updating [1, 2]. Both fair and unfair contest are taken into account. The results found here are to be compared with those achieved with deterministic updating that have been reported in former Chaps. 3 and 4.

In the updating mechanism described in Sect. 3.1, the evolution is guided by the deterministic imitation of the best paid neighbour, so that in the next generation, every player will adopt the parameters of his mate nearest-neighbour (including himself) that received the highest payoff. In this chapter, the generic player (i, j) will adopt the parameters of his mate neighbour (k, l) with a probability proportional to its payoff among their mate neighbours. Thus, denoting by $\mathcal{N}_{(i,j)}$ the mate neighbourhood of the cell (i, j), the probability that it is occupied by the parameters of the player (k, l) after the round played at time-step T is:

$$P[(\theta_{(i,j)}, \alpha_{(i,j)})^{(T+1)} = (\theta_{(k,l)}, \alpha_{(k,l)})^{(T)}] = \frac{p_{(k,l)}^{(T)}}{\displaystyle\sum_{(i^*, j^*) \in \mathcal{N}_{(i,j)}} p_{(i^*, j^*)}^{(T)}} \tag{6.1}$$

Because the convergence to a steady state is more problematic with a probabilistic component in the dynamics than with a deterministic one, the simulations in this chapter have been run up to $T = 500$ instead of up to $T = 200$ as it is usual in the rest of the book.

The probabilistic updating mechanism above described has been implemented in the classical context in [3], but differs from that considered in the study on the evolution of quantum strategies on networks performed in [4] dealing with the prisoner's dilemma and other symmetric games.

© Springer Nature Switzerland AG 2019
R. Alonso-Sanz, *Quantum Game Simulation*, Emergence, Complexity
and Computation 36, https://doi.org/10.1007/978-3-030-19634-9_6

6.1 The QBOS with Probabilistic Updating

In this section, in the QBOS players A and B are assigned the roles of *male* (♂
and *female* ♀ respectively. The BOS(R, r) game is studied here with fixed $r = 1$,
whereas R is parameterized to $R = 2$, $R = 4$ and $R = 6$. Mean quantum parameters
and mean-field payoff approaches are studied only for the (5,1) BOS parameters and
are sometimes presented in separate figures.

Figure 6.1 deals with the spatial quantum 2P-QBOS(5,1) with probabilistic updating and variable entanglement factor γ. So to say, it is the probabilistic version of the
deterministic 2P-QBOS(5,1) treated in Fig. 3.8. In both figures, both $\overline{\alpha}$ parameters
stabilize close to $\pi/4$ regardless of γ, particularly $\overline{\alpha}_B$, whereas both $\overline{\theta}$ parameters become fairly small, as soon as γ starts to increase. In both Figs. 6.1 and 3.8
$\overline{\theta}_A$ keeps at a low value, so that with maximum entanglement it only approaches
$\pi/16$. In contrast with this, $\overline{\theta}_A$ grows up to approximately around $\pi/4$ in Fig. 3.8,
whereas in Fig. 6.1 remains below $\pi/16$. Consequently, the mean-field payoffs for
not low entanglement very well correspond to the payoffs in a two-person game with
$\alpha_A + \alpha_B = \pi/2$ and $\theta_A = \theta_B = 0$.

According to Eqs. (2.8a) and (2.8d) in the conventional 2P-EWL quantum model,
if $\alpha_A + \alpha_B = \pi/2$ the general formulas of the diagonal elements of Π reduce to:

$$\pi_{11} = \cos^2 \omega_A \cos^2 \omega_B \cos^2 \gamma$$
$$\pi_{44} = (\sin\gamma \cos \omega_A \cos \omega_B + \sin \omega_A \sin \omega_B)^2 \qquad (6.2)$$

If also $\theta_A = \theta_B = 0$, Eqs. (6.2) simplify to $\pi_{11} = \cos^2 \gamma$, $\pi_{44} = \sin^2 \gamma$. Therefore,
in the QBOS game, if $\alpha_A + \alpha_B = \pi/2$ and $\theta_A = \theta_B = 0$ it is:

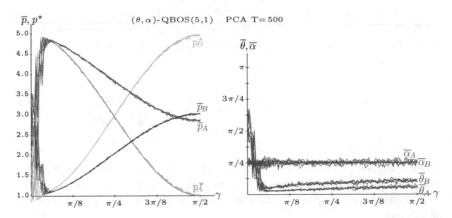

Fig. 6.1 The spatial 2P-QBOS(5,1) with probabilistic updating. Variable entanglement factor γ.
Five simulations at T = 500. Left frame: Mean actual payoffs and mean-field payoffs. Right frame:
Mean quantum parameters

$$p \begin{Bmatrix} \sigma^{\mathbb{1}} \\ \varphi \end{Bmatrix} = \begin{Bmatrix} R \\ r \end{Bmatrix} \cos^2 \gamma + \begin{Bmatrix} r \\ R \end{Bmatrix} \sin^2 \gamma \qquad (6.3)$$

Thus, the mean-field payoffs in Fig. 6.1 with small γ, very well correspond to the values given by the simple Eq. (6.3). In turn, the form of the graphs of the actual mean payoffs in the CA-simulations of Fig. 6.1 is a very deformed modification of the mean-field payoffs, which achieves payoffs equalization with a high γ value instead of with $\gamma = \pi/4$ as happens with the mean-field payoffs. The reason of the strong modification relies in the spatial effects emerging in the BOS-CA simulations, as will be commented for $\gamma = \pi/2$ in Fig. 6.3. It is remarkable that in the full entangling model, strategies with $\alpha_{\sigma^{\mathbb{1}}} + \alpha_{\varphi} = \pi/2$ and $\theta_{\sigma^{\mathbb{1}}} = \theta_{\varphi} = 0$ are in Nash equilibrium in the QBOS game [5].

Figure 6.2 shows an example of the dynamics and long-term patterns in the scenario of Fig. 6.1 dealing with a very low entanglement factor: $\gamma = \pi/16$. The far left panel of the figure shows the evolution up to $T = 500$ of the mean values of $\bar{\theta}$ and $\bar{\alpha}$, as well of the actual $(\bar{p})$ and mean-field payoffs $(p^{\star})$. The $\bar{\theta}$ values evolve initially in opposite direction from their mean values, close to $\pi/2 = 1.57$, $\bar{\theta}_{\sigma^{\mathbb{1}}}$ decreases up to 0.902 at $T = 6$, whereas $\bar{\theta}_{\varphi}$ grows up to 2.077. But the ulterior dynamics depletes this high $\bar{\theta}_{\varphi}$ value. At approximately by $T = 260$ both $\bar{\theta}$-parameters are fairly stabilized in a low value close to that reached at $T = 500$: $\bar{\theta}_{\varphi} = 0.196$, $\bar{\theta}_{\sigma^{\mathbb{1}}} = 0.098$. The dynamics of both $\bar{\alpha}$ parameter values is smoother compared to that of $\bar{\theta}$, so from the initial $\bar{\alpha} \simeq \pi/4 = 0.78$, at $T = 500$ it is: $\bar{\alpha}_{\sigma^{\mathbb{1}}} = 0.640$ and $\bar{\alpha}_{\varphi} = 0.814$. The parameter dynamics in Fig. 6.2 drive the $\bar{p}$ mean payoffs to distant values. The $\bar{p}_{\sigma^{\mathbb{1}}}$ parameter grows monotonically, whereas $\bar{p}_{\varphi}$ shows an early trend to grow which is soon corrected, so that from approximately $T = 260$, both mean payoffs become stabilized close the values reached at $T = 500$: $\bar{p}_{\varphi} = 1.098$, $\bar{p}_{\sigma^{\mathbb{1}}} = 4.819$. Thus, the

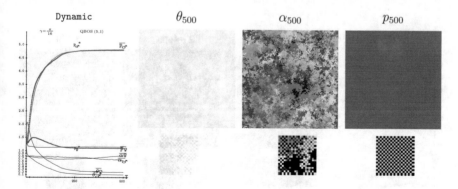

Fig. 6.2 A simulation in the 2P-QBOS(5,1) scenario of Fig. 6.1 with $\gamma = \pi/16$. Far Left: Evolving mean parameters and payoffs. Center: Parameter patterns at $T = 500$. Far Right: Payoff patterns at $T = 500$. Increasing grey levels indicate increasing pattern values

dynamics clearly favours the process to the male player, a kind of result contrasting with the bias towards the female player in the EWL model with full entangling.

Figure 6.2 also shows the snapshots of the parameter and payoff patterns at $T = 500$, both for the full lattice and its zoomed 23×23 central part. No particular spatial structures are appreciated at a first sight in the θ and α parameter lattices. Consequently, the mean-field payoffs p^* estimate fairly well the actual mean payoffs $\overline{p}$: p^*_{σ} estimates correctly $\overline{p}_{\sigma}$ all along the dynamic simulation, and p^*_{φ} only fails in the early period of $\overline{p}_{\varphi}$ increase, which is under-estimated by the mean-field approximation.

Figure 6.3 shows an example of the dynamics and long-term patterns in the scenario of Fig. 6.1 dealing with full entanglement: $\gamma = \pi/2$. The snapshots of the parameter and payoff patterns show that spatial structures emerge, so that the mean-field payoff approaches (which ignores the geometry of interactions) tend to be less indicative of the actual mean payoffs. Thus, the mean-field p^* values in its left panel of Fig. 6.3 do not reflect the actual mean payoffs of both kinds of players, they overestimate the female payoffs and underestimate the male payoffs.

As a general rule, as γ increases a kind of nucleation process emerges in the 2P-QBOS with probabilistic updating, particularly regarding the α parameter. This contrasts with the maze-like structures emerging in the 2P-QBOS with deterministic updating exemplified in Figs. 3.9 and 3.11.

Figure 6.4 shows the long-term payoffs in five simulations of two-parameter QBOS PCA with variable entanglement factor γ. The bias towards the female player (blue) in the model with maximal entangling (pointed out in the Introduction section) becomes apparent in Fig. 6.4 for high γ values, specifically for $\gamma > 3\pi/8$: Very close to this value with $R = 2$, beyond it with $R = 4$ and $R = 6$. In contrast to this, the initial increase of the entangling factor, from the classical $\gamma = 0$ context, leads to a dramatic bifurcation of the mean payoffs in favor to the male player (red), reaching a peak close to $(R, 1)$ with low γ values: When it approaches $\pi/8$ in the

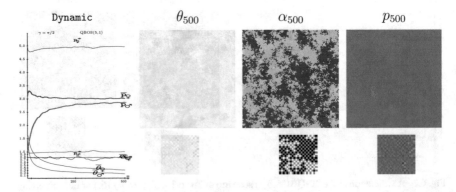

Fig. 6.3 A simulation in the 2P-QBOS(5,1) scenario of Fig. 6.1 with $\gamma = \pi/2$. Far Left: Evolving mean parameters and payoffs. Center: Parameter patterns at $T = 500$. Far Right: Payoff patterns at $T = 500$. Increasing grey levels indicate increasing pattern values

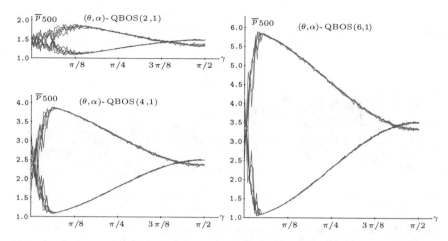

Fig. 6.4 Mean payoffs of the 2P-QBOS(R,1) PCA. Variable entanglement factor γ. Five simulations at $T = 500$. Upper left: $R = 2$. Lower left: $R = 4$. Right: $R = 6$. Red: male, Blue: female

$R = 2$ simulation, but before this value in the simulations with higher R. Before the initial *bifurcation*, the appearance of the graphs is rather noisy, particularly in the $R = 2$ simulation, but after reaching the mentioned peak in the bifurcation of the p-values, the mean payoffs of the five simulations become fairly equalized, so that they commence a smooth male-female approach as γ increases, reach fairly equal $\overline{p}$-values not far from the maximum feasible equalitarian payoff $p = (R + r)/2$, and finally, with very high values of the entangling factor, the $\overline{p}$-female over-rates the $\overline{p}$-male, though not at a high extent. Let us point out here that only maximal entangling is implemented in [4].

The main features of the graphs in the probabilistic updating scenario of Fig. 6.4 were already found in the deterministic updating context of Fig. 3.12. Maybe with the only difference being that the peaks of maximum bifurcation and equalization of the $\overline{p}$-values happen in the deterministic model closer to the $\gamma = \pi/8$ and $\gamma = 3\pi/8$ landmarks.

Three Parameters

Figure 6.5 deals with the spatial quantum 3P-QBOS(5,1) with probabilistic updating and variable entanglement factor γ. At variance with what happens in the 2P-QBOS simulations in Fig. 6.1, the plots in Fig. 6.5 exhibit a great variability. Also relevant in the comparison of both figures is the distinct baseline of oscillation of the quantum parameters: $\pi/2$ instead of a rather low level in the case of the $\overline{\theta}$ parameters, and $\pi/2$ instead of $\pi/4$ regarding the $\overline{\alpha}$ and $\overline{\beta}$ parameters. The value of the γ parameter seems no to affect the high variability of the plots of the mean-field payoffs approaches shown in the left frame of Fig. 6.5, maybe with the only exception of its reduction close to $\gamma = \pi/4$. As a rule, mean-field payoff approaches underestimate the actual mean payoffs, so that spatial effects become highly relevant in the simulations of

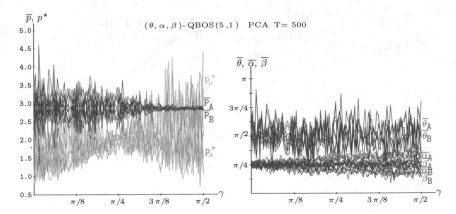

Fig. 6.5 The spatial 3P-QBOS(5,1) with probabilistic updating. Variable entanglement factor γ. Five simulations at T = 500. Left frame: Mean actual payoffs and mean-field payoffs. Right frame: Mean quantum parameters

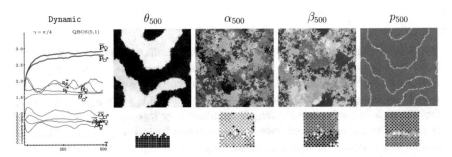

Fig. 6.6 A simulation in the 3P-QBOS(5,1) scenario of Fig. 6.5 with $\gamma = \pi/4$. Far Left: Evolving mean parameters and payoffs. Center: Parameter patterns at $T = 500$. Far Right: Payoff patterns at $T = 500$. Increasing grey levels indicate increasing pattern values

Fig. 6.5 as will be scrutinized with $\gamma = \pi/4$, in Fig. 6.6. No relevant differences are observable in the trait features in the graphs of the 3P-QBOS-CA simulation with probabilistic updating shown here in Fig. 6.5 compared to those with deterministic updating in Fig. 3.20, except regarding the behaviour with low entanglement.

An example of the dynamics and long-term patterns in the 3P-QBOS(5,1) scenario of Fig. 6.5 with $\gamma = \pi/4$ is shown in Fig. 6.6. Accordingly with the absence of bias favouring any of the types of players, in the three-parameter simulations both mean payoffs, and parameter values evolve in a fairly parallel way. In contrast with what happens with deterministic updating as shown in Fig. 3.22, no maze-like structures emerge in any of the patterns in Fig. 6.6. But the notable nucleation appreciated in the θ parameter pattern explains why the mean-field estimations p^*, do not follow the actual mean payoffs $\overline{p}$. Incidentally, the mean payoffs in Fig. 6.6 do not appear fully stabilized, thus longer simulation time should be implemented to achieve more accurate asymptotic results.

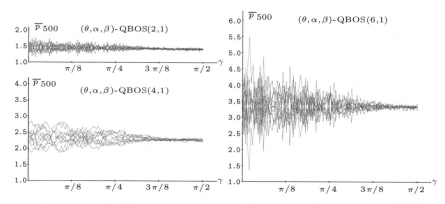

Fig. 6.7 Mean payoffs of the 3P-QBOS(R,1) PCA with variable entangling factor γ. Five simulations at $T = 500$. Upper left: $R = 2$. Lower left: $R = 4$. Right: $R = 6$. Red: male, Blue: female

Figure 6.7 shows the asymptotic mean payoffs in the QBOS(R,1)-CA with three quantum parameters and $R = 2$, $R = 4$, $R = 6$. At variance with what happens in the two parameter scenario of Fig. 6.4, the mean payoffs are not dramatically altered by the variation of γ. The overall effect of the increase of γ being a moderation in the variation of the $\overline{p}$ values corresponding to the five different simulations shown in every frame. Please, recall that in the three-parameter strategies model there is not any bias favoring the female player, so that there is not any tendency to any player over rating the other one. The spatial ordered structure induces a kind of self-organization effect, which allows to achieve fairly soon, approximately at $T = 20$, pairs of mean payoffs that are accessible only with correlated strategies in the two-person game. Recall that the maximum equalitarian payoff in the uncorrelated context is $p^+ = (R + r)/4$, whereas the payoffs in the simulations of Fig. 6.7 and in Fig. 6.5 with high entanglement factor reach values not far from the maximum feasible equalitarian payoff with correlated strategies: $p^= = (R + r)/2$. Thus, with $R = 6$ around 3.40, $R = 5$ around 2.90, with $R = 4$ around 2.40, and with $R = 2$ around 1.40.

Unfair Contests in the QBOS-PCA

Figure 6.8 shows the asymptotic mean payoffs in five simulations of an unfair two-parameter quantum male (red) versus classical female (blue) (2,1), (5,1) and (6,1)-BOS PCA with variable entanglement factor γ. The evolution of the mean payoffs of both players is that expected: As soon as γ takes off, the quantum player over rates the classical player. The bifurcation of both $\overline{p}$-values is very rapid, thus in the (5,1) and (6,1) scenarios before $\gamma = \pi/8$ the payoff of the classical player is lowered to the stable value 1, whereas that of the quantum player reaches an almost stable value close to R. With the (2,1) BOS parameters, the plots in the scenario of Fig. 6.8 do not present an immediate crisp bifurcation-like appearance: The noisy transition period stands up for γ values around $\pi/4$, beyond this value of the entangling factor, the quantum player definitely over rates the classical one, thus $p_{\sigma} \simeq 2.0$, $p_{\varphi} \simeq 1.0$.

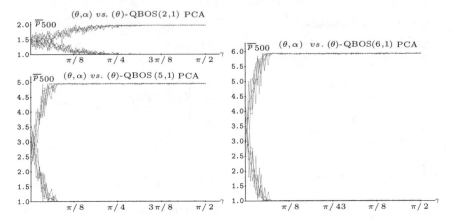

Fig. 6.8 Mean payoffs in an unfair QBOS(R,1) PCA two-parameter quantum male (red) versus classical female (blue). Variable entanglement factor γ. Five simulations at $T = 500$. Upper left: $R = 2$. Lower left: $R = 5$. Right: $R = 6$

In the unfair scenario of Fig. 6.8, but with the deterministic updating of strategies implemented in [6], the plots also show a bifurcation-like aspect, but not so clearly from so low values of γ. Thus, in the (5,1) and (6,1) scenarios the noisy transition up to (R, r) remains up to $\pi/4$, and with the (2,1) BOS parameters the plots for low γ values are highly noisy, with no clear predominance of any of the players. The case of (5,1)-BOS parameters is shown here in Fig. 4.6.

Figure 6.9 deals with a simulation of a quantum BOS(5,1) PCA in the unfair scenario of Fig. 6.8. In broad strokes, the evolution of the $\overline{p}$-curves in the left panel of Fig. 6.9 resembles that of Fig. 6.2: The $\overline{p}_{\sigma}$ parameter grows monotonically,

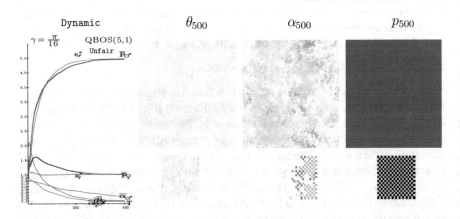

Fig. 6.9 A simulation in the QBOS(5,1) unfair scenario of Fig. 6.8 with $\gamma = \pi/16$. Far Left: Evolving mean parameters and payoffs. Center: Parameter patterns at $T = 500$. Far Right: Payoff patterns at $T = 500$. Increasing grey levels indicate increasing pattern values

whereas $\overline{p}_\varphi$ shows an early trend to grow which is soon corrected. From approximately $T = 260$ both mean payoffs become stabilized close to the values reached at $T = 500$: $\overline{p}_\varphi = 1.009$, $\overline{p}_\sigma = 4.956$. Thus, the dynamics clearly favours the process to the quantum male player facing the female player restricted to classical strategies: the male player reaches a mean p-value close to $R = 5$, dark cells in the far right panel, whereas the female is induced to get a mean value very close to $r = 1$, white cells in the far left panel. Both $\overline{\theta}$-parameter values plummet to zero, and $\overline{\alpha}_\sigma$ also decreases dramatically: $\overline{\alpha}_\sigma^{(500)} = 0.243$. As a result of this general drift to zero by every parameter as soon as γ takes off, the mean-field approach tends to operate properly regardless of the BOS parameters.

Reversing the type of unfairness studied in Fig. 6.8 (quantum-male versus classical-female), in Fig. 6.10 the unfair contest: classical-male versus quantum-female is taken into account. At variance with what happens in Fig. 6.8, where the quantum-male overrates the quantum-female in a very clear way, Fig. 6.10 shows a more variable dependence of γ, with $\gamma = \pi/4$ marking the turning point for the quantum female player over scoring the classical male player. Also remarkable in Fig. 6.10 is the perfect fit of the mean-field payoffs (p^*) to the actual mean payoffs ($\overline{p}$) as soon as γ takes off.

Figure 6.11 shows the asymptotic payoffs in five simulations of unfair three-parameter quantum male (red) versus classical female (blue) (2,1), (5,1) and (6,1)-BOS PCA with variable entanglement factor γ. Initially, for low γ, the general appearance of the curves in Fig. 6.11 is similar to that of the Fig. 6.8. In the case of $R = 2$, the bifurcation of the p values appears around $\pi/4$ in Fig. 6.11, thus unexpectedly much later than the almost immediate bifurcation in Fig. 6.8. For $R = 4$ and $R = 6$ the scenario is even more unexpected: around $\pi/16$ and up to around $\pi/8$, the payoff of the male players commences to decrease whereas that of the female players increase. Afterwards, the trends reverse again, so that both types of payoffs

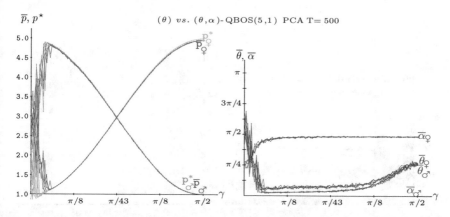

Fig. 6.10 The unfair QBOS(5,1) PCA classical male versus two-parameter quantum female with entanglement factor γ. Left: Actual mean payoffs ($\overline{p}$) and mean-field payoffs (p^*). Right: Mean parameter values. Five simulations at T $= 500$

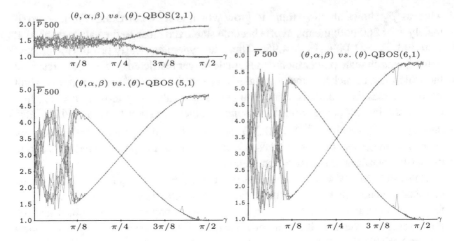

Fig. 6.11 Mean payoffs of an unfair three-parameter quantum male (red) versus classical female (blue) QBOS(R,1) PCA. Variable entanglement factor γ. Five simulations at $T = 500$. Upper left: $R = 2$. Lower left: $R = 5$. Right: $R = 6$

commence to approach and by $\gamma = \pi/4$ they equalize at a level $\overline{p} \simeq (R + r)/2$. After this middle-level of γ, the quantum player increasingly over rates the classical player. Nearly maximal entangling, the monotone evolution of the $\overline{p}$-values appears slightly disrupted, thus the (R,r) payoffs are not reached exactly. Even, coordination fails close to maximal entangling so that the payoff of the classical player goes slightly under $r = 1$, particularly with $R = 2$.

Figure 6.12 shows the asymptotic mean parameter and mean-field payoffs in the unfair QBOS(5,1) 3P scenario of Fig. 6.11. The main feature of the parameter plots is that of the plateaus reached in a wide γ-parameter interval, from approximately

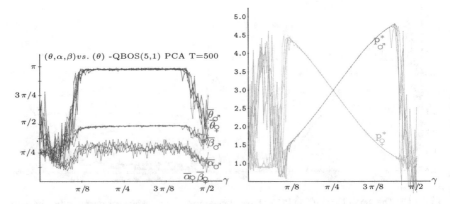

Fig. 6.12 Mean parameters (left) and mean-field payoffs (right) in the 3P-QBOS(5,1) unfair scenario of Fig. 6.11. Variable entanglement factor γ. Five simulations at T = 500

$\pi/8$ to beyond $3\pi/4$. Surprisingly, both the male and female player coincide in the θ-plateau level, $\overline{\theta} \simeq \pi$. The other parameters of the male (fully quantum) player stabilize in the mentioned γ-region as: $\overline{\alpha}_{\sigma} \simeq \pi/4$ and $\overline{\beta}_{\sigma} \simeq \pi/2$. Please, note that if $\theta = \pi$ the diagonal elements of U in (2.3) are zero, so that the α parameter turns out irrelevant. If both players have $\theta = \pi$, and it is $\overline{\beta}_{\sigma} = \pi/2, \overline{\beta}_{\female} = 0$, in the conventional (non-CA) 3P model the expected payoffs are[1]:

$$P \begin{Bmatrix} \sigma \\ \female \end{Bmatrix} = \begin{Bmatrix} R \\ r \end{Bmatrix} \sin^2 \gamma + \begin{Bmatrix} r \\ R \end{Bmatrix} \cos^2 \gamma \tag{6.4}$$

whose structure coincide with that given in (6.3), but with the position of sin and cos interchanged. The spatial effects become very relevant for the higher values of γ, as the plummeting of the male payoffs appearing in right panel of Fig. 6.12 is absent in the payoffs of the actual CA-simulations in the right panel of Fig. 6.11.

The forthcoming Figs. 6.13, 6.14, 6.15 and 6.16 consider the case of quantum male versus semi-quantum female contests, semi-quantum referring to players that may not implement one of the *quantum* parameters, either α or β, but have access to the other one, β or α respectively, in both cases with the θ parameter operative.

The changing predominance as effect of variable entanglement in the (4,1) and (6,1) scenarios of Fig. 6.13 is highly surprising, and that of the (θ, α)-player in the highest range on γ fully unexpected. This is not the case in Fig. 6.15, in which case what is very notable is the wide range of γ values in which the (θ, β)-player over rates the full quantum player, with no equalization at $\gamma = \pi/4$. With $(R=2, r=1)$, the full quantum player increasingly over rates the (θ, α)-player approximately from $\gamma = \pi/4$, whereas in the full quantum versus (θ, β) contest, the (θ, β)-player only slightly over-rates the full quantum player around the $[\pi/4, 3\pi/8]$ interval of γ.

Figure 6.14 shows the asymptotic mean parameters (left) and mean-field payoffs (right) in the QBOS(5,1) 3P unfair scenario of Fig. 6.13. The $\overline{\alpha}$ parameters appear fairly stabilized around $\pi/4$ in the left panel of Fig. 6.14, but the $\overline{\theta}$ parameters showa

---

[1] $\hat{U}_{\sigma} = \begin{pmatrix} 0 & i \\ i & 0 \end{pmatrix}$, $\hat{U}_{\female} = \begin{pmatrix} 0 & 1 \\ -1 & 0 \end{pmatrix}$, so that $\hat{U}_{\sigma} \otimes \hat{U}_{\female} = \begin{pmatrix} 0 & 0 & 0 & i \\ 0 & 0 & -i & 0 \\ 0 & i & 0 & 0 \\ -i & 0 & 0 & 0 \end{pmatrix}$. Consequently,

$\hat{U}_{\sigma} \otimes \hat{U}_{\female} \begin{pmatrix} \cos(\gamma/2) \\ 0 \\ 0 \\ i\sin(\gamma/2) \end{pmatrix} = \begin{pmatrix} -\sin(\gamma/2) \\ 0 \\ 0 \\ -i\cos(\gamma/2) \end{pmatrix}$, and $|\psi_f\rangle = \begin{pmatrix} -\cos(\gamma/2)i\sin(\gamma/2) - \sin(\gamma/2)\cos(\gamma/2) \\ 0 \\ 0 \\ i\sin^2(\gamma/2) - i\cos^2(\gamma/2) \end{pmatrix}$,

with associated $\Pi = \begin{pmatrix} \sin^2\gamma & 0 \\ 0 & \cos^2\gamma \end{pmatrix}$.

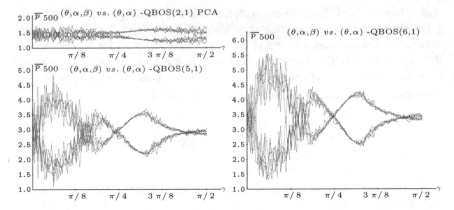

Fig. 6.13 Mean payoffs in an unfair three-parameter quantum male (red) versus (θ, α) semi-quantum female (blue) QBOS(R,1) PCA. Variable entanglement factor γ. Five simulations at $T = 500$. Upper left: $R = 2$. Lower left: $R = 5$. Right: $R = 6$

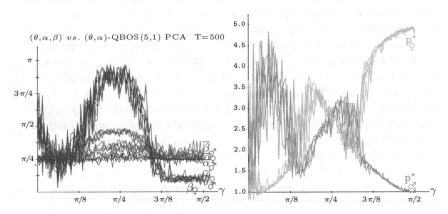

Fig. 6.14 Mean parameters (left) and mean-field payoffs (right) in the 3P-QBOS(5,1) unfair scenario of Fig. 6.13. Variable entanglement factor γ. Five simulations at $T = 500$

shape reminiscent of Gauss curve centered at $\gamma = \pi/4$. The strong bifurcation in the mean-field payoffs that occurs close to, but below, $3\pi/8$ in the right panel of Fig. 6.14 does not occur in the right panel Fig. 6.13. Thus, it is arguable that notable spatial effects emerge in the CA simulations with high γ parameter in the scenario of Fig. 6.13.

The (θ, β) parameterization of Fig. 6.15 was implemented in [7] and studied in [8] with strategies of the form: $\hat{U}(\theta, \beta) = \begin{pmatrix} \cos(\theta/2) & ie^{i\beta}\sin(\theta/2) \\ ie^{-i\beta}\sin(\theta/2) & \cos(\theta/2) \end{pmatrix}$. In [8] it is reported that in conventional (non-CA) (θ, β) EWL models with (2,1) BOS-parameters, the pair $\{\hat{F}, \hat{F}\}$, $\hat{F} = \hat{U}(0, \beta)$ is in Nash equilibrium for arbitrary β,

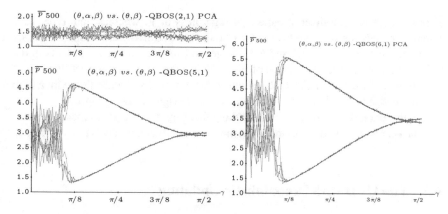

Fig. 6.15 Mean payoffs in an unfair three-parameter quantum male (red) versus (θ, β) semi-quantum female (blue) QBOS(R,1) PCA. Variable entanglement factor γ. Five simulations at $T = 500$. Upper left: $R = 2$. Lower left: $R = 5$. Right: $R = 6$

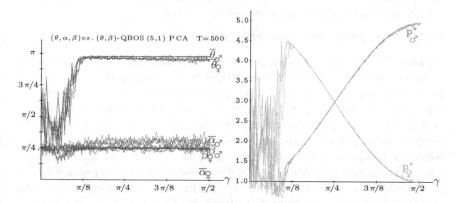

Fig. 6.16 Mean parameters (left) and mean-field payoffs (right) in the 3P-QBOS(5,1) unfair scenario of Fig. 6.15. Variable entanglement factor γ. Five simulations at $T = 500$

generating the payoffs: $p_{\male} = 2 - 2\sin^2\gamma$, $p_{\female} = 1 + 2\sin^2\gamma$, reversing the conclusions obtained with the usual (θ, α) parameterization, i.e., $\{\hat{B}, \hat{B}\}$, $\hat{B} = \hat{U}(0, \alpha)$ in NE for arbitrary α, with: $p_{\female} = 2 - 2\sin^2\gamma$, $p_{\male} = 1 + 2\sin^2\gamma$. This reversion may be appreciated in CA-simulations by comparing the plots in Fig. 6.15 to those in Fig. 6.4.

Figure 6.16 shows the asymptotic mean parameters (left) and the mean-field payoffs (right) in the unfair 3P-QBOS(5,1) scenario of Fig. 6.15. Both male and female

$\overline{\beta}$ parameters are stabilized around $\pi/4$ regardless of γ, whereas both types of $\overline{\theta}$ parameters become stabilized around π (which makes α irrelevant) from $\gamma \simeq \pi/8$.[2]

As in the case of Figs. 6.1 and 6.4, the general form of the actual payoffs in the CA-simulations in the (5,1) frame of Fig. 6.15 is a (very) deformed modification of that in the right frame of Fig. 6.16, which achieves payoffs equalization with a high γ value instead of with $\gamma = \pi/4$ as happens in the right frame of Fig. 6.16.

Let us conclude this section by remarking that Figs. 6.13 and 6.15 make apparent that the role of the α and β parameters in the three-parameter model is fairly different.

6.2 The QPD with Probabilistic Updating

Unfair QPD contests are studied in the first part of this section. In the fair QPD contests addressed in the second part of this section a scheme of interactions involving eight neighbours is implemented. Mean quantum parameters and mean-field payoff approaches are presented in separate figures.

Unfair Contests in the QPD-PCA

Figure 6.17 shows the asymptotic payoffs in five simulations of an unfair quantum two-parameter PD(5,3,2,1) PCA with variable entanglement factor γ. The behavior for low and high values of the entanglement factor is somehow that expected: with low values of γ the asymptotic payoff maintain the $P = 2$ punishment value, and with $\gamma > \pi/4$ the payoff of the quantum players (red-marked) out scores that of the classical player (blue-marked). But, what happens in the $(\gamma^*, \gamma^\bullet) = (0.524, \pi/4)$ interval is highly unexpected: the classical player out scores the quantum player. The peak emerging just after $\gamma^* = 0.524$ in particular is extremely surprising. For γ increasing from this value, up to $\gamma^\bullet = \pi/4$, the inversion in the ordering of the payoffs in depleted, up to reaching the same p-value, the reward $R = 3$, by $\gamma^\bullet = \pi/4$. It seems that the bias towards defection referred in the Sect. 2.1 overcomes the fact that the A player is a quantum player when entangling commences to operate but remains below its middle level. After the $\gamma^\bullet$ value, the expected ordering of the $\overline{p}$'s slowly appears, with a sort of noisy behaviour appearing after $\gamma = (\pi/4 + 3\pi/8)/2$. This discontinuity refers not only to the breaking of the increasing quantum over classical tendency, but also to the fact that the five different simulations become appreciable.

---

[2]If both players have $\theta = \pi$ and $\beta = \pi/4$, in the conventional EWL model it is: $\hat{U}_\sigma =$
$\hat{U}_\varphi = \dfrac{1}{\sqrt{2}} \begin{pmatrix} 0 & 1+i \\ -1+i & 0 \end{pmatrix}$, so that $\hat{U}_\sigma \otimes \hat{U}_\varphi = \dfrac{1}{2} \begin{pmatrix} 0 & 0 & 0 & 2i \\ 0 & 0 & -2 & 0 \\ 0 & -2 & 0 & 0 \\ -2i & 0 & 0 & 0 \end{pmatrix}$. Consequently, $\hat{U}_\sigma \otimes$

$\hat{U}_\varphi \begin{pmatrix} \cos(\gamma/2) \\ 0 \\ 0 \\ i\sin(\gamma/2) \end{pmatrix} = \begin{pmatrix} -\sin(\gamma/2) \\ 0 \\ 0 \\ -i\cos(\gamma/2) \end{pmatrix}$, which leads to the payoffs given in (6.4).

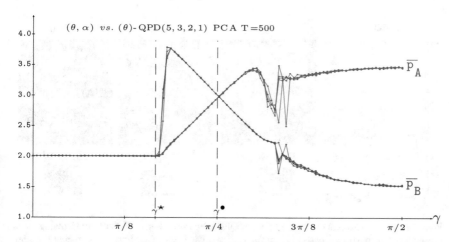

Fig. 6.17 Mean payoffs of an unfair quantum two-parameter (red) versus classical (blue) PD(5,3,2,1) PCA with variable entanglement factor γ. Five simulations at T $= 500$

It is remarkable that, with this exception, the p-values in the five simulations fairly coincide, also in the peak of the classical player.

Figure 6.18 shows the mean parameters and the mean-field payoffs in the unfair quantum two-parameter male versus classical female QPD scenario of Fig. 6.17. The mean-field payoffs estimations (p^*) given in the right frame of Fig. 6.18 fit fairly well to the actual ones shown in Fig. 6.17 for not too high values of γ. But when γ approaches $3\pi/8$ the p^* values notably differ from the $\overline{p}$ ones. This fact will be commented below, when dealing with Fig. 6.19.

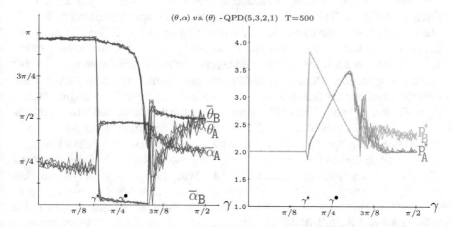

Fig. 6.18 Mean parameters and mean-field payoffs in the unfair quantum two-parameter (red) versus classical (blue) scenario of Fig. 6.17. Variable entanglement factor γ. Five simulations at T $= 500$

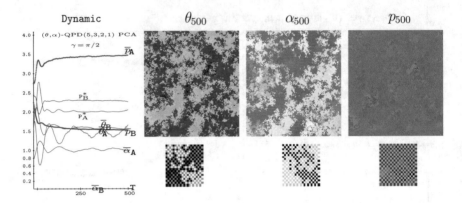

Fig. 6.19 A simulation in the $\gamma = \pi/2$ unfair QPD scenario of Fig. 6.17. Far Left: Evolving mean parameters and payoffs. Center: Parameter patterns. Far Right: Payoff patterns. Increasing grey levels indicate increasing pattern values

In a conventional unfair 2P-QPD game, the quantum player is well advised to play the *miracle* strategy $\hat{M} = \hat{U}(\pi/2, \pi/2) = \dfrac{1}{\sqrt{2}}\begin{pmatrix} i & 1 \\ -1 & -i \end{pmatrix}$ [9]. Provided that $\gamma > \pi/4$, the strategy $\hat{M}$ ensures to the quantum player to over-rate the classical player, even if the classical player chooses $\hat{D}$ (see Fig. 4.2). In the case of full entangling, $\hat{M}$ versus $\hat{D}$ will provide the payoff $\dfrac{\mathfrak{T}+P}{2}$ to the quantum player, whereas the classical player will get $\dfrac{S+P}{2}$. In the context of Fig. 6.17 it is: $\dfrac{\mathfrak{T}+P}{2} = 3.5$, and $\dfrac{S+P}{2} = 1.5$, numerical values of the payoffs close to those achieved by the $\overline{p}$ values for $\gamma = \pi/2$ in Fig. 6.18.

Figure 6.19 deals with a simulation in the $\gamma = \pi/2$ unfair QPD scenario of Fig. 6.17, where the B players (blue) are restricted to classical strategies, i.e., $\alpha_B = 0$. The far left panel of the figure shows the evolution up to $T = 500$ of the mean values of θ, α as well of the actual and mean-field payoffs. The parameter patterns in Fig. 6.19 show a kind of coarse-grain aspect, a kind of spatial heterogeneity that explains why the theoretical mean-field estimations of the payoffs very much differ from the actual ones, in this particular case even inverting the order of magnitude, i.e., $\overline{p}_A > \overline{p}_B$, but $p_A^* < \overline{p}_B$. The spatial structures of the parameter and payoff patterns here obtained with probabilistic CA-updating differ very much form that achieved with deterministic updating [10, 11], in which case rich maze like patterns emerge.

The trait features of the graphs in Fig. 6.17 are preserved in CA-simulations with the $(\gamma^* = \gamma^\bullet)$-PD-parameters (4,3,2,1). With the $(\gamma^* > \gamma^\bullet)$-PD-parameters (4,3,2,0), the actual $\overline{p} = P = 2$ remains stable for both types of players up to $\gamma^* = \pi/4$, at this value of γ a crisp bifurcation emerges, so that at $\gamma = \pi/2$ it is $\overline{p}_A = 3.0$ and $\overline{p}_B = 1.0$.

Unfair 3P Contests

Figure 6.20 shows the asymptotic payoffs in five simulations of an unfair quantum three-parameter PD(5,3,2,1) CA with variable entanglement factor γ. The general

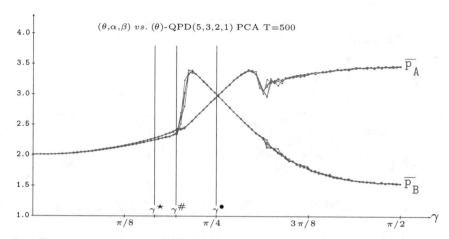

Fig. 6.20 Mean payoffs in an unfair three-parameter quantum (red) versus classical (blue) QPD(5,3,2,1) PCA with variable entanglement factor γ. Five simulations at T = 500

features of Fig. 6.17 participates of the trait features of the plots in Fig. 6.29 and in Fig. 6.17. Thus, for low values of γ the actual mean payoffs of both types of players evolve as in Fig. 6.29: Up to $\gamma = \gamma^*$ they coincide and increase its value smoothly, following the formula $p^{\#}(\gamma) = 2 + \sin^2(\gamma)$. In the $(\gamma^*, \gamma^{\#})$ interval the quantum player slightly over-rates the classical player, but by $\gamma = \gamma^{\#}$ both mean payoffs $\overline{p}$ diverge as in Fig. 6.17: The classical player over-rates the quantum player in the $(\gamma^{\#}, \gamma^{\bullet})$ interval, equalization achieved at level $\overline{p} = R = 3$ for $\gamma = \gamma^{\bullet}$, and $\overline{p}_A > \overline{p}_B$ for $\gamma > \gamma^{\bullet}$. The interval of γ for which the classical player over-rates the quantum player is narrower in Fig. 6.20 compared to that in Fig. 6.17 : $(\gamma^{\#}, \gamma^{\bullet})$ instead of $(\gamma^*, \gamma^{\bullet})$, with the classical player reaching a lower peak in his payoff. In Figs. 6.20 and 6.17 roughly the same $\overline{p}$-values are achieved by $\gamma = \pi/2$: $\overline{p}_A = 3.5$, $\overline{p}_B = 1.5$.

The mean-field payoffs in the scenario of Fig. 6.20, shown in the right frame of Fig. 6.21, prove the existence of heavy spatial effects for high values of γ: The mean-field payoffs are far distant from the actual mean payoffs, and with their ordering inverted.

The trait features of the graphs in Fig. 6.20 are preserved in CA-simulations with the PD-parameters (4,3,2,1), though with equal mean payoffs for both players reaching $\gamma = \gamma^{\#} > \gamma^{\bullet} = \gamma^*$. With the $(\gamma^* > \gamma^{\bullet})$-PD-parameters (4,3,2,0), equal mean payoffs remain up to approximately the center of the $(\gamma^*, \gamma^{\#})$ interval. In these scenarios, a fairly crisp bifurcation of the $\overline{p}$'s actual payoffs emerges after the equal $\overline{p}$'s behaviour, with a minimum γ interval in which the classical player over-rates the quantum player.

The plots in the simulations of an unfair quantum versus (θ, α)-semi-quantum three-parameter PD(5,3,2,1) CA shown in Fig. 6.22 preserve the trait features of those in Fig. 6.20, except regarding the tendency to equalization of the payoff of both types of players for the highest values of γ.

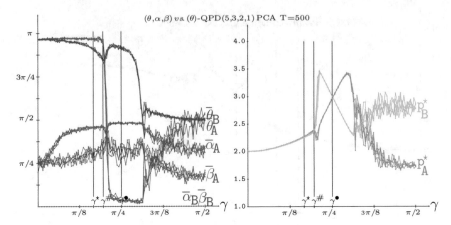

Fig. 6.21 Mean parameters and mean-field payoffs in the unfair quantum three-parameter (red) versus classical (blue) QPD scenario of Fig. 6.20. Variable entanglement factor γ. Five simulations at T = 500

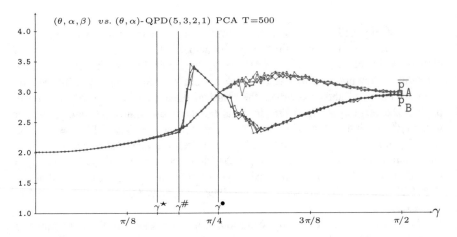

Fig. 6.22 Mean payoffs in an unfair three-parameter quantum (red) versus (θ, α)-semi-quantum (blue) PD(5,3,2,1) PCA with variable entanglement factor γ. Five simulations at T = 500

The mean-field payoffs in the scenario Fig. 6.22, shown in the right frame of Fig. 6.23, prove the existence of huge spatial effects for high values of γ: The mean-field payoffs rocket towards the extreme values (5,1), far distant from the actual mean payoffs. The discrepancy between the actual and the mean-field payoffs is still found with the (4,3,2,1)-PD payoffs, but not with the (4,3,2,0)-PD payoffs.

In the unfair quantum versus (θ, β)-semi-quantum simulations shown in Fig. 6.24 the plots appearance surprisingly looks like that in Fig. 6.29. Only a minor discrepancy between the actual mean payoffs and its mean-field approach may be appreciated in the left frame of Fig. 6.24 for high values of γ.

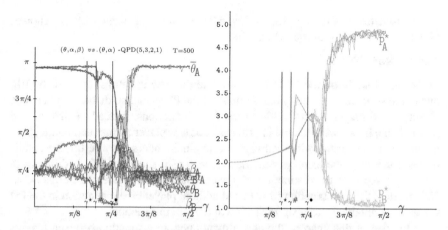

Fig. 6.23 Mean parameters and mean-field payoffs in the unfair quantum three-parameter (red) versus (θ, α)-semi-quantum (blue) scenario of Fig. 6.22. Variable entanglement factor γ. Five simulations at T = 500

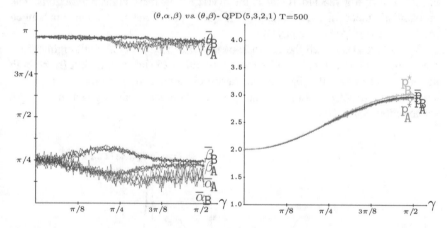

Fig. 6.24 The unfair three-parameter quantum (red) versus (θ, β)-semi-quantum (blue) PD(5,3,2,1) PCA. Variable entanglement factor γ. Five simulations at T = 500. Left: Mean quantum parameters. Right: Actual mean payoffs ($\overline{p}$) and mean-field payoffs (p^*),

The (θ, β) parameterization considered in Fig. 6.24 was implemented in [7] and studied in [8] with strategies of the form: $\hat{U}(\theta, \beta) = \begin{pmatrix} \cos(\theta/2) & ie^{i\beta}\sin(\theta/2) \\ ie^{-i\beta}\sin(\theta/2) & \cos(\theta/2) \end{pmatrix}$. In [8] it is reported that in conventional (non-CA) (θ, β) EWL models with (5,3,1,0) PD-parameters, the pair $\{\hat{D}', \hat{D}'\}$, $\hat{D}' = \frac{1}{\sqrt{2}}\begin{pmatrix} 0 & i-1 \\ i+1 & 0 \end{pmatrix}$, is in Nash equilibrium for all γ with equal payoffs for both players: $p_A = p_B = 1 + 2\sin^2\gamma$, whose form is reminiscent of that of the plots in Fig. 6.24.

The trait features of the plots in Fig. 6.24 remain unchanged with the choices (4,3,2,1) and (4,3,2,0) of the PD payoffs.

Eight Neighbours

In the initial study on the QPD-CA with deterministic updating reported in [10], due to the structural symmetry and fairness of the PD game, no distinction between player A and player B was made in the CA simulations. Instead of this, all the payers in the lattice were treated in [10] as, let us say, player A, interacting, i.e., both playing and comparing, with their eight nearest-neighbours. As a result, the details of the structure of the mean payoffs dependence of γ shown in [10] is different to those achieved with the distinction of players routinely implemented in this book. Nevertheless the results achieved with and without player type distinction in the PD game are qualitatively comparable as it will done in this section.

In the rest of this chapter, the above mentioned no distinction between players A and B is implemented in the probabilistic simulations of the QPD game. Thus, every player plays with his eight nearest-neighbours and with himself, so that the payoff $p_{i,j}^{(T)}$ of a given individual is the average over these eight interactions. The probabilistic mechanism for updating the quantum parameters remains unchanged involving said $p_{i,j}^{(T)}$ average payoffs.

Figure 6.25 deals with the two-parameter QPD(5,3,2,1) PCA with eight neighbours and variable entanglement factor γ. Figure 6.25 shows in its left frame both the actual mean payoffs ($\overline{p}$) and the mean-field approach (p^*) obtained by using strategies constructed from the mean quantum parameter values given in the right frame of the figure. Namely,

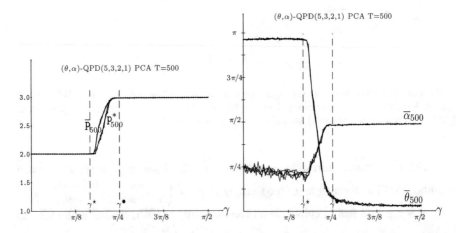

Fig. 6.25 The two-parameter QPD(5,3,2,1)-PCA with eight neighbours. Variable entanglement factor γ. Five simulations at T = 500. Left: Actual mean payoffs ($\overline{p}$) and mean-field payoffs (p^*). Right: Mean quantum parameters

$$U^* = \begin{pmatrix} e^{i\bar{\alpha}} \cos\frac{\bar{\theta}}{2} & \sin\frac{\bar{\theta}}{2} \\ \sin\frac{\bar{\theta}}{2} & e^{-i\bar{\alpha}} \cos\frac{\bar{\theta}}{2} \end{pmatrix} \tag{6.5}$$

The overall structure of the graphs in Fig. 6.25 regarding the 2P-QPD with no distinction between players and probabilistic updating is comparable to that in Fig. 3.1 regarding the 2P-QPD with distinction between players and deterministic updating. In both scenarios, mutual defection yields $P = 2$ below γ^*, $\{\hat{Q}, \hat{Q}\}$ yields $R = 3$ over $\gamma^\bullet$, and in the $(\gamma^*, \gamma^\bullet)$ phase-transition interval the actual mean payoffs and the mean-field payoffs differ, though not a great extent. In the deterministic updating with player distinction scenario of Fig. 3.1 it is $\gamma^* = \arcsin(1/2) = \pi/6$ and $\gamma^\bullet = \arcsin(1/\sqrt{2}) = \frac{\pi}{4}$. But in the probabilistic scenario of Fig. 6.25 the phase-transition commences slightly passed $\pi/6$ and culminates before $\pi/4$. This variation in the values of the γ-thresholds marking the phase-transition was also found in the 2P-QPD simulation with no distinction between players and deterministic updating reported in [10].

Figure 6.26 shows the dynamics up to T=100 in simulations in the QPD(5,3,2,1)-PCA scenario of Fig. 6.25 with $\gamma = 0$ (left) and $\gamma = \pi/2$ (right). As a result of the initial random assignment of the parameter values, it is initially in both frames: $\bar{\theta} \simeq \pi/2 = 1.57$, and $\bar{\alpha} \simeq \pi/4 = 0.78$. With $\gamma = 0$, $\bar{\alpha}$ remains unaltered around its initial $\pi/4$ value, whereas $\bar{\theta}$ rockets towards π, i.e., the parameters of the D strategy. With $\gamma = \pi/2$, $\bar{\alpha}$ rockets towards to $\pi/2$, and $\bar{\theta}$ plummets to zero, i.e., the parameters of the Q strategy. Please, note that in both cases the tendencies heavily emerge from the very beginning, despite the probabilistic nature of the updating of strategies.

Figure 6.27 deals with a simulation of a QPD(5,3,2,1)-PCA scenario of Fig. 6.25 for the entanglement factor $\gamma = 0.654$, i.e., in the middle of the $[\gamma^*, \gamma^\bullet]$ transition interval. The far left panel of the figure shows the evolution up to $T = 500$, of the

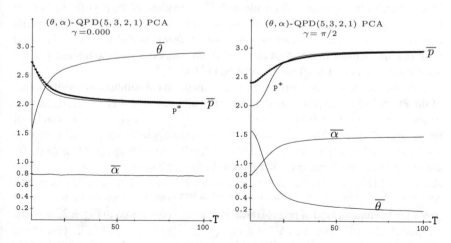

Fig. 6.26 Dynamics up to T=100 in simulations in the two-parameter QPD(5,3,2,1)-PCA scenario of Fig. 6.25. Left: $\gamma = 0$. Right: $\gamma = \pi/2$

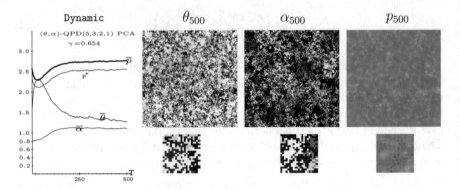

Fig. 6.27 A simulation in the two-parameter QPD(5,3,2,1)-PCA scenario of Fig. 6.25 with $\gamma = 0.654$. Far Left: Evolving mean parameters and payoffs. Center: Parameter patterns at $T = 500$. Far Right: Payoff patterns at $T = 500$. Increasing grey levels indicate increasing pattern values

mean values of θ, α as well as the actual payoffs ($\overline{p}$) and the mean-field payoffs (p^*). The $\overline{\alpha}$ parameter increases its value in the beginning of the simulation, but by T = 250 it has reached a plateau level of approximately 1.1. The $\overline{\theta}$ parameter notably increases its value at the first time-steps, but afterwards it decreases in a monotone way. Further evolution of this simulation maintains the smooth decrease of $\overline{\theta}$, accompanied by a small increase of $\overline{\alpha}$, so that by T = 1000 both parameter values become fairly equal, $\overline{\theta} \simeq \overline{\alpha} = 1.2$. The actual mean payoff $\overline{p}$ initially decreases its value, in parallel with the increase of $\overline{\theta}$, but soon recovers from the initial small depletion, in parallel with the $\overline{\theta}$ decrease, and becomes fairly stabilized by T = 300, with a very smooth further increase. It is $\overline{p} = 2.76$ at T = 500 and $\overline{p} = 2.79$ at T = 1000. The parameter patterns at $T = 500$ in Fig. 6.27 show a kind of fine-grain aspect, a kind of spatial heterogeneity that explains why the mean-field estimations of the payoffs differ from the actual mean payoffs, in this particular simulation with p^* below $\overline{p}$. The spatial structures of the parameter and payoff patterns shown in Fig. 6.27 notably differ form that much more uniform patterns achieved with deterministic updating in the simulations with eight neighbours reported in [10, 11].

The general form of the plots in Fig. 6.25 remains in CA-simulations with choices of the PD payoffs different from that primarily adopted here, i.e., (5,3,2,1). But the details of the transition from P to R vary notably accordingly to the hierarchy of the γ^* and $\gamma^\bullet$ values. Thus, the order $\gamma^* < \gamma^\bullet$ (that applies with (5,3,2,1)) may be inverted into $\gamma^\bullet < \gamma^*$, in which case, in the ($\gamma^\bullet$, γ^*) interval both $\{\hat{Q}, \hat{Q}\}$ and $\{\hat{D}, \hat{D}\}$ are in NE in conventional quantum games. In the limit case it is $\gamma^* = \gamma^\bullet$. In these alternative PD-scenarios, the $\overline{p}$-transition from P to R in CA-simulations occurs abruptly when γ increases. In these alternative PD-parameter scenarios the P to R transition becomes a kind of phase-transition that has been tested in Fig. 6.28 with the PD-parameters (4,3,2,0), i.e., $\gamma^* = \arcsin(1/2) > \gamma^\bullet = \arcsin(1/\sqrt{2})$ (P-marked plots in the right frame of Fig. 6.28), and with (4,3,2,1), i.e., $\gamma^* = \gamma^\bullet = \arcsin(1/2)$ (P-marked plots in the left frame of Fig. 6.28). In the former case the abrupt transition

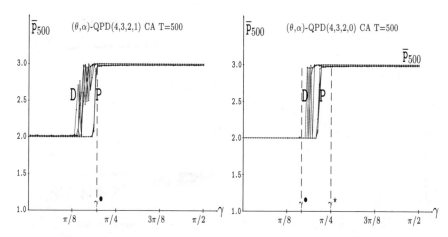

Fig. 6.28 Mean payoffs of the two-parameter QPD-CA with eight neighbours. Variable entanglement factor γ. Five simulations at T = 500. Left: (4,3,2,1) payoffs. Right: (4,3,2,0) payoffs. P: Probabilistic updating, D: Deterministic updating

from $P = 2$ to $R = 3$ occurs close to the middle of the $(\gamma^\bullet, \gamma^*)$ interval, and in the latter case, as foreseeable, close to $\gamma^* = \gamma^\bullet$. The D-marked plots in both frames of Fig. 6.28 show the transition from $P = 2$ to $R = 3$ with deterministic updating of the strategy parameters. In both scenarios, the transition is more noisy than with probabilistic updating, and occurs at lower γ values, particularly in the $\gamma^* = \gamma^\bullet$ scenario (left frame).

Figure 6.29 shows the results achieved in the scenario of Fig. 6.25, but in the three-parameter strategies model. At variance with what happens with two parameters in Fig. 6.25, the actual mean payoffs $(\overline{p})$ increase fairly monotonically under the three-parameter strategies model (left frame of Fig. 6.29), again, from a mean payoff equal to punishment $P = 2$ up to the reward $R = 3$. The mean-field payoff approach (p^*) fits almost perfectly to the actual payoff $(\overline{p})$. The right frame of Fig. 6.29 shows that: (i) the $\overline{\beta}$ parameter does not vary appreciably from its initial assignment $\pi/4$, regardless of γ, (ii) the $\overline{\theta}$ parameter does not depend very much of γ, reaching values close to π, and (iii) $\overline{\alpha}$ smoothly decreases from $\pi/4$ as γ increases.

The overall structure of the graphs in Fig. 6.29 regarding the 3P-QPD with no distinction between players and probabilistic updating is comparable to that in Fig. 3.17 regarding the 3P-QPD with distinction between players and deterministic updating, albeit in Fig. 6.29 both the mean payoffs and the quantum parameter graphs show a much more crisp aspect. Anyhow, in both scenarios no discontinuity is apparent at $\gamma^\# = \arcsin(\sqrt{1/3}) = 0.615$, so that, one may conjecture that the dynamics in this scenario resorts somehow to the mixed strategies in NE described in [12]. In any case, for the (5,3,2,1) PD-parameters in Fig. 6.29, the payoffs induced by the strategy pair $(\hat{\mathfrak{X}}, \hat{\mathfrak{X}})$, i.e., $p^{\hat{\mathfrak{X}},\hat{\mathfrak{X}}}(\gamma) = 2 + \sin^2 \gamma$, is perfectly verified for the actual mean payoffs and their mean-field approaches. Thus for example for the two extreme values of

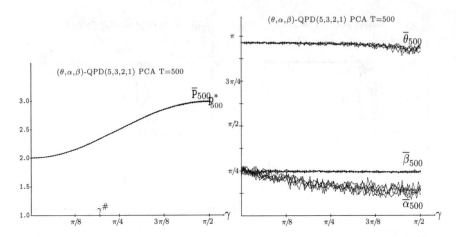

Fig. 6.29 A three-parameter QPD(5,3,2,1)-PCA with eight neighbours. Variable entanglement factor γ. Five simulations at T $= 500$. Left: Actual mean payoffs $\overline{p}$, and mean-field payoffs p^*. Right: Mean parameter values

the entangling factor: $\overline{p}(0) = p^{\hat{x},\hat{x}}(0) = P = 2$, $p^{\#}(\pi/2) = R = 3$, and fairly for its middle value $\overline{p}(\pi/4) = p^{\hat{x},\hat{x}}(\pi/4) = (P + R)/2 = 2.5$.

The results just reported in Fig. 6.29 apply also to CA-simulations with $(\gamma^* = \gamma^{\bullet})$ in the PD with payoffs (4,3,2,1), and with $(\gamma^* > \gamma^{\bullet})$ in the PD with payoffs (4,3,2,0). In these alternative PD-scenarios, the $\overline{\alpha}$ mean parameter remains close to $\pi/4$, as $\overline{\beta}$ does, but no other distinctive effects have been found.

References

1. Alonso-Sanz, R.: A quantum battle of the sexes cellular automaton with probabilistic updating. Int. J. Parallel Emergent Distrib. Syst. **31**(4), 305–218 (2016)
2. Alonso-Sanz, R.: A quantum prisoner's dilemma cellular automaton with probabilistic updating. J. Cell. Autom. **11**(2–3), 145–166 (2015)
3. Alonso-Sanz, R.: Self-organization in the spatial battle of the sexes with probabilistic updating. Phys. A **390**, 2956–2967 (2011)
4. Li, Q., Iqbal, A., Chen, M., Abbott, D.: Quantum strategies win in a defector-dominated population. Phys. A **391**, 3316–3322 (2012)
5. Du, J., Xu, X., Li, H., Zhou, X., Han, R.: Nash Equilibrium in Quantum Games (2000). http://cds.cern.ch/record/468453/files/0010050.pdf
6. Alonso-Sanz, R.: Variable entangling in a quantum battle of the sexes cellular automaton. In: ACRI-2014. LNCS, vol. 8751, pp. 125–135 (2014)
7. Du, J.F., Xu, X.D., Li, H., Zhou, X., Han, R.: Entanglement enhanced multiplayer quantum games. Phys. Lett. A **302**, 222–233 (2002)
8. Flitney, A.P., Hollengerg, L.C.L.: Nash equilibria in quantum games with generalized two-parameter strategies. Phys. Lett. A **363**, 381–388 (2007)
9. Eisert, J., Wilkens, M., Lewenstein, M.: Quantum games and quantum strategies. Phys. Rev. Lett. **83**(15), 3077–3080 (1999)

10. Alonso-Sanz, R.: Variable entangling in a quantum prisoner's dilemma cellular automaton. Quantum Inf. Process. **14**, 147–164 (2015)
11. Alonso-Sanz, R.: A quantum prisoner's dilemma cellular automaton. Proc. R. Soc. A **470**, 20130793 (2014)
12. Du, J.F., Li, H., Xu, X.D., Zhou, X., Han, R.: Phase-transition-like behaviour of quantum games. J. Phys. A Math. Gen. **36**(23), 6551–6562 (2003)

Chapter 7
Quantum Noise

The disrupting effect of quantum noise on the dynamics of the spatial formulation of quantum games with variable entanglement is studied in this chapter. The Prisoner's Dilemma and the Samaritan's Dilemma [1] are considered in this study. It is concluded here that quantum noise induces in these games the need for higher entanglement in order to make possible the emergence of the strategy pair that renders the social welfare solution. The effect of quantum noise in quantum-relativistic games will be studied in the last section of next Chap. 8.

7.1 The Density Matrix Formalism

Quantum mechanics may be formulated in terms of a mathematical tool known as *density operator* or *density matrix* instead of state vectors. This alternate formulation is mathematically equivalent to the state vector approach, but provides a more convenient language for dealing with some commonly encountered scenarios in quantum mechanics. Whether one uses the density operator language or the state vector language is in principle a matter of taste, since both give the same results; however it is sometimes much easier to approach the problem from one point of view rather than the other [2].

The density matrix for a system whose state vector is known to be exactly $|\Psi\rangle$ turns out to be: $\rho = |\Psi\rangle\langle\Psi|$. Thus, in the EWL model, with initial state,

$$|\psi_i\rangle = \hat{J}|00\rangle = \cos\frac{\gamma}{2}|00\rangle_{A,B} + i\sin\frac{\gamma}{2}|11\rangle_{A,B} = (\cos\frac{\gamma}{2}, 0, 0, i\sin\frac{\gamma}{2})' \quad (7.1)$$

© Springer Nature Switzerland AG 2019
R. Alonso-Sanz, *Quantum Game Simulation*, Emergence, Complexity
and Computation 36, https://doi.org/10.1007/978-3-030-19634-9_7

the initial density matrix is,

$$
\rho_i = |\psi_i\rangle\langle\psi_i| = \begin{pmatrix} \cos^2\gamma/2 & 0\ 0 & -i\cos\gamma/2\sin\gamma/2 \\ 0 & 0\ 0 & 0 \\ 0 & 0\ 0 & 0 \\ i\cos\gamma/2\sin\gamma/2 & 0\ 0 & \sin^2\gamma/2 \end{pmatrix} \tag{7.2}
$$

After the application of the players strategies and the $\hat{J}^\dagger$ gate, the state of the game evolves to $|\psi_f\rangle = \hat{J}^\dagger(\hat{U}_A \otimes \hat{U}_B)\hat{J}|00\rangle$, with associated density matrix,

$$
\rho_f = |\psi_f\rangle\langle\psi_f| = \hat{J}^\dagger(\hat{U}_A \otimes \hat{U}_B)\rho_i(\hat{U}_A \otimes \hat{U}_B)^\dagger \hat{J} \tag{7.3}
$$

Finally, the elements of Π are obtained as the diagonal elements of the final density matrix: $\Pi = \begin{pmatrix} \rho_{11} & \rho_{22} \\ \rho_{33} & \rho_{44} \end{pmatrix}$.

When using the density matrix formalism, it is useful to express the entangling operator as $J = \cos\frac{\gamma}{2}I + i\sin\frac{\gamma}{2}P$, with the permutation matrix $P = P' = \begin{pmatrix} 0 & 0 & 0 & 1 \\ 0 & 0 & -1 & 0 \\ 0 & -1 & 0 & 0 \\ 1 & 0 & 0 & 0 \end{pmatrix}$. Thus, $J^\dagger = \cos\frac{\gamma}{2}I - i\sin\frac{\gamma}{2}P$, so that $J^\dagger\delta_i J = (\cos\frac{\gamma}{2}I - i\sin\frac{\gamma}{2}P)$ $\delta_i(\cos\frac{\gamma}{2}I + i\sin\frac{\gamma}{2}P)$. Therefore, the final density matrix may be calculated as indicated in Eq. (7.4), where $\delta_i = (\hat{U}_A \otimes \hat{U}_B)\rho_i(\hat{U}_A \otimes \hat{U}_B)^\dagger$

$$
\rho_f^{CC} = \cos^2\frac{\gamma}{2}\delta_i + \sin^2\frac{\gamma}{2}P\delta_i + i\sin\frac{\gamma}{2}\cos\frac{\gamma}{2}(\delta_i P - P\delta_i) \tag{7.4}
$$

7.2 Amplitude-Damping Noise

Real-world quantum information processing systems inevitably interact with the environment. The environmental noise may destroy the quantum properties and be harmful for information processing. Thus, in general, it is important to take quantum noise into consideration in order to provide a detailed description of the system dynamics. However, let us remark that Yin et al. [3] have demonstrated that noise has a negligible effect in a the breakthrough experiment, where the feasibility of transmitting entangled photons over large distances using a satellite-based distribution was demonstrated.

In the presence of noise, the shared state is corrupted before the players apply their (unaltered) strategies $\hat{U}_A$ and $\hat{U}_B$. If we assume that the environment thermalizes sufficiently fast, the noise effect on a single qubit can be described through Kraus operators using Completely Positive Trace-Preserving maps (usually known as CPTP), which distort the density matrix as $\mathcal{N}(\rho) = \sum_i K_i\rho K_i^\dagger$, and fulfill the completeness relation $\sum_i K_i^\dagger K_i = I$ where the Kraus operators K_i^μ that describe the noise effect.

We assume that both qubits of the shared state are affected by the same kind of noise. As a result, the Kraus operators transform the initial density matrix into,

$$\rho_i^* = \sum_{i=1}^{2}\sum_{j=1}^{2}(K_i \otimes K_j)\rho_i(K_i \otimes K_j)^\dagger \tag{7.5}$$

We will consider here the effect of amplitude-damping noise type that can be characterized by the Kraus operators:

$$K_1 = \begin{pmatrix} 1 & 0 \\ 0 & \sqrt{1-\mu} \end{pmatrix}, \quad K_2 = \begin{pmatrix} 0 & \sqrt{\mu} \\ 0 & 0 \end{pmatrix} \tag{7.6}$$

where $\mu \in [0, 1]$ represents the strength of noise.

It is, $K_1 \otimes K_1 = \begin{pmatrix} 1 & 0 & 0 & 0 \\ 0 & \sqrt{1-\mu} & 0 & 0 \\ 0 & 0 & \sqrt{1-\mu} & 0 \\ 0 & 0 & 0 & 1-\mu \end{pmatrix}$, $K_1 \otimes K_2 = \begin{pmatrix} 0 & \sqrt{\mu} & 0 & 0 \\ 0 & 0 & 0 & 0 \\ 0 & 0 & 0 & \sqrt{\mu}\sqrt{1-\mu} \\ 0 & 0 & 0 & 0 \end{pmatrix}$, $K_2 \otimes K_1 = \begin{pmatrix} 0 & 0 & \sqrt{\mu} & 0 \\ 0 & 0 & 0 & \sqrt{\mu}\sqrt{1-\mu} \\ 0 & 0 & 0 & 0 \\ 0 & 0 & 0 & 0 \end{pmatrix}$, $K_2 \otimes K_2 = \begin{pmatrix} 0 & 0 & 0 & \mu \\ 0 & 0 & 0 & 0 \\ 0 & 0 & 0 & 0 \\ 0 & 0 & 0 & 0 \end{pmatrix}$.

With maximum noise, the direct products of the Kraus matrices become, $K_1 \otimes K_1 = \begin{pmatrix} 1 & 0 & 0 & 0 \\ 0 & 0 & 0 & 0 \\ 0 & 0 & 0 & 0 \\ 0 & 0 & 0 & 0 \end{pmatrix}$, $K_1 \otimes K_2 = \begin{pmatrix} 0 & 1 & 0 & 0 \\ 0 & 0 & 0 & 0 \\ 0 & 0 & 0 & 0 \\ 0 & 0 & 0 & 0 \end{pmatrix}$, $K_2 \otimes K_1 = \begin{pmatrix} 0 & 0 & 1 & 0 \\ 0 & 0 & 0 & 0 \\ 0 & 0 & 0 & 0 \\ 0 & 0 & 0 & 0 \end{pmatrix}$, $K_2 \otimes K_2 = \begin{pmatrix} 0 & 0 & 0 & 1 \\ 0 & 0 & 0 & 0 \\ 0 & 0 & 0 & 0 \\ 0 & 0 & 0 & 0 \end{pmatrix}$. Consequently it is $\rho_i[\mu = 1.0] = \begin{pmatrix} 1 & 0 & 0 & 0 \\ 0 & 0 & 0 & 0 \\ 0 & 0 & 0 & 0 \\ 0 & 0 & 0 & 0 \end{pmatrix}$, which leads to $\Pi^{C,C} = \Pi^{Q,Q} = \begin{pmatrix} \cos^2 \frac{\gamma}{2} & 0 \\ 0 & \sin^2 \frac{\gamma}{2} \end{pmatrix}$. Permuting the columns, the rows, and the rows and columns of said $\Pi^{C,C}$, provides the $\Pi^{C,D} = \Pi^{Q,D}$, $\Pi^{D,C} = \Pi^{D,Q}$, and $\Pi^{D,D} = \begin{pmatrix} \sin^2 \frac{\gamma}{2} & 0 \\ 0 & \cos^2 \frac{\gamma}{2} \end{pmatrix}$ with full noise. Noticeably, these joint probabilities with full noise relate to the noiseless ones given in Eqs. 2.7 with γ replaced by $\gamma/2$.

The Kraus operators (7.6) transform the initial density matrix given by Eq. (7.2) into,

$$\rho_i^*(\mu) = \begin{pmatrix} \frac{1}{2}[1+\mu^2+(1-\mu^2)\cos\gamma] & 0 & 0 & -\frac{i}{2}(1-\mu)\sin\gamma \\ 0 & (1-\mu)\mu\sin^2\frac{\gamma}{2} & 0 & 0 \\ 0 & 0 & (1-\mu)\mu\sin^2\frac{\gamma}{2} & 0 \\ \frac{i}{2}(1-\mu)\sin\gamma & 0 & 0 & (1-\mu)^2\sin^2\frac{\gamma}{2} \end{pmatrix} \tag{7.7}$$

The particular case of the density matrix for $\mu = 1/2$ is given in Eq. (7.8). The expressions of Π for the (Q,Q) and (Q,D) contests with $\mu = 1/2$ are given in Eqs. (7.9) and (7.10). The components of ρ_f calculated according to the method proposed in Eq. (7.4) for said (Q,Q) and (Q,D) contests are given in the footnotes[1] and[2]. The entries of $\Pi^{DD}[\mu = 0.5]$ coincide with that of $\Pi^{CC}[\mu = 0.5]$ in Eq. (7.9), but with the diagonal elements interchanged, i.e., $\pi_{11}^{DD} = \sin^2 \frac{\gamma}{2}, \pi_{22}^{DD} = 4 - 3\sin^2 \frac{\gamma}{2}$.

$$^1\rho_i^* P = \frac{1}{4}\begin{pmatrix} 4-3\sin^2\frac{\gamma}{2} & 0 & 0 & -i\sin\gamma \\ 0 & \sin^2\frac{\gamma}{2} & 0 & 0 \\ 0 & 0 & \sin^2\frac{\gamma}{2} & 0 \\ i\sin\gamma & 0 & 0 & \sin^2\frac{\gamma}{2} \end{pmatrix}\begin{pmatrix} 0 & 0 & 0 & 1 \\ 0 & 0 & -1 & 0 \\ 0 & -1 & 0 & 0 \\ 1 & 0 & 0 & 0 \end{pmatrix} =$$

$$\begin{pmatrix} -i\frac{1}{4}\sin\gamma & 0 & 0 & \cos^2\frac{\gamma}{2}+\frac{1}{4}\sin^2\frac{\gamma}{2} \\ 0 & 0 & -\frac{1}{4}\sin^2\frac{\gamma}{2} & 0 \\ 0 & -\frac{1}{4}\sin^2\frac{\gamma}{2} & 0 & 0 \\ \frac{1}{4}\sin^2\frac{\gamma}{2} & 0 & 0 & i\frac{1}{4}\sin\gamma \end{pmatrix},$$

$$P\rho_i^* = \begin{pmatrix} i\frac{1}{4}\sin\gamma & 0 & 0 & \frac{1}{4}\sin^2\frac{\gamma}{2} \\ 0 & 0 & -\frac{1}{4}\sin^2\frac{\gamma}{2} & 0 \\ 0 & -\frac{1}{4}\sin^2\frac{\gamma}{2} & 0 & 0 \\ \cos^2\frac{\gamma}{2}+\frac{1}{4}\sin^2\frac{\gamma}{2} & 0 & 0 & -i\frac{1}{4}\sin\gamma \end{pmatrix}$$

$$\rho_i^* P - P\rho_i^* = \begin{pmatrix} -i\frac{1}{2}\sin\gamma & 0 & 0 & \cos^2\frac{\gamma}{2} \\ 0 & 0 & 0 & 0 \\ 0 & 0 & 0 & 0 \\ -\cos^2\frac{\gamma}{2} & 0 & 0 & i\frac{1}{2}\sin\gamma \end{pmatrix}, \quad P\rho_i^* P = \begin{pmatrix} \frac{1}{4}\sin^2\frac{\gamma}{2} & 0 & 0 & i\frac{1}{4}\sin\gamma \\ 0 & \frac{1}{4}\sin^2\frac{\gamma}{2} & 0 & 0 \\ 0 & 0 & \frac{1}{4}\sin^2\frac{\gamma}{2} & 0 \\ -i\frac{1}{4}\sin\gamma & 0 & 0 & \cos^2\frac{\gamma}{2}+\frac{1}{4}\sin^2\frac{\gamma}{2} \end{pmatrix}$$

$\rho_i^*(2,2) = \rho_i^*(3,3) = \frac{1}{4}(\cos^2\frac{\gamma}{2}\sin^2\frac{\gamma}{2}+\sin^2\frac{\gamma}{2}\sin^2\frac{\gamma}{2}) = \frac{1}{4}\sin^2\frac{\gamma}{2}$

$\rho_i^*(1,1) = \cos^2\frac{\gamma}{2}(\cos^2\frac{\gamma}{2}+\frac{1}{4}\sin^2\frac{\gamma}{2})+\sin^2\frac{\gamma}{2}\frac{1}{4}\sin^2\frac{\gamma}{2}+\sin^2\frac{\gamma}{2}\cos^2\frac{\gamma}{2} = \cos^2\frac{\gamma}{2}(\cos^2\frac{\gamma}{2}+\frac{1}{4}\sin^2\frac{\gamma}{2})+$
$\sin^2\frac{\gamma}{2}(\frac{1}{4}\sin^2\frac{\gamma}{2}+\cos^2\frac{\gamma}{2}) = \frac{1}{4}\sin^2\frac{\gamma}{2}+\cos^2\frac{\gamma}{2}$

$\rho_i^*(4,4) = \cos^2\frac{\gamma}{2}\frac{1}{4}\sin^2\frac{\gamma}{2}+\sin^2\frac{\gamma}{2}(\cos^2\frac{\gamma}{2}+\frac{1}{4}\sin^2\frac{\gamma}{2})-\sin^2\frac{\gamma}{2}\cos^2\frac{\gamma}{2} = \cos^2\frac{\gamma}{2}\frac{1}{4}\sin^2\frac{\gamma}{2}+\sin^2\frac{\gamma}{2}\frac{1}{4}\sin^2\frac{\gamma}{2} = \frac{1}{4}\sin^2\frac{\gamma}{2}$.

$$^2\delta_i = (\hat{Q}\otimes\hat{D})\rho_i^*(\hat{Q}\otimes\hat{D})^\dagger = \begin{pmatrix} 0 & 1 & 0 & 0 \\ -1 & 0 & 0 & 0 \\ 0 & 0 & 0 & -1 \\ 0 & 0 & 1 & 0 \end{pmatrix}\frac{1}{4}\begin{pmatrix} 0 & -4+3\sin^2\frac{\gamma}{2} & i\sin\gamma & 0 \\ \sin^2\frac{\gamma}{2} & 0 & 0 & 0 \\ 0 & 0 & 0 & \sin^2\frac{\gamma}{2} \\ 0 & -i\sin\gamma & -\sin^2\frac{\gamma}{2} & 0 \end{pmatrix} =$$

$$\frac{1}{4}\begin{pmatrix} \sin^2\frac{\gamma}{2} & 0 & 0 & 0 \\ 0 & 4-3\sin^2\frac{\gamma}{2} & -i\sin\gamma & 0 \\ 0 & i\sin\gamma & \sin^2\frac{\gamma}{2} & 0 \\ 0 & 0 & 0 & \sin^2\frac{\gamma}{2} \end{pmatrix}, \quad \delta_i P = \frac{1}{4}\begin{pmatrix} 0 & 0 & 0 & \sin^2\frac{\gamma}{2} \\ 0 & i\sin\gamma & -4+3\sin^2\frac{\gamma}{2} & 0 \\ 0 & -\sin^2\frac{\gamma}{2} & -i\sin\gamma & 0 \\ \sin^2\frac{\gamma}{2} & 0 & 0 & 0 \end{pmatrix}$$

$$P\delta_i = \frac{1}{4}\begin{pmatrix} 0 & 0 & 0 & \sin^2\frac{\gamma}{2} \\ 0 & -i\sin\gamma & -\sin^2\frac{\gamma}{2} & 0 \\ 0 & -4+3\sin^2\frac{\gamma}{2} & i\sin\gamma & 0 \\ \sin^2\frac{\gamma}{2} & 0 & 0 & 0 \end{pmatrix},$$

$$\delta_i P - P\delta_i = \frac{1}{4}\begin{pmatrix} 0 & 0 & 0 & 0 \\ 0 & 2i\sin\gamma & -4+4\sin^2\frac{\gamma}{2}+i\sin\gamma & 0 \\ 0 & 4-4\sin^2\frac{\gamma}{2} & -2i\sin\gamma & 0 \\ 0 & 0 & 0 & 0 \end{pmatrix}$$

$$P\delta_i P = \frac{1}{4}\begin{pmatrix} \sin^2\frac{\gamma}{2} & 0 & 0 & 0 \\ 0 & \sin^2\frac{\gamma}{2} & i\sin\gamma & 0 \\ 0 & -i\sin\gamma & 4-3\sin^2\frac{\gamma}{2} & 0 \\ 0 & 0 & 0 & \sin^2\frac{\gamma}{2} \end{pmatrix}, \quad \rho_{11} = \rho_{44} = \frac{1}{4}(\cos^2\frac{\gamma}{2}\sin^2\frac{\gamma}{2}+\sin^2\frac{\gamma}{2}\sin^2\frac{\gamma}{2}) =$$

$\frac{1}{4}\sin^2\frac{\gamma}{2}$

$\rho_{22} = \frac{1}{4}\left(\cos^2\frac{\gamma}{2}(4-3\sin^2\frac{\gamma}{2})+\sin^2\frac{\gamma}{2}\sin^2\frac{\gamma}{2}+i\sin\frac{\gamma}{2}\cos\frac{\gamma}{2}2i\sin\gamma\right) = \frac{1}{4}(4-4\sin^2\frac{\gamma}{2}-$
$3\sin^2\frac{\gamma}{2}\cos^2\frac{\gamma}{2}+\sin^2\frac{\gamma}{2}\sin^2\frac{\gamma}{2}-4\sin^2\frac{\gamma}{2}\cos^2\frac{\gamma}{2} = \frac{1}{4}(4-4\sin^2\frac{\gamma}{2}+\sin^4\frac{\gamma}{2}-7\sin^2\frac{\gamma}{2}(1-$
$\sin^2\frac{\gamma}{2}))$

$\rho_{33} = \frac{1}{4}\left(\cos^2\frac{\gamma}{2}\sin^2\frac{\gamma}{2}+\sin^2\frac{\gamma}{2}(4-3\sin^2\frac{\gamma}{2})+i\sin\frac{\gamma}{2}\cos\frac{\gamma}{2}(-2i\sin\gamma)\right) = \frac{1}{4}(\sin^2\frac{\gamma}{2}-$
$\sin^4\frac{\gamma}{2}+4\sin^2\frac{\gamma}{2}-3\sin^4\frac{\gamma}{2}+4\sin^2\frac{\gamma}{2}-4\sin^4\frac{\gamma}{2})$.

$$\rho_i^*[\mu = 1/2] = \frac{1}{4}\begin{pmatrix} 4 - 3\sin^2\frac{\gamma}{2} & 0 & 0 & -i\sin\gamma \\ 0 & \sin^2\frac{\gamma}{2} & 0 & 0 \\ 0 & 0 & \sin^2\frac{\gamma}{2} & 0 \\ i\sin\gamma & 0 & 0 & \sin^2\frac{\gamma}{2} \end{pmatrix} \tag{7.8}$$

$$\Pi^{\hat{Q},\hat{Q}}[\mu = 1/2] = \Pi^{\hat{C},\hat{C}} = \frac{1}{4}\begin{pmatrix} 4 - 3\sin^2\frac{\gamma}{2} & \sin^2\frac{\gamma}{2} \\ \sin^2\frac{\gamma}{2} & \sin^2\frac{\gamma}{2} \end{pmatrix} \tag{7.9}$$

$$\Pi^{\hat{Q},\hat{D}}[\mu = 1/2] = \frac{1}{4}\begin{pmatrix} \sin^2\frac{\gamma}{2} & 4 - 11\sin^2\frac{\gamma}{2} + 8\sin^4\frac{\gamma}{2} \\ 9\sin^2\frac{\gamma}{2} - 8\sin^4\frac{\gamma}{2} & \sin^2\frac{\gamma}{2} \end{pmatrix} \tag{7.10}$$

According to Eqs. (7.9)–(7.10), the payoffs of the $\hat{Q}$, $\hat{D}$ and $\hat{C}$ strategy combinations in the $QPD(\mathfrak{T}, R, P, S)$ with $\mu = 0.5$ noise are given in Eqs. (7.11).

$$p_{A,B}^{\hat{Q},\hat{Q}}(0.5) = p_{A,B}^{\hat{C},\hat{C}}(0.5) = R - \frac{1}{4}(3R - \mathfrak{T} - S - P)\sin^2\frac{\gamma}{2} \tag{7.11a}$$

$$p_{A,B}^{\hat{D},\hat{D}}(0.5) = P + \frac{1}{4}(\mathfrak{T} + S + R - 3P)\sin^2\frac{\gamma}{2} \tag{7.11b}$$

$$p_A^{\hat{Q},\hat{D}}(0.5) = p_B^{\hat{D},\hat{Q}}(0.5) = S + \frac{1}{4}(R + P + 9\mathfrak{T} - 11S)\sin^2\frac{\gamma}{2} - 2(\mathfrak{T} - S)\sin^4\frac{\gamma}{2} \tag{7.11c}$$

$$p_A^{\hat{D},\hat{Q}}(0.5) = p_B^{\hat{Q},\hat{D}}(0.5) = \mathfrak{T} - \frac{1}{4}(11\mathfrak{T} - R - P - 9S)\sin^2\frac{\gamma}{2} + 2(\mathfrak{T} - S)\sin^4\frac{\gamma}{2} \tag{7.11d}$$

$$p_{A,B}^{\hat{Q},\hat{C}}(0.5) = p_{A,B}^{\hat{C},\hat{Q}}(0.5) = R - \frac{1}{4}(11R - 9P - S - \mathfrak{T})\sin^2\frac{\gamma}{2} + 2(R - P)\sin^4\frac{\gamma}{2} \tag{7.11e}$$

The payoff dependence on γ of the $\hat{C}$, $\hat{D}$, and $\hat{Q}$, strategy interactions in the QPD(5,3,2,1) game with $\mu = 1/2$ noise and full noise are shown in Fig. 7.1 (left and right frames respectively). The graphs with full noise are the (5,3,2,1)-particular realization of the general formulas: $p_{A,B}^{Q,Q} = p_{A,B}^{C,C} = R\cos^2\frac{\gamma}{2} + P\sin^2\frac{\gamma}{2}$, $p_{A,B}^{D,D} = R\sin^2\frac{\gamma}{2} + P\cos^2\frac{\gamma}{2}$, $p_A^{QD} = p_A^{CD} = p_B^{DC} = p_B^{DQ} = \mathfrak{T}\sin^2\frac{\gamma}{2} + S\cos^2\frac{\gamma}{2}$, and $p_A^{DC} = p_A^{DQ} = p_B^{CD} = p_B^{QD} = S\sin^2\frac{\gamma}{2} + \mathfrak{T}\cos^2\frac{\gamma}{2}$. In both frames. the $\gamma^\star$ landmark is given at the intersection of p^{DD} and p_A^{QD}, whereas $\gamma^\bullet$ landmark is given at the intersection of p_A^{QD} and p^{QQ}.

The payoff dependence on γ of the $\hat{C}$, $\hat{D}$, and $\hat{Q}$, strategy interactions in the QPD(4,3,2,1) game with $\mu = 1/2$ noise and full noise are shown in Fig. 7.2 (left and right frames respectively). In Fig. 7.2, at variance with what happens in Fig. 7.1, p^{QQ} does not intersect p_B^{QD} and p_B^{QD} in the same point with $\mu = 0.5$ noise.

Let us remark that only amplitude-damping quantum noise will be under scrutiny in this study. The effect of other types of quantum noise such as those describing

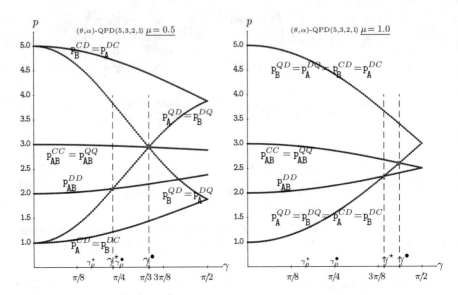

Fig. 7.1 Payoffs of the $\hat{Q}$, $\hat{D}$ and $\hat{C}$ strategy interactions in a QPD(5,3,2,1) with variable entanglement factor γ and amplitude damping noise of factor μ. Left: $\mu = 0.5$, Right: $\mu = 1.0$

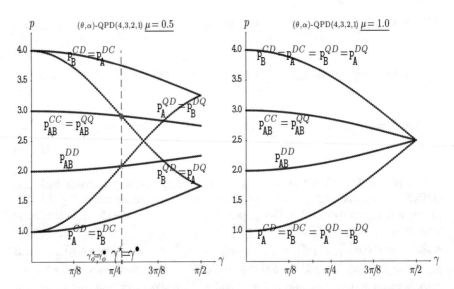

Fig. 7.2 Payoffs of the $\hat{Q}$, $\hat{D}$ and $\hat{C}$ strategy interactions in a QPD(4,3,2,1) with variable entanglement factor γ and amplitude damping noise of factor μ. Left: $\mu = 0.5$, Right: $\mu = 1.0$

bit and phase flip channels, depolarizing channels, or phase damping channels [2, 4] will be taken into account in future work. Particular attention deserves that of phase-damping, structurally close to amplitude-damping as both damping noises have the same K_1 Kraus operator, deferring in $K_2 = \begin{pmatrix} 0 & 0 \\ 0 & \sqrt{\mu} \end{pmatrix}$ with phase damping. Mirroring the proximity of these two noise types, with full phase-damping noise, it turns out that Π^{QQ} becomes diagonal as with amplitude damping, but with $\pi_{22}^{QQ} = \frac{1}{2}\sin^2\gamma$ instead of $\pi_{22}^{QQ} = \sin^2\frac{\gamma}{2}$. With $\mu = 0.5$ phase damping, Π^{QQ} also turns out diagonal with $\pi_{22}^{QQ} = \frac{1}{4}\sin^2\gamma$.

7.3 The QPD with Noise

Figure 7.3 deals with the results obtained in QPD(5,3,2,1)-CA simulations with $\mu = 0.5$ noise. The general form of the graphs of $(\overline{p}_A, \overline{p}_B)$ shown in the simulations free of noise in Fig. 3.1 is still recognizable with $\mu = 0.5$ noise in Fig. 7.3, albeit it is significantly altered. Particularly relevant is that both the $\gamma^\star$ and $\gamma^\bullet$ thresholds increase their value with noise, so that mutual defection persists for larger γ values and mutual Q demands higher γ.

Recall from the noiseless QPD(5,3,2,1) simulations reported in Fig. 3.1 that the CA imitation dynamics enables the emergence of NE according to the scheme: $\{\hat{D}, \hat{D}\} \rightarrow \{\{\hat{D}, \hat{Q}\}, \{\hat{Q}, \hat{D}\}\} \rightarrow \{\hat{Q}, \hat{Q}\}$ as γ increases, with thresholds in the QPD with (5,3,2,1) parameters being $\gamma^\star = \arcsin(\sqrt{1/4}) = \pi/6$, and $\gamma^\bullet = \arcsin(\sqrt{1/2}) = \pi/4$. These thresholds in noiseless ($\mu = 0.0$) simulations will be referred here for convenience with 0.0 subindex. Thus,

$$\gamma_{0.0}^\star = \arcsin\left(\sqrt{\frac{P-S}{\mathfrak{T}-S}}\right) \quad , \quad \gamma_{0.0}^\bullet = \arcsin\left(\sqrt{\frac{\mathfrak{T}-R}{\mathfrak{T}-S}}\right) \qquad (7.12)$$

The γ thresholds in the $QPD(\mathfrak{T}, R, P, S)$ game with $\mu = 0.5$ noise are given in Eqs. (7.13). The $\gamma_{0.5}^\star$ landmark is obtained at the intersection of p^{DD} given in Eq. (7.11b) and p_A^{QD} given in Eq. (7.11c).[3] The $\gamma_{0.5}^\bullet$ landmark is obtained at the intersection of p_B^{QD} given in Eq. (7.11d) and p^{QQ} given in Eq. (7.11a).[4]

---

[3] $P + \frac{1}{4}(\mathfrak{T} + S + R - 3P)\sin^2\frac{\gamma}{2} = S + \frac{1}{4}(R + P + 9\mathfrak{T} - 11S)\sin^2\frac{\gamma}{2} - 2(\mathfrak{T} - S)\sin^4\frac{\gamma}{2} \Rightarrow 2(S - \mathfrak{T})\sin^4\frac{\gamma}{2} + \frac{1}{4}(8\mathfrak{T} + 4P - 12S)\sin^2\frac{\gamma}{2} + S - P = 0$. It is $(2\mathfrak{T} + P - 3S)^2 - 8(S - \mathfrak{T})(S - P) = 4\mathfrak{T}^2 + P^2 + 9S^2 + 4\mathfrak{T}P - 12\mathfrak{T}S - 6PS - 8S^2 + 8PS + 8\mathfrak{T}S - 8\mathfrak{T}P = 4\mathfrak{T}^2 + P^2 + S^2 - 4\mathfrak{T}P - 4\mathfrak{T}S + 2PS = (2\mathfrak{T} - P - S)^2$. Thus, $\sin^2\frac{\gamma}{2} = \dfrac{-(2\mathfrak{T} + P - 3S) + (2\mathfrak{T} - P - S)}{4(S - \mathfrak{T})}$ $= \dfrac{-P + S}{2(S - \mathfrak{T})} = \dfrac{P - S}{2(\mathfrak{T} - S)}$.

[4] $\mathfrak{T} - \frac{1}{4}(11\mathfrak{T} - R - P - 9S)\sin^2\frac{\gamma}{2} + 2(\mathfrak{T} - S)\sin^4\frac{\gamma}{2} = R - \frac{1}{4}(3R - \mathfrak{T} - S - P)\sin^2\frac{\gamma}{2} \Rightarrow 2(\mathfrak{T} - S)\sin^4\frac{\gamma}{2} - \frac{1}{4}(12\mathfrak{T} - 4R - 8S)\sin^2\frac{\gamma}{2} + \mathfrak{T} - R = 0$. It is $(3\mathfrak{T} - R - 2S)^2 - 8(\mathfrak{T} - S)(\mathfrak{T} - R) = 9\mathfrak{T}^2 + R^2 + 4S^2 - 6\mathfrak{T}R - 12\mathfrak{T}S + 4RS - 8\mathfrak{T}^2 + 8\mathfrak{T}R + 8\mathfrak{T}S - 8SR = \mathfrak{T}^2 + R^2 + 4S^2 + 2$

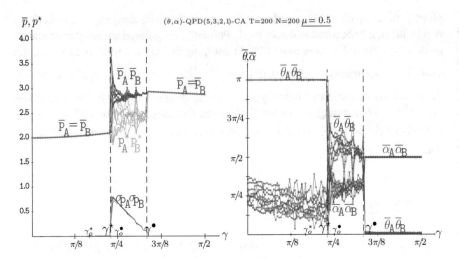

Fig. 7.3 The QPD(5,3,2,1)-CA with $\mu = 0.5$ noise factor. Variable entanglement factor γ. Five simulations at T= 200. Left: Mean payoffs ($\overline{p}$) and mean-field payoffs ($p^\star$). Right: Mean quantum parameters

$$\gamma_{0.5}^\star = 2\arcsin\left(\sqrt{\frac{P-S}{2(\mathfrak{T}-S)}}\right) \quad , \quad \gamma_{0.5}^\bullet = 2\arcsin\left(\sqrt{\frac{\mathfrak{T}-R}{2(\mathfrak{T}-S)}}\right) \qquad (7.13)$$

In the QPD(5,3,2,1) simulations of Fig. 7.3 it is $\gamma_{0.5}^\star = 2\arcsin\left(\sqrt{\frac{2-1}{2(5-1)}} = \frac{1}{8}\right) =$ $0.723 > \gamma_{0.0}^\star = \pi/6 = 0.524$, whereas $\gamma_{0.5}^\bullet = 2\arcsin\left(\sqrt{\frac{5-3}{2(5-1)}} = \frac{1}{4} = \frac{1}{2}\right) = 2\frac{\pi}{6} =$ $\pi/3 = 2\gamma_{0.0}^\bullet > \gamma_{0.0}^\bullet = \pi/4$. The payoffs equalization induced with mutual Q for $\gamma \geq \gamma_{0.5}^\bullet$ shows a small negative slope, as from Eq. (7.11a) it is $p^{QQ} = 3 + \frac{1}{4}\sin^2\frac{\gamma}{2}(-1)$, which varies from $p^{QQ}(\gamma = 0.0) = R = 3.0$ up to $p^{QQ}(\gamma = \pi/2) = 3 + \frac{1}{4}\frac{1}{2}$ $(-1) = 2.875$. In Fig. 7.3 $p_{A,B}^{Q,Q}$ is shown from $\gamma = \pi/3$ with $\sin(\gamma = \pi/6) = 1/2$, so that $p^{QQ}(\gamma = \pi/3) = 3 + \frac{1}{4}\frac{1}{4}(-1) = 2.937$, i.e., the small decrease in $p_{A,B}^{Q,Q}$ is just from 2.937 down to 2.875. The standard deviation (σ) of both players rockets to circa 0.75 at $\gamma_{0.5}^\star$ and monotonically decreases down to zero at $\gamma_{0.5}^\bullet$. Similarly to what happens in the noiseless simulations of Fig. 3.1 in the ($\gamma_{0.0}^\star, \gamma_{0.0}^\bullet$) interval.

The general form of the plots in Fig. 7.3 remains in CA-simulations of the PD with choices of the payoffs different from that adopted in most of the simulations in this book, i.e., (5,3,2,1) so that $\gamma_{0.0}^\star < \gamma_{0.0}^\bullet$. But the details of the transition from

---

$\mathfrak{T}R - 4\mathfrak{T}S - 4RS = (\mathfrak{T} + R - 2S)^2$. Thus, $\sin^2\frac{\gamma}{2} = \frac{(3\mathfrak{T} - R - 2S) - (\mathfrak{T} + R - 2S)}{4(\mathfrak{T} - S)} =$

$\frac{\mathfrak{T} - R}{2(\mathfrak{T} - S)}$.

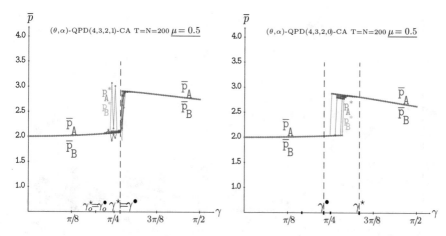

Fig. 7.4 Mean payoffs and mean-field payoffs in QPD-CA games with $\mu = 0.5$ noise. Variable entanglement factor γ. Five simulations at T= 200. Left: (4,3,2,1)-PD parameters. Right: (4,3,2,0)-PD parameters

$\{\hat{D}, \hat{D}\}$ to $\{\hat{Q}, \hat{Q}\}$ vary notably accordingly to the hierarchy of the γ^* and $\gamma^\bullet$ values. Figure 7.4 shows one example of PD-payoffs with $\gamma^*_{0.0} = \gamma^\bullet_{0.0}$, that of (4,3,2,1), and one example of PD-payoffs with $\gamma^\bullet_{0.0} < \gamma^*_{0.0}$, that of (4,3,2,0). In the $\gamma^* = \gamma^\bullet = \arcsin(\sqrt{1/3}) = 0.615$ scenario of the left frame of Fig. 7.4 the transition from (D,D) to (Q,Q) becomes an abrupt phase transition that happens at $\gamma^{*,\bullet}_{0.0} = 0.615$ in noiseless simulations, and that is delayed in simulations with $\mu = 0.5$ noise up to $\gamma^{*,\bullet}_{0.5} =$
$2\arcsin(\sqrt{\dfrac{P - S}{2(\mathfrak{T} - S)}} = \dfrac{\mathfrak{T} - R}{2(\mathfrak{T} - S)} = \dfrac{1}{6}) = 0.841$, greater but close to $\pi/4 = 0.785$
(accordingly with the threshold value in the left frame of Fig. 7.2), as it is shown in the left frame of Fig. 7.4. In the (4,3,2,0)-QPD context of the right frame of Fig. 7.4, both $\{\hat{Q}, \hat{Q}\}$ and $\{\hat{D}, \hat{D}\}$ are in NE in the $(\gamma^\bullet, \gamma^*)$ interval in conventional PD quantum games free of noise, as a result in the $\mu = 0.5$-(4,3,2,0)-QPD simulations in the right frame of Fig. 7.4 the mean-field approach fits almost perfectly to the actual mean payoffs regardless of the γ value.

Figure 7.5 deals with the results obtained in QPD(5,3,2,1)-CA simulations with full noise. In this scenario, mutual defection persists up the high threshold γ^* producing $p^{D,D}_{A,B} = R \sin^2 \frac{\gamma}{2} + P \cos^2 \frac{\gamma}{2} = 2 + (R - P) \sin^2 \frac{\gamma}{2}$. The threshold γ^* emerges at the intersection of $p^{D,D}$ and $p^{Q,D}_A = p^{D,Q}_B = S \sin^2 \frac{\gamma}{2} + \mathfrak{T} \cos^2 \frac{\gamma}{2}$, giving $\gamma^*{*_{1.0}} = 2\arcsin(\sqrt{\dfrac{P - S}{\mathfrak{T} + P - R - S}} = \dfrac{1}{3}) = 1.231$. Beyond γ^*, both $\overline{\theta}_A$ and $\overline{\theta}_B$
parameters notably decrease, as shown in the right frame of Fig. 7.5, and as a result both mean payoffs increase their value compared to that of mutual defection (recall that cooperation is featured by $\theta = 0$), so that at $\gamma = \pi/2$ it is $\overline{p}_A \simeq \overline{p}_B \simeq 3.0$, i.e., the payoff of mutual cooperation. Anyhow, the effect of amplitude-damping quantum noise in the 2P-QPD before the γ^* landmark turns out to be fairly surprising because

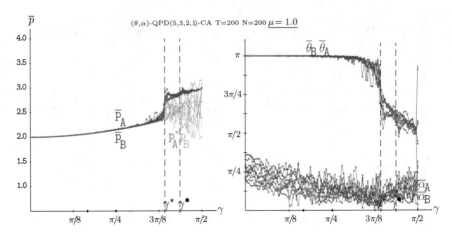

Fig. 7.5 The QPD(5,3,2,1)-CA with full noise. Variable entanglement factor γ. Five simulations at T= 200. Left: Mean payoffs ($\overline{p}$) and mean-field payoffs ($p^\star$). Right: Mean quantum parameters

the payoff of both players increases with the entanglement, instead of remaining constant at the P value as happens in noiseless simulations.

In the QHD(3,2,−1,0) simulation with $\mu = 0.5$ noise of Fig. 7.6, the $\gamma_{0.5}^\bullet$ threshold is notably higher than that in its noiseless counterpart in Fig. 3.5. In the QHD(3,2,−1,0) it is $\gamma_{0.5}^\bullet = 2\arcsin\left(\sqrt{\dfrac{3-2}{2(3-0)}} = \dfrac{1}{6}\right) = 0.8411$ and $p^{QQ}(\mu = 0.5) = 2 + \dfrac{1}{4}\sin^2\dfrac{\gamma}{2}$

$(3 + 0 - -1 - -6) = 2 - \sin^2\dfrac{\gamma}{2}$, which varies from $p^{QQ}(\gamma = 0.0) = R = 2.0$ up

to $p^{QQ}(\gamma = \pi/2) = 2 - \dfrac{1}{4} = 1.50$.

Unfair QPD Contests

In the unfair simulations that follow, a type of players is restricted to classical strategies $\tilde{U}(\theta, 0)$, whereas the other type of players may use quantum $\hat{U}(\theta, \alpha)$ ones.

Figure 7.7 deals with QPD(5,3,2,1)-CA contests involving quantum (θ, α)-player A (red) versus classical θ-player B (blue). The $\mu = 0.5$ and full noise scenarios are covered in this figure. The noiseless reference simulation to assess the effect of noise in the unfair scenario of Fig. 7.7 was studied in Fig. 4.1. In such a noiseless QPD-CA unfair simulation it was stated that the left threshold $\gamma^\star = \pi/6$ remains unaltered compared with that of the fair simulations, but the right threshold $\gamma^\bullet = 0.955$ is greater than that of the fair simulations. Also it was concluded that: (i) mutual defection remains up to $\gamma^\star$, (ii) the (Q,D) strategy pair emerges isolated in the $(\gamma^\star, \gamma^\bullet)$ transition interval, (iii) the miracle strategy $\hat{M} = \hat{U}(\pi/2, \pi/2)$ seems to play some role with $\gamma > \gamma^\bullet$.

The top frames of Fig. 7.7 deal with $\mu = 0.5$ simulations. Both $\gamma^\star$ and $\gamma^\bullet$ thresholds drift to the right compared to those in noiseless unfair simulations, so that mutual defection persists for higher values of γ and the (Q,D) pair appears later with noise. Again, as in the noiseless scenario, the left threshold $\gamma^\star$ in the $\mu = 0.5$

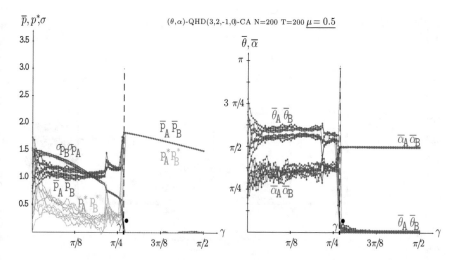

Fig. 7.6 The QHD(3,2,-1,0)-CA with $\mu = 0.5$ noise factor. Variable entanglement factor γ. Five simulations at T= 200. Left: Mean and standard deviations of the payoffs $(\overline{p}, \sigma)$ and mean-field payoffs $(p^\star)$. Right: Mean quantum parameters

unfair scenario coincide with that in the fair contest, but the right threshold $\gamma^\bullet$ is greater in the unfair game, in fact it coincides with the lower threshold in the fair game with full noise, i.e., $\gamma^\star_{1.0} = 2\arcsin(\sqrt{\dfrac{P-S}{\mathfrak{T}+P-R-S}} = \dfrac{1}{3}) = 1.231 > \pi/3$. In the $(\gamma^\star, \gamma^\bullet)$ interval, with players adopting the pair (Q,D) with $\mu = 0.5$, according to Eqs. (7.11c)–(7.11d) the payoffs intersection occurs at $\gamma^= = 2\arcsin\frac{1}{2} = \pi/3$,[5] giving the equalized payoff $p^= = \frac{1}{16}(R + P + 7(S + \mathfrak{T})) = 2.937$ in Fig. 7.7. Incidentally, $\gamma^= = 2\arcsin\frac{1}{2}$ coincides with $\gamma^\bullet_{0.5}$ in the QPD(5,3,2,1) fair game with $\mu = 0.5$, as may be ascertained in the left frame of Fig. 7.1. Please, note that $\gamma^=$ does not depend on the PD payoffs, so that, for example, it applies in simulations with (4,3,2,1) and (4,3,2,0) PD-payoffs, giving $p^= = 2.500$ and $p^= = 2.062$ respectively. After the right $\gamma^\bullet$ threshold, the payoffs of both players become approached compared to those achieved in the noiseless simulations, again with the payoff of the quantum player outperforming that of the classical player. Much as expected.

The bottom frames of Fig. 7.7 show the results in the unfair scenario with $\mu = 1.0$ noise. Mutual defection keeps in the full noise unfair scenario up to approximately the same threshold as in the fair simulations, in turn coincident with the right threshold in unfair $\mu = 0.5$ simulations, i.e., $\gamma^\star_{1.0} = \gamma^\bullet_{0.5}$. With higher entanglement the payoff of the quantum player outperforms that of the classical player in a short range of γ, just up to their equalization at $\gamma^\bullet = 2\arcsin(\sqrt{\dfrac{R-S}{\mathfrak{T}+R-P-S}} = \dfrac{2}{5}) = 1.369$.

---

[5] $p^{QD}_A = p^{QD}_B \Rightarrow S(4 - \sin^2\frac{\gamma}{2}(11 - 8\sin^2\frac{\gamma}{2})) + \mathfrak{T}(\sin^2\frac{\gamma}{2}(9 - 8\sin^2\frac{\gamma}{2})) = \mathfrak{T}(4 - \sin^2\frac{\gamma}{2}(11 - 8\sin^2\frac{\gamma}{2})) + S(\sin^2\frac{\gamma}{2}(9 - 8\sin^2\frac{\gamma}{2})) \Rightarrow 4 - \sin^2\frac{\gamma}{2}(11 - 8\sin^2\frac{\gamma}{2}) = \sin^2\frac{\gamma}{2}(9 - 8\sin^2\frac{\gamma}{2}) \Rightarrow 4 = \sin^2\frac{\gamma}{2}(20 - 16\sin^2\frac{\gamma}{2}) \Rightarrow 4\sin^4\frac{\gamma}{2} - 5\sin^2\frac{\gamma}{2} - 1 = 0 \Rightarrow \sin^2\frac{\gamma}{2} = \frac{1}{2}.$

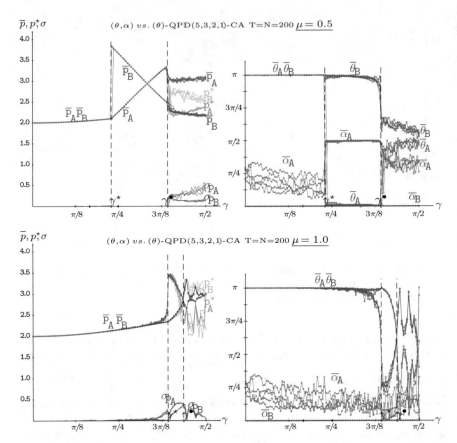

Fig. 7.7 The QPD(5,3,2,1)-CA unfair quantum (θ, α)-player A versus classical θ-player B with μ factor noise. Variable entanglement γ. Five simulations at T= 200. Top: $\mu = 0.5$. Bottom: $\mu = 1.0$. Left: Mean and variance payoffs $(\overline{p}, \sigma)$, and mean-field payoffs (p^*). Right: Mean quantum parameters

For $\gamma > \gamma^\bullet$, unexpectedly the classical player outperforms the quantum player, though in a small extent, due to the drift to high $\overline{\theta}_B$ accompanied with that to low $\overline{\theta}_A$.

The mean-field payoffs estimations fit fairly perfectly to the actual ones up to γ reaching $\gamma^\bullet$ in the simulations of the two scenarios of Fig. 7.7. Accordingly with this, the standard deviation (σ) of both players remains null up to $\gamma^\bullet$. But with $\gamma > \gamma^\bullet$, the mean-field payoff estimations p^* notably differ from the actual $\overline{p}$ ones in the noiseless simulations and in a lesser extent in the simulations with noise. The divergence between both actual and mean-field payoffs is not only quantitative but qualitative, in the sense that $\overline{p}_A > \overline{p}_B$ whereas $p_A^* < p_B^*$. The explanation of this divergence relies in spatial effects as illustrated in Fig. 7.8 for the unfair $\mu = 0.5$ scenario with $\gamma = \pi/2$. The parameter patterns at $T = 100$ in this figure, show a kind of maze-like aspect, particularly crisp in the θ pattern (much as in the noiseless case [5]). Again,

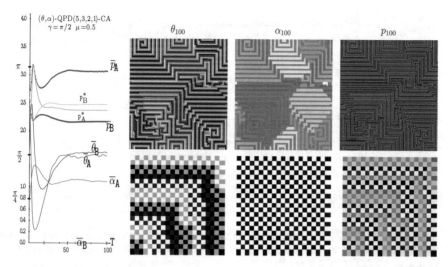

Fig. 7.8 A simulation in the $\gamma = \pi/2$ unfair, $\mu = 0.5$ scenario of Fig. 7.7. Far Left: Evolving mean parameters and payoffs. Center: parameter patterns. Far Right: Payoff pattern. The upper row of patterns shows the whole 200 × 200 lattice. The lower row of patterns zooms its 20 × 20 central part. Increasing grey levels indicate increasing values in the patterns

this spatial heterogeneity explains why the mean-field p^* differ from the actual payoffs $\overline{p}$. This divergence becomes apparent from the initial iterations in the far left frame of Fig. 7.8 where, after a short initial transition time, it is $\overline{p}_A > p_B^* > p_A^* > \overline{p}_B$.

Figure 7.9 deals with the case of an unfair game where the player B is allowed to resort only to the parameter α instead to θ as in the unfair simulations just before considered. In such a scenario, the (Q,Q) pair emerges in NE in the noiseless game (already studied in Fig. 4.4) and in the game with $\mu = 0.5$ noise. Beyond the threshold γ^* (far left and center frames of Fig. 7.9). Before the mentioned threshold the full quantum player A defects, whereas the player B sets his free parameter to its maximum $\overline{\alpha}_B = \pi/2$. As a result, the (D,Q) pair emerges inducing $\Pi_{0.0}^{D,Q} = \begin{pmatrix} 0 & \sin^2 \gamma \\ \cos^2 \gamma & 0 \end{pmatrix}$, and $\Pi_{0.5}^{D,Q}$ obtained interchanging the off-diagonal elements of $\Pi_{0.5}^{Q,D}$ given in (7.10). It is $\gamma_{0.0}^* = \pi/4 < \gamma_{0.5}^* = \pi/3$, therefore the advantage of player A over player B remains for a greater γ in the $\mu = 0.5$ scenario compared to the noiseless one. In the full noise scenario of the far right frame of Fig. 7.9, the full quantum player A defects for all γ, whereas the α-player B drifts to $\overline{\alpha}_B = \pi/4$, so that $\Pi_{1.0}^{D,U(0,\pi/4)} = \begin{pmatrix} 0 & \sin^2 \frac{\gamma}{2} \\ \cos^2 \frac{\gamma}{2} & 0 \end{pmatrix}$, and consequently $\overline{p}_A = 5\cos^2 \frac{\gamma}{2} + \sin^2 \frac{\gamma}{2} > \overline{p}_B = \cos^2 \frac{\gamma}{2} + 5\cos^2 \frac{\gamma}{2}$. Incidentally, in the full noise scenario the strategy $\mathbb{Q} = U(0, \alpha)$ when confronting the defection strategy D generates a Π that does not depend on α. Thus in the $(\hat{D}_A, \hat{\mathbb{Q}}_B)$ game the Π just

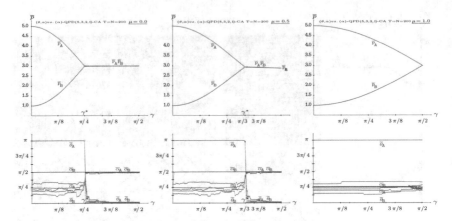

Fig. 7.9 The QPD(5,3,2,1)-CA unfair quantum (θ, α)-player A (red) versus (α)-player B (blue) with μ factor noise. Variable entanglement γ. Five simulations at T= 200. Far left: Noiseless. Center: $\mu = 0.5$., Far right: $\mu = 1.0$

aforementioned for the particular case $\alpha_B = \pi/4$, i.e., the Π of the (D, Q) pair. This is so because $(\hat{D}_A \otimes \hat{Q}_B)\rho_{1.0}^*(\hat{D}_A \otimes \hat{Q}_B) = \begin{pmatrix} 0 & 0 & e^{i\alpha_A}e^{-i\alpha_A} & 0 \\ 0 & 0 & 0 & 0 \\ 0 & 0 & 0 & 0 \\ 0 & 0 & 0 & 0 \end{pmatrix} = \begin{pmatrix} 0 & 0 & 1 & 0 \\ 0 & 0 & 0 & 0 \\ 0 & 0 & 0 & 0 \\ 0 & 0 & 0 & 0 \end{pmatrix}.$

Three-Parameter Strategies

Figure 7.10 shows the results achieved in three parameter, QPD(5,3,2,1)-CA simulations in the two noise scenarios: $\mu = 0.5$ noise (upper) and full noise (bottom). Please, recall that the noiseless 3P-QPD(5,3,2,1) simulations were reported in Fig. 3.17. The $\mu = 0.5$ graphs in Fig. 7.10 very much resemble those in Fig. 3.17. Therefore, also in the $\mu = 0.5$ simulations here the $(\hat{\mathfrak{X}}, \hat{\mathfrak{X}})$ pair roughly dominates the scene.

In the 3P noiseless CA simulations in the Fig. 3.17 the $\gamma^{\#}$ landmark appeared to mark the appearance of instability of the $\bar{\theta}$ parameters, which was reflected in an erratic behaviour of the mean-field payoffs frame. In the $\mu = 0.5$ 3P-CA simulations in the upper frames in Fig. 7.10, the instability in the $\bar{\theta}$ parameters and the consequent erratic behaviour of the mean-field payoffs appears beyond $\gamma^{\#}$. In the 3P simulations with full noise in the lower frames of Fig. 7.10 both actual mean payoffs increase according to the equation $p = 2 + \sin^2 \frac{\gamma}{2}$ (induced by the $(\hat{\mathfrak{X}}, \hat{\mathfrak{X}})$ pair with full noise) up to $\gamma^* = 1.231$, i.e., the same landmark as in the 2P simulations with full noise in Fig. 7.5. The same general considerations are to be made for $\gamma > 1.231$ in Fig. 7.10 as those made in Fig. 7.5.

Figure 7.11 deals with the results achieved in 3P-QPD(5,3,2,1)-CA unfair simulations, i.e., the 3P version of the 2P unfair simulations in Fig. 7.7. The structure of the graphs of the actual mean payoffs (left frames) of Fig. 7.11 notably resembles that of Fig. 7.7: equal payoffs for both players with not high γ, a transition γ-interval

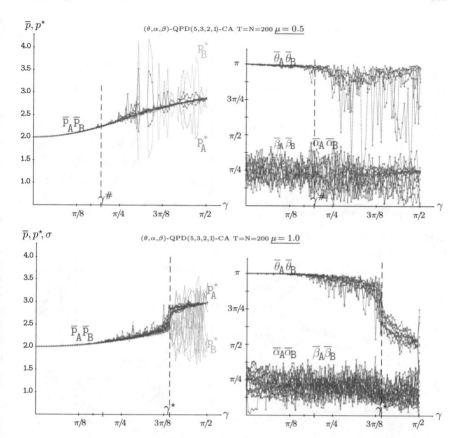

Fig. 7.10 The three-parameter QPD(5,3,2,1)-CA with μ noise factor. Variable entanglement factor γ. Five simulations at T= 200. Upper: $\mu = 0.5$, Bottom: $\mu = 1.0$. Left: Actual mean payoffs across the lattice $\overline{p}$, and mean-field payoffs p^*. Right: Mean quantum parameters

inducing (Q,D), and prevalence of the actual mean payoff of the quantum player over that of the classical player for high γ.

The 3P unfair simulations in Fig. 7.11 with $\mu = 0.5$ (upper frames) very much resemble the corresponding noiseless simulations reported in Fig. 4.11: below γ^* the player B adopts the $\hat{D} = U(\pi, 0, 0)$ strategy and the player A adopts the $\hat{\mathfrak{D}} = U(\pi, \alpha, \pi/2)$ strategy, whose associated Π is that given in Eq. (7.10) but with its rows interchanged. Therefore, below γ^* the common payoff for both players is: $p_{0.5}^{\mathfrak{D},D} = P + \frac{1}{4}\sin^2\frac{\gamma}{2}(\mathfrak{T} + 9R - 11P + 8(P - R)\sin^2\frac{\gamma}{2})$, which in the (5,3,2,1)-PD game becomes $p_{0.5}^{\mathfrak{D},D} = 2 + \frac{1}{4}\sin^2\frac{\gamma}{2}(11 - 8\sin^2\frac{\gamma}{2})$. As a result of noise, $p_{0.5}^{\mathfrak{D},D} < p_{0.0}^{\mathfrak{D},D} = 2 + \sin^2\gamma, \forall \gamma > 0$. The $\gamma_{0.5}^*$ threshold emerges at the intersection of $p_{0.5}^{\mathfrak{D},D}$ and $p_A^{Q,D} = 1 + \frac{1}{4}\sin^2\frac{\gamma}{2}(39 - 31\sin^2\frac{\gamma}{2})$, resulting $\gamma_{0.5}^* = 0.837$. The right threshold turns out to be $\gamma_{0.5}^\bullet = 2\gamma_{0.0}^* = 1.230$, i.e., the same γ^* of the 2P unfair full noise simulations.

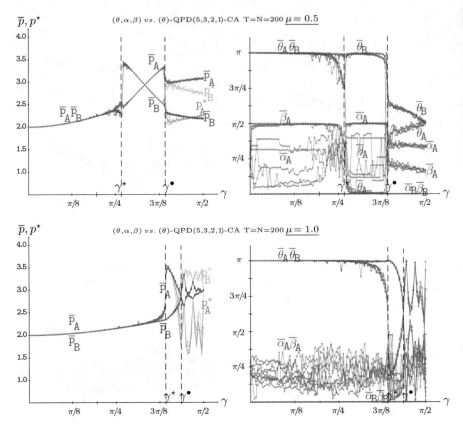

Fig. 7.11 The QPD(5,3,2,1)-CA unfair quantum (θ, α, β)-player A versus classical θ-player B with μ noise factor. Variable entanglement γ. Five simulations at T= 200. Upper: $\mu = 0.5$. Lower: $\mu = 1.0$. Left: Actual mean payoffs $\overline{p}$, and mean-field payoffs p^*. Right: Mean quantum parameters

7.4 The QSD with Noise

The payoff dependence on γ in the Samaritan's Dilemma (SD) game of Q versus the pure strategies C and D with $\mu = 1/2$ may be checked in Fig. 7.12 under the label $\mu = 1/2$. In the (Q,Q) scenario in particular, the payoffs become: $p_A^{QQ} = 3 - \frac{11}{4}\sin^2\frac{\gamma}{2}$, and $p_B^{QQ} = 2 - \frac{2}{4}\sin^2\frac{\gamma}{2}$, which equalize at $1 = \frac{9}{4}\sin^2\frac{\gamma}{2}$, i.e., at $\gamma = 2\arcsin\left(\sqrt{\frac{4}{9}}\right) = 1.460$, giving $p_{A,B} = 1.777$. Beyond $\gamma = 1.460$ it is $p_A^{QQ} < p_B^{QQ}$, as with full entangling where $p_A^{QQ} = 13/8 < p_B^{QQ} = 14/8$. The payoff dependences on γ in the SD of the pure strategies with full noise may be checked in Fig. 7.12 under the label $\mu = 1$. In particular, in the (Q, Q) context, $p_A^{QQ} = 3\cos^2\frac{\gamma}{2}$, $p_B^{QQ} = 2\cos^2\frac{\gamma}{2}$.

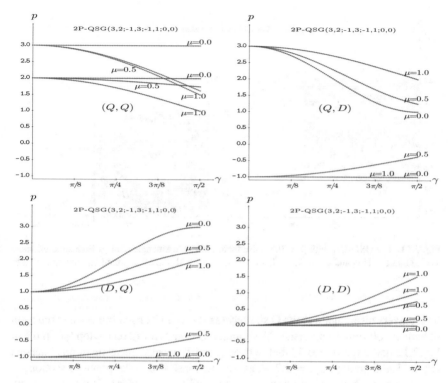

Fig. 7.12 Payoffs in the QSD with variable entanglement factor γ. Top-Left: $(\hat{Q}, \hat{Q})$, Top-Right: $(\hat{Q}, \hat{D})$, Bottom-Left: $(\hat{D}, \hat{Q})$, Bottom-Right: $(\hat{D}, \hat{D})$. Three levels of noise (μ) are considered in each frame. Red: payoffs of player A (Samaritan), Blue: payoffs of player B (beneficiary)

Recall that only amplitude-damping noise is taken into account in this study. We only will point out here that with $\mu = 0.5$ phase damping, as Π^{QQ} turns out diagonal with $\pi_{22}^{QQ} = \frac{1}{4}\sin^2\gamma$, in the QSD game it is $p_A^{QQ} > p_B^{QQ}$, $\forall\gamma$, at variance with what happens with amplitude damping where both payoffs intersect as indicated above.

Figure 7.13 deals with the results obtained in a QSD-CA simulation with $\mu = 0.5$ noise. The general form of the graphs of $(\overline{p}_A, \overline{p}_B)$ for not high entanglement shown in the simulations free of noise shown in Fig. 3.6 remains in Fig. 7.13 up to nearly $\gamma = \pi/4$ with $\mu = 0.5$ noise. From approximately $\gamma = \pi/4$, $\overline{\theta}_A$ gradually tends to zero, so that, as in the noiseless simulation, the mean payoff of the player A is notably stabilized. The pair $\{\hat{Q}, \hat{D}\}$ at $\mu = 0.5$ generates the joint probabilities: $\pi_{11} = \pi_{44} = \frac{1}{4}\sin^2\frac{\gamma}{2}$ $\pi_{12} = \cos^2\frac{\gamma}{2} - \frac{1}{4}\sin^2\frac{\gamma}{2}(7 - 8\sin^2\frac{\gamma}{2})$, $\pi_{21} = \frac{1}{4}\sin^2\frac{\gamma}{2}(9 - 8\sin^2\frac{\gamma}{2})$. Thus, $p_B^{QD} = 3 - \frac{22}{4}\sin^2\frac{\gamma}{2} + 4\sin^4\frac{\gamma}{2}$. It was $p_B^{QQ} = 2 - \frac{2}{4}\sin^2\frac{\gamma}{2}$ so that both payoffs equalizes as $1 - 5\sin^2\frac{\gamma}{2} + 4\sin^4\frac{\gamma}{2} = 0$, therefore at $\gamma^{\bullet} = 2\arcsin(1/2) = \pi/3$. Analogously, $p_A^{DQ} = -1 + \frac{3}{4}\sin^2\frac{\gamma}{2}$, and $p_A^{QQ} = 3 - \frac{11}{4}\sin^2\frac{\gamma}{2}$ so that $p_A^{QQ} > p_A^{DQ}$, $\forall\gamma$. In consequence, the pair $\{\hat{Q}, \hat{Q}\}$ at $\mu = 0.5$ is in NE for $\gamma > \gamma^{\bullet} = \pi/3$. As a reflect of this fact, in Fig. 7.13 from $\gamma > \pi/3 = 1.047$

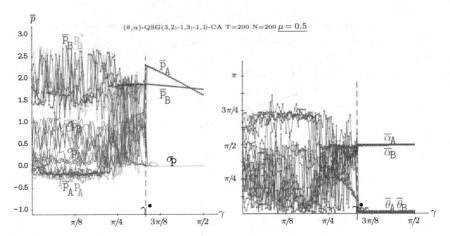

Fig. 7.13 The QSD-CA with $\mu = 0.5$ noise factor. Variable entanglement γ. Five simulations at T= 200. Left: Mean and standard deviation of the payoffs. Right: Mean quantum parameter values

both players resort again to the Q strategy (as shown in its right frame), and the form of both payoffs emerging in the CA simulations is that shown on the top left frame of Fig. 7.12 under the $\mu = 0.5$ label, where the fairly locally linear functions intersect at $\gamma = 1.460$ giving $p_{A,B} = 1.777$, as was demonstrated in the previous section.

Despite the structural proximity of both amplitude and phase-damping quantum noises (only K_2 varies), the pair (Q,Q) is not in NE with phase-damping noise for high values of γ. Consequently, in QSG-CA simulations with $\mu = 0.5$ phase-damping (not the shown here) the discontinuity observed with $\mu = 0.5$ amplitude-damping noise in Fig. 7.13 when γ approaches $\pi/3$ does not emerge in the payoffs graph. The parameter graph shows in this scenario how the Samaritan player drifts to the Q strategy ($\overline{\theta}_A \to 0$, $\overline{\alpha}_A \to \pi/2$) when γ increases beyond $\pi/3$, but the beneficiary player does not accompany the Samaritan in this trend.

Figure 7.14 deals with the results obtained in the QSD-CA scenario of Fig. 7.13 but with full $\mu = 1.0$ noise. Full noise seems to impede the emergence of the (Q,Q) pair, so that the oscillations of both payoffs not far from (–0.2,1.5) shown in Figs. 3.6 and 7.13 for low γ, remains here for higher entanglement.

In Fig. 7.14, with high entanglement the (θ, α) parameters of both players tend to approach their middle levels ($\pi/2$, $\pi/4$), leading to $\Pi = \frac{1}{4} \begin{pmatrix} 1 - \sin\gamma & 1 \\ 1 & 1 + \sin\gamma \end{pmatrix}$, that induces $p_A(\gamma) = \frac{1}{4}(1 - 3\sin\gamma)$, $p_B(\gamma) = \frac{1}{4}(6 - 2\sin\gamma)$. These payoffs smoothly decrease as γ increases: p_A from 0.25 down to –0.5, p_B from 1.5 down to 1.0. The mean-field approaches (p^*) shown in left frame of Fig. 7.14 follow fairly well the $p_A(\gamma)$ and $p_B(\gamma)$ given by said equations at high entanglement. In fact, at maximum $\gamma = \pi/2$, it is $p_A^* \simeq -0.5$, $p_B^* \simeq 1.0$. But the actual mean payoffs ($\overline{p}$) tend to provide higher values of the payoffs. Thus for example, at $\gamma = \pi/2$ in the left frame of Fig. 7.14 turn out to be over the expected (–0.5, 1.0). This is due to spatial

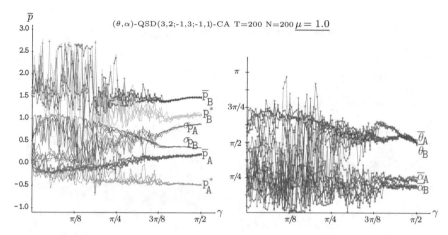

Fig. 7.14 The QSD-CA with full noise. Variable entanglement factor γ. Five simulations at T= 200. Left: Mean payoffs. Right: Mean quantum parameter values

Fig. 7.15 Quantum parameter (θ, α) and payoff (p) patterns at $T = 200$ in a simulation with $\gamma = \pi/2$ in the QSD-CA with full noise scenario of Fig. 7.14

effects (foreseeable since the standard deviations of the payoffs are not negligible) that make it difficult to estimate the actual mean payoffs from the mean parameters. An example of spatial structure at $\gamma = \pi/2$ is given in Fig. 7.15, where the parameter values (and the payoffs in consequence) show a kind of *maze*-like aspect.

Unfair QSD Contests

The simulations in the $\alpha_B = 0$ unfair scenario with $\mu = 0.5$ noise (not shown here) produce roughly the same results as in the noiseless unfair scenario of Fig. 4.5, albeit the payoffs of the charity player show a notably smaller variation around 1.5, and those of the beneficiary player become positive at a greater extent as in Fig. 4.5. Spatial effects are not relevant neither in Fig. 4.5, nor in the $(\alpha_B = 0, \mu = 0.5)$ simulations, so that the mean-field payoffs estimations fit well to the actual ones in these unfair scenarios.

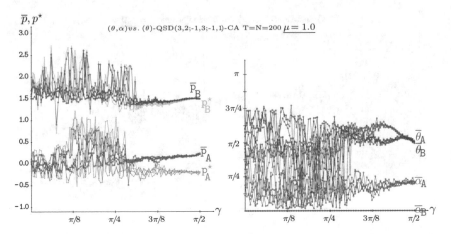

Fig. 7.16 The QSD-CA unfair quantum (θ, α)-player A (red) versus a classical θ-player B (blue) with full noise. Variable entanglement γ. Five simulations at T = 200. Left: Actual mean payoffs $\overline{p}$ and mean-field payoffs p^*. Right: Mean quantum parameter values

Figure 7.16 shows the results in the $\alpha_B = 0$ unfair scenario of Fig. 4.5, but with $\mu = 1.0$ noise. The results regarding both the payoffs and mean parameters shown in Fig. 7.16 are qualitatively similar to those achieved in the fair, $\mu = 1.0$ scenario of Fig. 7.14. Spatial effects arise with high entanglement, but oddly they seem to affect only to the charity player A. Consequently, the parameter and the payoffs spatial structures with high entanglement and full noise show a much fuzzified *maze*-like aspect compared to those in the fair scenario of Fig. 7.15.

Figure 7.17 deals with the case of an unfair game where the beneficiary player is allowed to resort only to the parameter α instead to θ as in the unfair simulations just before considered. In such a scenario, the (Q,Q) pair emerges in NE regardless of γ in the noiseless and $\mu = 0.5$ scenarios, as the beneficiary player can not resort to strategies such as D (or Loaf) which demands $\theta > 0$. As a result, in the noiseless far left frame of Fig. 7.17 it is $\overline{p}_A = 3$, and $\overline{p}_B = 2$, and in the $\mu = 0.5$-noise center frame the $(\overline{p}_A, \overline{p}_B)$ payoffs fit the expressions given in the quantum noise introductory Sect. 7.2. In the full noise scenario of the far right frame, where $\overline{\theta}_A = \overline{\theta}_B = 0$, it is
$\Pi = \begin{pmatrix} \cos^2 \frac{\gamma}{2} & 0 \\ 0 & \sin^2 \frac{\gamma}{2} \end{pmatrix}$, so that $\overline{p}_A = 3\cos^2 \frac{\gamma}{2}, \overline{p}_B = 2\cos^2 \frac{\gamma}{2}$ regardless of the values of the α parameters. This is so because $\theta = 0$ makes $\hat{U}$ diagonal and it turns out that

$$(\hat{U}_A \otimes \hat{U}_B)\rho_{1.0}^*(\hat{U}_A \otimes \hat{U}_B) = \begin{pmatrix} e^{i(\alpha_A+\alpha_B)}e^{-i(\alpha_A+\alpha_B)} & 0 & 0 & 0 \\ 0 & 0 & 0 & 0 \\ 0 & 0 & 0 & 0 \\ 0 & 0 & 0 & 0 \end{pmatrix} = \begin{pmatrix} 1 & 0 & 0 & 0 \\ 0 & 0 & 0 & 0 \\ 0 & 0 & 0 & 0 \\ 0 & 0 & 0 & 0 \end{pmatrix}.$$

Three-Parameter QSD

In the three-parameter strategies QSD-CA simulations with $\mu = 0.5$ noise (not shown here) roughly similar results as in the noiseless 3P-QSD scenario of Fig. 3.19 are achieved. Nevertheless, the form of the graphs of the actual mean and mean-field

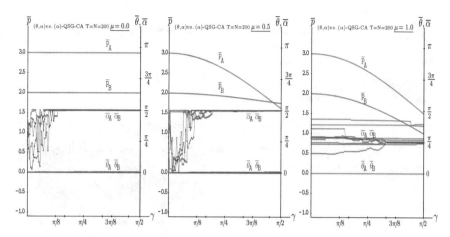

Fig. 7.17 The QSD-CA unfair quantum (θ, α)-player A (red) versus a (α)-player B (blue). Variable entanglement γ. Five simulations at T $= 200$. Far Left: $\mu = 0.0$, Centre: $\mu = 0.5$, Far Right: $\mu = 1.0$

payoffs are altered, much advancing the features observed in the 3P-QSD simulations with full noise described in Fig. 7.18. This is particularly true in what respect to a notable decreasing in the erratic behaviour of the mean-field estimations, particularly those of player B.

Figure 7.18 deals with the results achieved in the scenario of Fig. 3.19, but with full noise. The actual mean payoffs of both players monotonically increase their values as the entanglement increases, in the case of player A from approximately zero up to approximately 1.5, in the case of player B from approximately 1.5 up to approximately 2.0. Thus, player B overrates player A all along the γ variation, much as in Fig. 3.19, but with full noise in a more crisp manner. At variance with what happens in Fig. 3.19, the mean-field payoff estimations fit fairly well the actual mean payoffs with full noise, so that the standard deviations of the parameters have not been shown in Fig. 7.18. In fact, the actual mean payoffs in Fig. 7.18 may be very well fitted by the equations: $p_A = \sin \gamma$, $p_B = 1.5 + \sqrt{\sin \gamma}$. Note that the functions $-0.5 + \sin \gamma$ and $1.0 + \sqrt{\sin \gamma}$ have been plotted in the left frame of Fig. 7.18 as references. Again at variance with what happens in Fig. 3.19, some trends become apparent in the mean parameter graphics with full noise, as the mean parameter patterns of player A and $\overline{\theta}_B$ drift towards $\pi/2$.

In three-parameter strategies QSD-CA unfair simulations with $\mu = 0.5$ noise (not shown here), although the general form of the graphs of the payoffs and parameters versus γ is similar to that in the noiseless scenario of Fig. 3.19, the graphs are altered advancing the features observed in the 3P-QSD unfair simulations with full noise described below in Fig. 7.19. In particular, with $\mu = 0.5$ noise the actual mean payoffs of both players increase their values as γ increases at a lower degree compared to those achieved in the noiseless simulations.

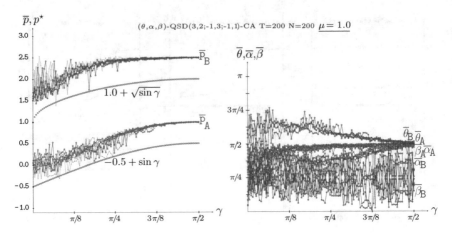

Fig. 7.18 The three-parameter QSD-CA with full noise. Variable entanglement γ. Five simulations at T = 200. Left: Mean payoffs $\overline{p}$ and mean-field payoffs p^*. Right: Mean quantum parameters

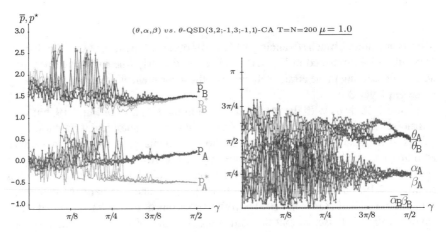

Fig. 7.19 The unfair three-parameter QSD-CA with full noise. Variable entanglement γ. Five simulations at T = 200. Left: Mean payoffs $\overline{p}$ and mean-field payoffs p^*. Right: Mean quantum parameters

In the $\mu = 1.0$ unfair 3P-QSD scenario of Fig. 7.19, where $\alpha_B = \beta_B = 0$, if $\theta_A = \theta_B = \pi/2$ and $\alpha_A = \beta_A = \pi/4$ it is: $\Pi = \frac{1}{4} \begin{pmatrix} 1 - \sin\gamma & 1 + \sin\gamma \\ 1 - \sin\gamma & 1 + \sin\gamma \end{pmatrix}$, with associated payoffs: $p_B = 1.5$, $p_A = 0.25 - 1.5\sin\gamma$. The latter smoothly decreases from $p_A(\gamma = 0) = 0.25$, down to $p_A(\gamma = \pi/2) = -0.50$. The right frame of Fig. 7.19 indicates that the just described quantum parameters scenario applies with high entanglement in the simulations of this figure, and in consequence the mean-field payoff estimations in its left frame correspond with the theoretical values approximately from $\gamma = \pi/4$. Spatial effects minimally influence the actual mean payoffs of the beneficiary player A in the CA simulations, whereas they support those of the charity player A so that he achieves no negative payoffs.

References

1. Alonso-Sanz, R., Situ, H.: A quantum Samaritan's dilemma cellular automaton. R. Soc. Open Sci. **4**(6), 863–160669 (2017)
2. Nielsen, M.A., Chuang, I.L.: Quantum Computation and Quantum Information. Cambridge University Press, Cambridge (2000)
3. Yin, J., et al.: Satellite-based entanglement distribution over 1200 kilometers. Science **356**, 1140 (2017)
4. Huang, Z.M., Qiu, D.: Quantum games under decoherence. Int. J. Theor. Phys. **55**(2), 965–992 (2016)
5. Alonso-Sanz, R.: Variable entangling in a quantum prisoner's dilemma cellular automaton. Quantum Inf. Process. **14**, 147–164 (2015)

Chapter 8
Quantum Relativistic Games

This chapter studies the spatial quantum relativistic formulation of the iterated Prisoner's Dilemma [1] and the Battle of the Sexes [2] games with variable entangling. The noiseless and with quantum noise scenarios will be scrutinized. The density matrix formalism described in Chap. 7 will be used in this chapter.

8.1 The Unruh Effect

In the Quantum-Relativistic (QR) games we deal with in this chapter, in addition to the prerogatives stated in the EWL Sect. 2.1, we also assume that the player A moves with a uniform acceleration, and the player B remains stationary [3, 4]. QR games with both players accelerated are to come for scrutiny in subsequent studies.

As in previous chapters, both players share the same initial entangled state given in (7.1), composed of two qubits (one for each player, as indicated by the subscripts) at a point in flat Minkowski spacetime.

Additionally, in order to describe the interaction between the players, we must introduce Rindler coordinates, which define two causally disconnected regions, therefore named I and II. An observer in one region has no access to the information that is leaked from the other through decoherence effects. Thus, we assume the players capable of measuring only a single Minkowski mode in region I for the sake of simplicity, with a highly monochromatic detector. As demonstrated in other works [5, 6], a given Minkowski mode of a particular frequency spreads over all positive Rindler frequencies $((\omega c)/a)$ that peaks about the Minkowski frequency. The constants ω, c and a are the Dirac frequency of the particle, the speed of light and the acceleration of player A. We also define the dimensionless acceleration parameter r, given by $r = (e^{-2\pi\omega c/a} + 1)^{-1/2}$, which varies in the range $[0, \pi/4]$ as the acceleration a varies in the range $[0, \infty]$. In the player A accelerated frame, the vacuum state is a two-mode squeezed state given in terms of Rindler modes by,

© Springer Nature Switzerland AG 2019
R. Alonso-Sanz, *Quantum Game Simulation*, Emergence, Complexity
and Computation 36, https://doi.org/10.1007/978-3-030-19634-9_8

$$|0\rangle_A = \cos r |0\rangle_I |0\rangle_{II} + \sin r |1\rangle_I |1\rangle_{II} \tag{8.1}$$

while the single particle state is given, also in terms of Rindler modes, by:

$$|1\rangle_A = |1\rangle_I |0\rangle_{II} \tag{8.2}$$

Equation (8.1) indicates the appearance in our system of what is called the Unruh effect, where the noninertial observer that moves with constant acceleration in region I sees a thermal state instead of the vacuum state. Substituting (8.1) and (8.2) into (7.1), we obtain the following entangled initial state:

$$|\psi\rangle_{B,I,II} = \cos\frac{\gamma}{2}\cos r |0\rangle_B |0\rangle_I |0\rangle_{II} + \cos\frac{\gamma}{2}\sin r |0\rangle_B |1\rangle_I |1\rangle_{II} + i\sin\frac{\gamma}{2}|1\rangle_B |1\rangle_I |0\rangle_{II} \tag{8.3}$$

Since player A has no access to information from region II, we trace out the modes related to it, obtaining:

$$\rho_{B,I} = Tr_{II}(|\psi\rangle\langle\psi|)$$
$$= \langle 0|_{II}|\psi\rangle_{B,I,II}\langle\psi|_{B,I,II}|0\rangle_{II} + \langle 1|_{II}|\psi\rangle_{B,I,II}\langle\psi|_{B,I,II}|1\rangle_{II}$$
$$= (\cos\frac{\gamma}{2}\cos r|00\rangle + i\sin\frac{\gamma}{2}|11\rangle)(\cos\frac{\gamma}{2}\cos r\langle 00| - i\sin\frac{\gamma}{2}\langle 11|) + \cos^2\frac{\gamma}{2}\sin^2 r|01\rangle\langle 01|$$

$$= \begin{pmatrix} \cos^2\frac{\gamma}{2}\cos^2 r & 0 & 0 & -i\cos\frac{\gamma}{2}\sin\frac{\gamma}{2}\cos r \\ 0 & \cos^2\frac{\gamma}{2}\sin^2 r & 0 & 0 \\ 0 & 0 & 0 & 0 \\ i\cos\frac{\gamma}{2}\sin\frac{\gamma}{2}\cos r & 0 & 0 & \sin^2\frac{\gamma}{2} \end{pmatrix} \tag{8.4}$$

Remarkably, with no entanglement it is $\rho_{B,I}(\gamma = 0) = \begin{pmatrix} \cos^2 r & 0 & 0 & 0 \\ 0 & \sin^2 r & 0 & 0 \\ 0 & 0 & 0 & 0 \\ 0 & 0 & 0 & 0 \end{pmatrix}$,
which only coincides with the matrix obtained from Eq. (7.2) for $\gamma = 0$ in the inertial case, i.e., $r = 0$ so that $\cos r = 1$, $\sin r = 0$.

Therefore, the final density matrix introduced in Eq. (7.3) turns out to be that given in Eq. (8.5),

$$\rho_f = \hat{J}^\dagger(\hat{U}_A \otimes \hat{U}_B)\rho_{B,I}(\hat{U}_A \otimes \hat{U}_B)^\dagger \hat{J} \tag{8.5}$$

The probability distribution Π for $\gamma = 0$ and arbitrary r given in (8.6) may be expressed as: $\Pi = \mathbf{xy'}\cos^2 r + \mathbf{x(1-y)'}\sin^2 r = \mathbf{xy'}\cos 2r + \mathbf{x1'}\sin^2 r$. In this $\gamma = 0$ scenario, the equalization of the θ parameter values, so $x = y = 1/2$, leads to the equalization of the probabilities in Π, so that both players get the arithmetic mean of the payoff values regardless of r. If both players use pure strategies, i.e., $x \in \{0, 1\}$, $y \in \{0, 1\}$ in Eq. (8.6), the probabilities are simplified as: $\Pi^{CC} = \begin{pmatrix} \cos^2 r & \sin^2 r \\ 0 & 0 \end{pmatrix}$, Π^{CD} is obtained permuting the columns of Π^{CC}, Π^{DC} permuting the rows of Π^{CC}, and Π^{DD} permuting the rows and the columns of Π^{CC}.

$$\Pi_{\gamma=0} = \begin{pmatrix} xy\cos^2 r + x(1-y)\sin^2 r & x(1-y)\cos^2 r + xy\sin^2 r \\ (1-x)y\cos^2 r + (1-x)(1-y)\sin^2 r & (1-x)(1-y)\cos^2 r + (1-x)y\sin^2 r \end{pmatrix} \quad (8.6)$$

Equation (8.7) gives the probability distribution for mutual cooperation for arbitrary γ and r parameters.[1] As with $\gamma = 0$, permuting the columns, the rows, and the rows and columns of Π^{CC}, generate Π^{CD}, Π^{DC}, and Π^{DD}. As in the inertial frame, it is: $\Pi^{QQ} = \Pi^{CC}, \forall \gamma, r$. Π^{CC} for $\gamma = \pi/2$ is given in Eq. (8.8).

$$\Pi^{CC}(\gamma, r) = \begin{pmatrix} (\cos^2 \frac{\gamma}{2} \cos r + \sin^2 \frac{\gamma}{2})^2 & \cos^4 \frac{\gamma}{2} \sin^2 r \\ \frac{1}{4} \sin^2 \gamma \sin^2 r & \frac{1}{4} \sin^2 \gamma (1 - \cos r)^2 \end{pmatrix} \quad (8.7)$$

$$\Pi^{CC}(\gamma = \pi/2, r) = \frac{1}{4} \begin{pmatrix} (1 + \cos r)^2 & \sin^2 r \\ \sin^2 r & (1 - \cos r)^2 \end{pmatrix} \quad (8.8)$$

8.2 A Quantum Relativistic Prisoner's Dilemma

From Eq. (8.7) it is: $p_A < p_B$ if (C, C) or (C, D), and $p_A > p_B$ if (D, C) or (D, D). Let us prove the first inequality in the case of mutual cooperation. In order to compare the values of p_A^{CC} and p_B^{CC} in the PD, only π_{12} and π_{21} are relevant. These probabilities share the factor $\sin^2 r$, therefore $\cos^4 \frac{\gamma}{2}$ is to be compared to $\frac{1}{4} \sin^2 \gamma$. It turns out

$${}^{1}\rho_{B,1}P = \begin{pmatrix} \cos^2 \frac{\gamma}{2} \cos^2 r & 0 & 0 & -i\cos\frac{\gamma}{2}\sin\frac{\gamma}{2}\cos r \\ 0 & \cos^2\frac{\gamma}{2}\sin^2 r & 0 & 0 \\ 0 & 0 & 0 & 0 \\ i\cos\frac{\gamma}{2}\sin\frac{\gamma}{2}\cos r & 0 & 0 & \sin^2\frac{\gamma}{2} \end{pmatrix} \begin{pmatrix} 0 & 0 & 0 & 1 \\ 0 & 0 & -1 & 0 \\ 0 & -1 & 0 & 0 \\ 1 & 0 & 0 & 0 \end{pmatrix} =$$

$$\begin{pmatrix} -i\cos\frac{\gamma}{2}\sin\frac{\gamma}{2}\cos r & 0 & 0 & \cos^2\frac{\gamma}{2}\cos^2 r \\ 0 & 0 & -\cos^2\frac{\gamma}{2}\sin^2 r & 0 \\ 0 & 0 & 0 & 0 \\ \sin^2\frac{\gamma}{2} & 0 & 0 & i\cos\frac{\gamma}{2}\sin\frac{\gamma}{2}\cos r \end{pmatrix},$$

$$P\rho_{B,1} = \begin{pmatrix} i\cos\frac{\gamma}{2}\sin\frac{\gamma}{2}\cos r & 0 & 0 & \sin^2\frac{\gamma}{2} \\ 0 & 0 & 0 & 0 \\ 0 & -\cos^2\frac{\gamma}{2}\sin^2 r & 0 & 0 \\ \cos^2\frac{\gamma}{2}\cos^2 r & 0 & 0 & -i\cos\frac{\gamma}{2}\sin\frac{\gamma}{2}\cos r \end{pmatrix}$$

$$\rho P - P\rho = \begin{pmatrix} -i\sin\gamma\cos r & 0 & 0 & \cos^2\frac{\gamma}{2}\cos^2 r - \sin^2\frac{\gamma}{2} \\ 0 & 0 & 0 & 0 \\ 0 & \cos^2\frac{\gamma}{2}\sin^2 r & 0 & 0 \\ \sin^2\frac{\gamma}{2} - \cos^2\frac{\gamma}{2}\cos^2 r & 0 & 0 & i\sin\gamma\cos r \end{pmatrix}, \quad P\rho_{B,1}P =$$

$$\begin{pmatrix} \sin^2\frac{\gamma}{2} & 0 & 0 & i\cos\frac{\gamma}{2}\sin\frac{\gamma}{2}\cos r \\ 0 & 0 & 0 & 0 \\ 0 & 0 & \cos^2\frac{\gamma}{2}\sin^2 r & 0 \\ -i\cos\frac{\gamma}{2}\sin\frac{\gamma}{2}\cos r & 0 & 0 & \cos^2\frac{\gamma}{2}\cos^2 r \end{pmatrix}$$

$\rho(2,2) = \cos^2\frac{\gamma}{2}\cos^2\frac{\gamma}{2}\sin^2 r = \cos^4\frac{\gamma}{2}\sin^2 r$, $\rho(3,3) = \sin^2\frac{\gamma}{2}\cos^2\frac{\gamma}{2}\sin^2 r = (\frac{1}{2}\sin\gamma)^2\sin^2 r$

$\rho(4,4) = \cos^2\frac{\gamma}{2}\sin^2\frac{\gamma}{2} + \sin^2\frac{\gamma}{2}\cos^2\frac{\gamma}{2}\cos^2 r + i\frac{1}{2}\sin\gamma i\sin\gamma\cos r = (\frac{1}{2}\sin\gamma)^2 + (\frac{1}{2}\sin\gamma)^2\cos^2 r - \frac{1}{2}\sin^2\gamma\cos r = \frac{1}{4}\sin^2\gamma(1 + \cos^2 r - 2\cos r)$.

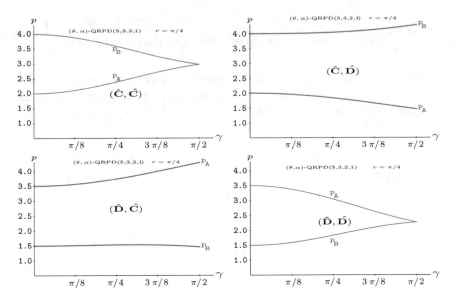

Fig. 8.1 Payoffs of the pure strategies in a QRPD(5,3,2,1) with variable entanglement factor γ at $r = \pi/4$. Top-Left: $(\hat{C}, \hat{C})$, Top-Right: $(\hat{C}, \hat{D})$, Bottom-Left: $(\hat{D}, \hat{C})$, Bottom-Right: $(\hat{D}, \hat{D})$

that, $\cos^4 \frac{\gamma}{2} \in [1, 1/4]$ and $\frac{1}{4} \sin^2 \gamma \in [0, 1/4]$, and consequently $\pi_{12} > \pi_{21}$, which implies $p_A^{CC} > p_B^{CC}$.

Figure 8.1 shows the payoffs of the pure strategies in the QRPD with (5,3,2,1) parameters at $r = \pi/4$ for variable γ. At this maximum level of r it is $\Pi_{\gamma=0}^{CC} = \Pi_{\gamma=0}^{CD} = \frac{1}{2}\begin{pmatrix} 1 & 1 \\ 0 & 0 \end{pmatrix}$, and $\Pi_{\gamma=0}^{DC} = \Pi_{\gamma=0}^{DD} = \frac{1}{2}\begin{pmatrix} 0 & 0 \\ 1 & 1 \end{pmatrix}$, and consequently: $p_A^{CC} = p_A^{CD} = \frac{R+S}{2}, p_A^{DC} = p_A^{DD} = \frac{\mathfrak{T}+S}{2}, p_B^{CC} = p_B^{CD} = \frac{\mathfrak{T}+P}{2}, p_B^{DC} = p_B^{DD} = \frac{S+P}{2}$. Still at $r = \pi/4$, if $\gamma = \pi/2$ it is $p_A^{CD} = p_B^{DC}, p_B^{CD} = p_A^{DC}$. In the non-inertial frame it is also: $\Pi^{QQ} = \Pi^{CC}, \forall \gamma, r$.

The Spatialized QRPD with Noise

As customary in the spatial simulations of this book, in the QRPD-CA simulations in this section, the A and B players alternate in the site occupation in a 2D lattice, interact in the cellular automata manner and the evolution is ruled by the deterministic imitation of the best paid mate neighbour, included the r parameter value in the case of the accelerated player A. Also as usual, the simulations in this section are run up to $T = 200$ in a $N = 200$ lattice with periodic boundary conditions, and five different initial random assignment of the (θ, α, r) parameter values are implemented in every studied scenario.

Figure 8.2 deals with the QRPD(5,3,2,1)-CA with variable entanglement factor γ. Said figure shows in its left frame the mean payoffs across the lattice ($\overline{p}$) of both

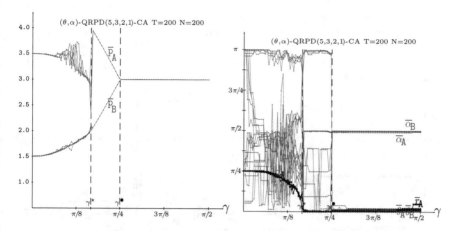

Fig. 8.2 The two parameter QRPD(5,3,2,1)-CA with variable entanglement factor γ. Five simulations at T = 200. Left: Mean payoffs. Right: Mean quantum parameters and r_A values

player types, and in its right frame the quantum mean parameter and mean r_A values across the lattice in five simulations.

Let us recall here that in the QPD inertial game, only $\{\hat{D}, \hat{D}\}$ are in NE for γ under $\gamma^\star$, for values of the entanglement factor over $\gamma^\bullet$, only $\{\hat{Q}, \hat{Q}\}$ are in NE, and that in the $(\gamma^*, \gamma^\bullet)$ interval, both $\{\hat{Q}, \hat{D}\}$ and $\{\hat{D}, \hat{Q}\}$ are in NE [7, 8]. The critical values that delimit the NE-regions given in Eqs. 2.9.

The right frame of Fig. 8.2 shows how the CA imitation dynamics enables the emergence of an effect of γ in the context of the QRPD-CA simulations qualitatively comparable to that just described in the conventional QPD inertial game regarding NE: Mutual defection below $\gamma^* = \arcsin(1/2) = \pi/6$, phase-transition in the $(\gamma^*, \gamma^\bullet)$ interval, and $\{\hat{Q}, \hat{Q}\}$ over $\gamma^\bullet = \arcsin(1/\sqrt{2}) = \dfrac{\pi}{4}$. In the $[0, \gamma^*]$ γ-interval it is $\bar{\theta}_A \simeq \bar{\theta}_B \simeq \pi$, so that the erratic behaviour of the α parameters in it is irrelevant as $\cos\pi/2 = 0$ annihilates the influence of α in Eq. (2.3). In the particular case of $\gamma = 0$ it is $\bar{\theta}_A = \bar{\theta}_B = \pi$ and $\bar{r} = \pi/4$, i.e., the (D,D) scenario of Fig. 8.1, yielding the $p_A = 3.5$, $p_B = 1.5$ payoffs in the left frame. In the $[\gamma^\bullet, \pi/2]$ γ-interval it is $\bar{\alpha}_A \simeq \bar{\alpha}_B \simeq \pi/2$ and $\bar{r} = 0$, so that $\{\hat{Q}, \hat{Q}\}$ yields $R = 3$ for both player types in the left frame. For very low values of γ, roughly up to $\gamma = \pi/16$, the $\bar{r}$ parameter keeps fairly stable at $\pi/4$, but for higher γ the value of $\bar{r}$ decreases generating an appreciable turbulence in the $\overline{p}_A$ payoff in the left frame before γ^*, but at $\gamma = \gamma^*$ it is already $\bar{r} = 0.0$. In the $[\gamma^*, \gamma^\bullet]$ interval, it is $\bar{\theta}_A \simeq \pi$ (that makes irrelevant $\bar{\alpha}_A$), $\bar{\alpha}_B \simeq \pi/2$, $\bar{\theta}_A \simeq \pi$, so the parameters of the $\{\hat{D}, \hat{Q}\}$ game, with associated $\Pi^{DQ} = \begin{pmatrix} \frac{1}{4}\sin^2\gamma\sin^2 r & \frac{1}{4}\sin^2\gamma(\cos r + 1)^2 \\ (-\cos^2(\gamma/2)(\cos r + 1) + 1)^2 & \cos^4(\gamma/2)\sin^2 r \end{pmatrix}$, that at $r = 0$

reduces to $\Pi^{DQ}(r = 0) = \begin{pmatrix} 0 & \sin^2\gamma \\ \cos^2\gamma & 0 \end{pmatrix}$, that generate payoffs that behave in a

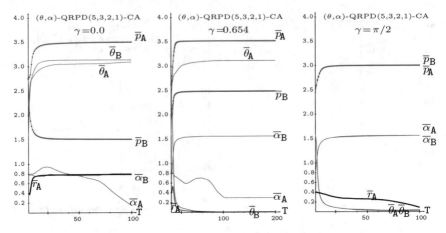

Fig. 8.3 Dynamics in three simulations in the QRPD(5,3,2,1)-CA scenario of Fig. 8.2. Left: $\gamma = 0$, Center: $\gamma = (\gamma^* + \gamma^{\bullet})/2$, Right: $\gamma = \pi/2$

nearly linear way in the $[\gamma^*, \gamma^{\bullet}]$ interval. It is $\Pi_{\gamma=0}^{DQ}(r = 0) = \begin{pmatrix} 0 & 1 \\ 0 & 0 \end{pmatrix}$, $\Pi_{\gamma=\pi/2}^{DQ}(r = 0) = \begin{pmatrix} 0 & 0 \\ 1 & 0 \end{pmatrix}$, $\Pi_{\gamma=\pi/4}^{DQ}(r = 0) = \begin{pmatrix} 0 & 1/2 \\ 1/2 & 0 \end{pmatrix}$. In the later case it is: $p_A = p_B = \frac{\tau + s}{2}$, in the scenario of Fig. 8.2 $p_A = p_B = 3.0$, as emerging for $\gamma = \pi/4$.

Figure 8.3 shows the dynamics in simulations in the QRPD(5,3,2,1)-CA scenario of Fig. 8.2 with $\gamma = 0$ (left), and $\gamma = (\gamma^* + \gamma^{\bullet})/2$ (center), and $\gamma = \pi/2$ (right). As a result of the initial random assignment of the parameter values, it is initially in both frames: $\bar{\theta} \simeq \pi/2 = 1.570$, $\bar{\alpha} \simeq \pi/8 = 0.785$, and $\bar{r} \simeq \pi/8 = 0.196$. With $\gamma = 0$, (i) $\bar{\alpha}_B$ remains fairly unaltered, whereas $\bar{\alpha}_A$ decreases in the long term, (ii) both $\bar{\theta}$'s rocket to π, (iii) $\bar{r}$ rockets to $\pi/4$. With $\gamma = \pi/2$, both $\bar{\alpha}$'s rocket towards $\pi/2$, both $\bar{\theta}$'s plummet to zero, i.e., the parameters of the Q strategy, and $\bar{r}$ decreases towards zero. Please, note that the tendencies heavily emerge from the very beginning, despite the full range of parameters initially accessible in the CA local interactions. In the central frame with $\gamma = (\gamma^* + \gamma^{\bullet})/2$, the tendencies operate in a dramatically fast manner ($\bar{\alpha}_B \to \pi/2, \bar{\theta}_A \to \pi, \bar{\theta}_B \to 0$), with the exception of $\bar{\alpha}_A$ that only stabilizes at $T = 100$ to $\bar{\alpha}_A = 0.294$, not far from $\pi/8 = 0.393$.

The general form of the plots in Fig. 8.2 remains in CA-simulations with choices of the PD parameters different from that primarily adopted here, i.e., (5,3,2,1). But the details of the transition from $\{\hat{D}, \hat{D}\}$ to $\{\hat{Q}, \hat{Q}\}$ vary notably accordingly to the hierarchy of the γ^* and $\gamma^{\bullet}$ values. Thus, the order $\gamma^* < \gamma^{\bullet}$ (that applies with (5,3,2,1)) may be inverted into $\gamma^{\bullet} < \gamma^*$, in which case, in the $(\gamma^{\bullet}, \gamma^*)$ interval both $\{\hat{Q}, \hat{Q}\}$ and $\{\hat{D}, \hat{D}\}$ are in NE in conventional inertial quantum games. In the limit case it is $\gamma^* = \gamma^{\bullet}$. In these alternative PD-scenarios, the transition from $\{\hat{D}, \hat{D}\}$ to $\{\hat{Q}, \hat{Q}\}$ in CA-simulations occurs abruptly when γ increases, in the form of a kind of phase-transition that has been tested in Fig. 8.4 with the PD-parameters (4,3,2,0), i.e., $\gamma^* = \arcsin(1/2) > \gamma^{\bullet} = \arcsin(1/\sqrt{2})$, right frame of Fig. 8.4, and with (4,3,2,1),

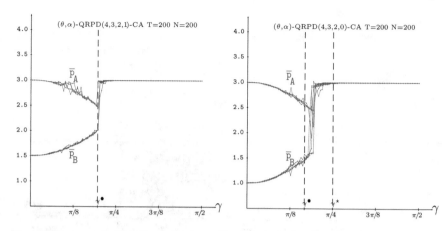

Fig. 8.4 Mean payoffs ($\overline{p}$) in five simulations at T = 200 of a QRPD-CA with variable entanglement factor γ. Left: (4,3,2,1)-PD parameters. Right: (4,3,2,0)-PD parameters

i.e., $\gamma^* = \gamma^{\bullet} = \arcsin(1/2)$, left frame of Fig. 8.4. In the (4, 3, 2, 1)-PD case the abrupt transition from $\{\hat{D}, \hat{D}\}$ to $\{\hat{Q}, \hat{Q}\}$ occurs roughly at $\gamma^{\bullet}$, in the (4,3,2,0)-PD case over but close to $\gamma^{\bullet}$.

Remarkably, the transition: $\{\hat{D}, \hat{D}\} \rightarrow \{\{\hat{D}, \hat{Q}\} \cup \{\hat{Q}, \hat{D}\}\} \rightarrow \{\hat{Q}, \hat{Q}\}$ as γ increases found here in CA simulations in the noninertial frame, generalizes that found in the inertial frame [9, 10]. If contrary to the accelerated player A, inertial player B scenario assumed in this work, the reverse is adopted, i.e., player B accelerated, player A inertial, the emergence of the pair $\{\hat{Q}, \hat{Q}\}$ beyond $\gamma^{\bullet}$ remains operative as Fig. 8.5 shows. Before this γ-threshold, the accelerated player, now B, also overrates the inertial one, as in Fig. 8.2, which in this case implies $\{\hat{C}, \hat{C}\}$ before γ^*, and $\{\hat{C}, \hat{D}\}$ in the $(\gamma^*, \gamma^{\bullet})$ interval.

Unfair Contests

Let us assume the unfair situation: A type of players is restricted to classical strategies $\tilde{U}(\theta, 0)$, whereas the other type of players may use quantum $\hat{U}(\theta, \alpha)$ ones [11].

Figure 8.6 deals with five simulations of an accelerated (θ, α)-player A (red) versus an inertial θ-player B (blue) in a QRPD(5,3,2,1)-CA with variable entanglement factor γ. Its left frame shows the asymptotic mean payoffs across the lattice ($\overline{p}$) of both player types, and in its right frame the mean parameter values. The effect of γ up to close $\gamma^{\bullet}$ is somehow comparable to that in Fig. 8.2, i.e., that of mutual defection at $r = \pi/4$; but near $\gamma^{\bullet}$ a kind of discontinuity emerges so that the trend to $\overline{p}_A$ and $\overline{p}_B$ to approach ceases, and is replaced by a trend to the separation of both mean payoffs, so that at $\gamma = \pi/2$ both payoffs are separated nearly as at $\gamma = 0$. As a result, the player with two parameters overrates that with only the θ parameter regardless of γ.

In the conventional unfair inertial QPD game, the so-called *miracle* strategy $\hat{M} = \hat{U}(\pi/2, \pi/2)$ ensures that the quantum player over-rates the classical player provided that $\gamma > \pi/4$, even if the classical player chooses $\hat{D}$ (see Fig. 4.2). In the case of full

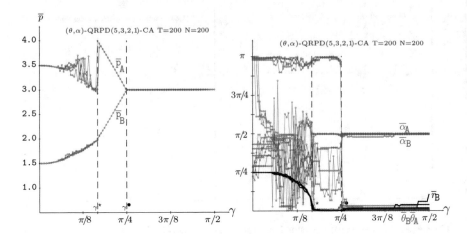

Fig. 8.5 The QRPD(5,3,2,1)-CA with variable entanglement factor γ. Player A stationary and player B accelerated. Five simulations at T = 200. Left: Mean payoffs. Right: Mean quantum parameters and r_A values

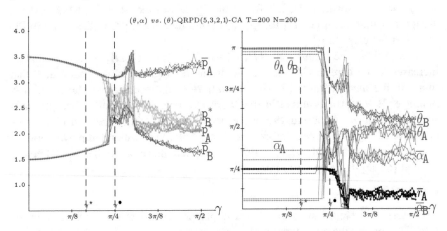

Fig. 8.6 The QRPD(5,3,2,1)-CA unfair quantum accelerated (θ, α)-player A (red) versus inertial θ-player B (blue) game. Variable entanglement factor γ. Five simulations at T = 200. Left: Actual mean payoffs ($\overline{p}$) and mean-field payoffs (p^*). Right: Mean quantum parameters and r_A values

entangling, $\hat{M}$ versus $\hat{D}$ will provide the payoff $\frac{\mathfrak{T}+P}{2}$ to the quantum player, whereas the classical player will get $\frac{S+P}{2}$. In the context of Fig. 8.6 it is: $\frac{\mathfrak{T}+P}{2} = 3.5$, and $\frac{S+P}{2} = 1.5$, numerical values of the payoffs close to those achieved by the $\overline{p}$ values for $\gamma = \pi/2$ in the left frame of Fig. 8.6.

The left frame of Fig. 8.6 also shows the mean-field payoffs (p^*) achieved in a single hypothetical two-person game with players adopting the mean parameters

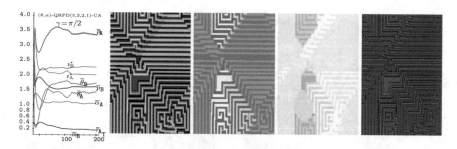

Fig. 8.7 A simulation in the $\gamma = \pi/2$ unfair scenario of Fig. 8.6. Far Left: Evolving mean parameters and payoffs. Center: (θ, α, r) parameter patterns at $T = 200$. Far Right: Payoff pattern at $T = 200$. Increasing grey levels indicate increasing parameter values

appearing in the spatial dynamic simulation, those given in the right panel of Fig. 8.6. Namely, with $r = \bar{r}$ and U_A^* and U_B^* as given in Eqs. 3.1.

The mean-field payoffs estimations given in the left frame of Fig. 8.6 fit fairly perfectly to the actual ones up to γ near to γ^*. But for high γ values the mean-field p^* values notably differ from the actual $\bar{p}$ ones. This is so even qualitatively, in the sense that for $\gamma > \gamma^\bullet$ it is $\bar{p}_A > \bar{p}_B$ whereas $\bar{p}_A^* < \bar{p}_B^*$. The explanation of this divergence relies in spatial effects (fairly absent in fair contests), as commented below when dealing with Fig. 8.7.

Figure 8.7 deals with a simulation in the $\gamma = \pi/2$ unfair scenario of Fig. 8.6. A PD(5,3,2,1)-CA where the B players are restricted to classical strategies, i.e., $\alpha_B = 0$. The far left panel of the figure shows the evolution up to $T=100$ of the mean values across the lattice of θ, α and $\bar{r}$ as well of the actual and theoretical mean payoffs. As a result of the random assignment of the parameter values (as in Fig. 8.3) it is initially: $\bar{\theta}_A \simeq \bar{\theta}_B \simeq \pi/2 = 1.57$, $\bar{\alpha}_A \simeq \pi/4 = 0.78$, and $\bar{r} \simeq \pi/8 = 0.196$. But at variance with what happens in the right frame of Fig. 8.3, the $\bar{\theta}$ parameters do not plummet to zero, and $\bar{r}$ remains at levels not close to zero. The parameter patterns at $T = 200$ shown in Fig. 8.7 show a kind of maze-like aspect, particularly crisp in the θ pattern, a kind of spatial heterogeneity that explains why the theoretical mean-field estimations of the payoffs very much differ from the actual ones. The divergence of the mean-field p^* values from the actual $\bar{p}$ ones becomes apparent from the initial iterations in the left frame of Fig. 8.7, so that after a short transition time, it is $\bar{p}_A > \bar{p}_B^* > p_A^* > \bar{p}_B$.

Figure 8.8 deals with five simulations of an accelerated θ-player A (red) versus an inertial (θ, α)-player B (blue) in a QRPD(5,3,2,1)-CA with variable entanglement factor γ, thus the scenario of Fig. 8.6 but with reversed unfairness. The structure of the graphics of the asymptotic mean payoffs in the left frame of Fig. 8.8 resembles that of Fig. 8.6 up to $\gamma = \gamma^\bullet$, but beyond this threshold they differ so that the linearity which is lost at $\gamma^\bullet$ in Fig. 8.6 remains in Fig. 8.8 at some higher γ extent. For higher γ values, the (θ, α)-player B overrates the θ-player B in Fig. 8.8, reversing what happens in Fig. 8.6. The mean-field payoffs estimations given in the left frame of Fig. 8.8 fit in a noisy way to the actual ones up to γ^*, perfectly well to over but close

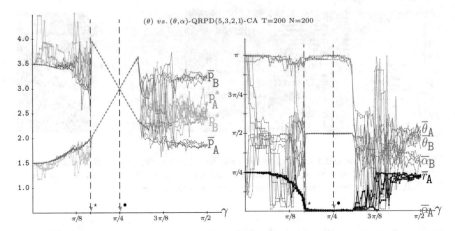

Fig. 8.8 The QRPD(5,3,2,1)-CA unfair accelerated θ-player A (red) versus an inertial (θ, α)-player B (blue) game. Variable entanglement factor γ. Five simulations at T = 200. Left: Actual mean payoffs $(\overline{p})$ and mean-field payoffs (p^*). Right: Mean quantum parameters and r_A values

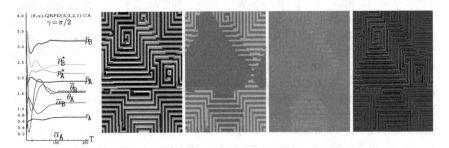

Fig. 8.9 A simulation in the $\gamma = \pi/2$ unfair scenario of Fig. 8.8. Far Left: Evolving mean quantum parameters, payoffs and r_A values. Center: (θ, α, r) parameter patterns at $T = 200$. Far Right: Payoff pattern $T = 200$. Increasing grey levels indicate increasing parameter values

$\gamma^{\bullet}$, but for high γ values the mean-field p^* values notably differ from the actual $\overline{p}$ ones, much as already observed in Fig. 8.6.

Figure 8.9 deals with a simulation in the $\gamma = \pi/2$ unfair scenario of Fig. 8.8. A PD(5,3,2,1)-CA where the A players are restricted to classical strategies, i.e., $\alpha_A = 0$. The far left panel of the figure shows the evolution up to T=100 of the mean values across the lattice of θ, α and $\overline{r}$ as well of the actual and theoretical mean payoffs. Figure 8.9 keeps the trait features of the graphs in Fig. 8.7. The $\overline{\theta}$ parameters do not plummet to zero, and $\overline{r}$ remains at levels not close to zero; the parameter patterns show a kind of maze-like aspect, particularly crisp in the θ pattern, which explains why the theoretical mean-field estimations of the payoffs very much differ from the actual ones, again from the initial iterations, so that after a short transition time, it is $\overline{p}_B > \overline{p}_B^* > p_A^* > \overline{p}_A$.

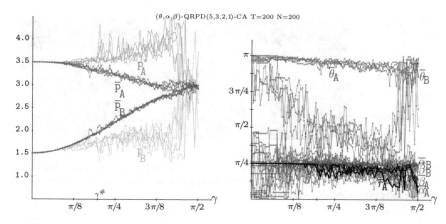

Fig. 8.10 The three-parameter QRPD(5,3,2,1)-CA with variable entanglement factor γ. Five simulations at T = 200. Left: Actual mean payoffs ($\overline{p}$) and mean-field payoffs (p^*). Right: Mean quantum parameters and r_A values

Three-Parameter Strategies

This section deals with the full space of strategies SU(2), operating with three parameters as given in Eq. (2.3) in the QRPD game.

Figure 8.10 shows the results achieved in the scenario of Fig. 8.2, but in the three-parameter strategies model. The graph of the actual mean payoffs (left frame) is reminiscent of that of the $(\hat{D}, \hat{D})$ game at $r = \pi/2$, as in Fig. 8.2, but at variance with this, all along the γ variation. No pair of pure strategies in NE exists in the conventional (non-CA) inertial model with three parameters at $\gamma = \pi/2$, but there are an infinite number of mixed strategies in NE [12]. In [8] pure strategies in NE are described in the 3P scenario below the threshold $\gamma^{\#} = \arcsin(\sqrt{\frac{P-S}{\mathfrak{T}+P-R-S}})$, which seems not to play any particular role in the payoff frame of the noninertial model of Fig. 8.10 (this also happens in the inertial frame studied in [9]). But the right frame of this figure shows that it is $\overline{r} = \pi/4$ up to approximately $\gamma = \gamma^{\#}$, and beyond this γ-threshold $\overline{r}$ shows a much more trembling graph, albeit not far from $\overline{r} = \pi/4$.

A second feature of the payoffs in the left frame of Fig. 8.10 is that of the dramatic deviation of the mean-field estimation from the actual mean payoffs beyond $\gamma = \pi/8$, with increasing overestimation of the A player payoff, and underestimation of the B player payoff. This is due to heavy spatial effects exemplified at $\gamma = 3\pi/8$ in Fig. 8.11, where the far left frame shows that the deviation of the p^*'s values from their corresponding $\overline{p}'s$, i.e., the spatial effects, commence very early in the dynamics, and the parameter patterns show, again, a kind of maze-like aspect.

Figure 8.12 shows the payoffs at $T = 200$ in a simulation of unfair QRPD(5,3,2,1)-CA games with variable γ, where one of the players is allowed to tune three parameters, whereas only θ is allowed for the other one. The graphs in the two alternative scenarios of Fig. 8.12 are much alike as those shown in the left frames of Figs. 8.6 and 8.8. So that in the scenario of the left frame, the 3P-player over scores the

Fig. 8.11 A simulation in the $\gamma = 3\pi/8$ scenario of Fig. 8.10. Far Left: Evolving mean parameters and payoffs up to $T = 100$. Right: $(\theta, \alpha, \beta, r)$ parameter patterns at $T = 200$. Increasing grey levels indicate increasing parameter values

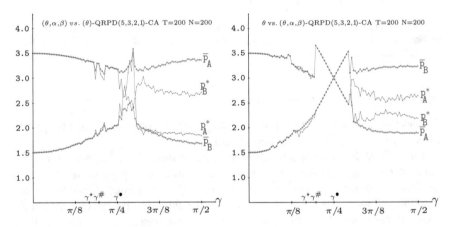

Fig. 8.12 Actual mean payoffs across the lattice $\overline{p}$, and mean-field payoffs p^* at $T = 200$ in two unfair simulations of the QRPD(5,3,2,1)-CA game. Variable entanglement factor γ. Left: (θ, α, β)-accelerated player A (red) versus θ-inertial player B (blue). Right: θ-accelerated player A (red) versus (θ, α, β)-inertial player B (blue)

θ-player regardless of γ, whereas in the scenario of the right frame the 3P-player over scores the θ-player only for $\gamma > \gamma^{\bullet}$. Again, the divergence between the actual ($\overline{p}$) and mean-field (p^*) payoffs for high γ values relies in heavy spatial effects.

It is concluded in this section that in the quantum relativistic PD with high levels of the entanglement factor (specifically from $\gamma > \gamma^*$) the accelerated player tends to brake, so that the results in CA simulations in the relativistic and inertial scenarios tend to resemble for high γ. In fact, for γ over the $\gamma^{\bullet}$ threshold the accelerated player becomes fully braked, so that the results in both scenarios coincide. In contests fair regarding the quantum component of the game, the transition from mutual defection to mutual Q when increasing the entangling factor in CA simulations is roughly marked for the same thresholds as in the inertial quantum games. In the unfair quantum versus classical scenario, the effect of entangling in CA simulations appears to be much more complex, as spatial effects arise with high entanglement levels, though the player that may use two-parameter quantum strategies overscores the one restricted to classical strategies. In simulations with players enabled to access three parameters, the accelerated player over scores the inertial one in CA simulations where high spatial effects arise.

8.3 A Quantum Relativistic Battle of the Sexes

In this section, player A will be assigned the male ($\circlearrowleft$) role whereas player B will be assigned that of the female ($\circlearrowleft$) role in a quantum relativistic Battle of the Sexes (QRBOS) game.

From Eq. (8.6) it is known that if both players use pure strategies, i.e., $x \in \{0, 1\}$, $y \in \{0, 1\}$, $\Pi(\gamma = 0)$ gets simple forms. In particular, $\Pi^{FF}(\gamma = 0) = \begin{pmatrix} \cos^2 r & \sin^2 r \\ 0 & 0 \end{pmatrix}$. Π^{FB} is obtained permuting the columns of Π^{FF}, Π^{BF} permuting the rows of Π^{FF}, and Π^{BB} permuting the rows and the columns of Π^{FF}.

The probability distribution for mutual F and arbitrary γ and r parameters is given in Eq. (8.7). As with $\gamma = 0$, permuting the columns, the rows, and the rows and columns of Π^{FF}, generates Π^{FB}, Π^{BF}, and Π^{BB}. As a result, in the QRBOS [13, 14] it is: $p_{\circlearrowleft} > p_{\circlearrowleft}$ if (F, F) or (F, B), and $p_{\circlearrowleft} < p_{\circlearrowleft}$ if (B, F) or (B, B).

Figure 8.13 shows the payoffs of the pure strategies in the QRBOS with (5,1) parameters at $r = \pi/4$ for variable γ. At this maximum level of r it is $\Pi^{FF}_{\gamma=0} = \Pi^{FB}_{\gamma=0} = \frac{1}{2}\begin{pmatrix} 1 & 1 \\ 0 & 0 \end{pmatrix}$, and $\Pi^{BF}_{\gamma=0} = \Pi^{BB}_{\gamma=0} = \frac{1}{2}\begin{pmatrix} 0 & 0 \\ 1 & 1 \end{pmatrix}$, and consequently: $p^{FF}_{\circlearrowleft} = p^{FB}_{\circlearrowleft} = p^{BF}_{\circlearrowleft} = p^{BB}_{\circlearrowleft} = \frac{R}{2} = 2.5$, $p^{FF}_{\circlearrowleft} = p^{FB}_{\circlearrowleft} = p^{BF}_{\circlearrowleft} = p^{BB}_{\circlearrowleft} = \frac{\rho}{2} = 0.5$.

The Spatialized QRBOS

As customary in the spatial simulations of this book, in the QRBOS-CA simulations in this section, the male player A and the female player B alternate in the site occupation in a 2D lattice, interact in the cellular automata manner and the evolution is ruled by the deterministic imitation of the best paid mate neighbour. Also as usual, the simulations in this section are run up to $T = 200$ in a $N = 200$ lattice with periodic boundary conditions, and five different initial random assignment of the (θ, α, r) parameter values are implemented in every studied scenario.

Figure 8.14 shows the dynamics up to $T = 100$ in simulations in the QRBOS(5,1)-CA scenario with $\gamma = 0$ (left), $\gamma = \pi/4$ (center), and $\gamma = \pi/2$ (right). As a result of the initial random assignment of the parameter values, it is initially in every frame: $\bar{\theta} \simeq \pi/2 = 1.570$, $\bar{\alpha} \simeq \pi/8 = 0.785$, and $\bar{r}_{\circlearrowleft} \simeq \pi/8 = 0.196$. With $\gamma = 0$, $\bar{r}_{\circlearrowleft}$ drifts to zero (i.e., to the inertial scenario), as well a both $\bar{\theta}_{\circlearrowleft}$ and $\bar{\theta}_{\circlearrowleft}$, which means that both players choose F and consequently $\bar{p}_{\circlearrowleft} = R = 5$, $\bar{p}_{\circlearrowleft} = \rho = 1$ in the long term. The stabilization of the mean values of both α parameters in the $\gamma = 0$ scenario (where the $\alpha_{\circlearrowleft}$ and $\alpha_{\circlearrowleft}$ parameters are in fact irrelevant) indicates

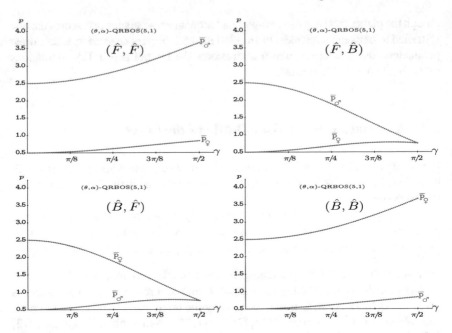

Fig. 8.13 Payoffs of the interactions of the pure strategies in a QRBOS(5,1) game at $r = \pi/4$. Variable entanglement factor γ. Top-Left: $(\hat{F}, \hat{F})$, Top-Right: $(\hat{F}, \hat{B})$, Bottom-Left: $(\hat{B}, \hat{F})$, Bottom-Right: $(\hat{B}, \hat{B})$,

the complete stabilization of the system, once the (5,1) payoffs are fixed all around the lattice. With $\gamma = \pi/4$, $\bar{r}_{\sigma}$ holds a non-zero value (though at a small value), both $\bar{\theta}_{\sigma}$ and $\bar{\theta}_{\varphi}$ reach similar values, whereas $\bar{\theta}_{\varphi}$ outperforms $\bar{\theta}_{\sigma}$. With $\gamma = \pi/2$, the dynamics of the $\bar{\alpha}$ and $\bar{\theta}$ parameters are similar to those with $\gamma = \pi/4$, whereas $\bar{r}_{\sigma}$ reaches a value that is higher than the low one shown with $\gamma = \pi/4$. It is remarkable, that the just commented tendencies heavily emerge from the very beginning as a rule, despite the full range of parameters initially accessible in the CA local interactions. Please, note that the simulations in Fig. 8.14 are shown up to $T = 100$, not up to $T = 200$ as implemented here to make sure that the dynamics achieve stable states. In this way, the fluctuations and apparent lack of convergence observable in some parameters in the center and right frames of Fig. 8.14 vanish long before $T = 200$.

Figure 8.14 also shows the dynamics of the mean-field payoffs (p^*) achieved in a single hypothetical two-person game with players adopting the mean parameters appearing in the spatial dynamic simulation. Namely, with $r_{\sigma} = \bar{r}_{\sigma}$ and,

$$U^*_{\sigma} = \begin{pmatrix} e^{i\bar{\alpha}_{\sigma}} \cos \bar{\omega}_{\sigma} & \sin \bar{\omega}_{\sigma} \\ -\sin \bar{\omega}_{\sigma} & e^{-i\bar{\alpha}_{\sigma}} \cos \bar{\omega}_{\sigma} \end{pmatrix}, \; U^*_{\varphi} = \begin{pmatrix} e^{i\bar{\alpha}_{\varphi}} \cos \bar{\omega}_{\varphi} & \sin \bar{\omega}_{\varphi} \\ -\sin \bar{\omega}_{\varphi} & e^{-i\bar{\alpha}_{\varphi}} \cos \bar{\omega}_{\varphi} \end{pmatrix} \quad (8.9)$$

Figure 8.15 shows the patterns in the $\gamma = \pi/2$ scenario of Fig. 8.14, where increasing grey levels indicate increasing values. From left to right: (θ, α, r) parameter

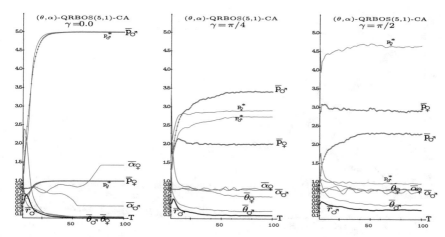

Fig. 8.14 Dynamics up to $T = 100$ in three simulations in the QRBOS(5,1)-CA. Left: $\gamma = 0$, Center: $\gamma = \pi/4$, Right: $\gamma = \pi/2$

Fig. 8.15 Patterns at $T = 200$ in a simulation in the $\gamma = \pi/2$ scenario of Fig. 8.14. From left to right: (θ, α, r) parameter patterns, and payoff pattern. Increasing grey levels indicate increasing values

patterns, and payoff pattern. The parameter patterns in Fig. 8.15 show a kind of maze-like aspect, particularly crisp in the α pattern, a kind of spatial heterogeneity that explains why the mean-field estimations of the payoffs differ from the actual ones in the $\gamma \neq 0$ scenarios of Fig. 8.14. The divergence of the mean-field payoffs and the actual ones becomes apparent from the initial iterations in the right frame ($\gamma = \pi/2$) of Fig. 8.14, leading to notably distant p^* and $\overline{p}$ values in the long term. The actual versus mean-field divergence in the central frame of Fig. 8.14, i.e., with $\gamma = \pi/4$, becomes apparent latter compared to the $\gamma = \pi/2$ case. The mean-field payoff estimates of both player types with $\gamma = \pi/4$ (central frame) slowly approach, so that at $T = 200$ their values equalize, as becomes apparent in Fig. 8.16, but in Fig. 8.15 they are still different at $T = 100$.

Figure 8.16 deals with the results obtained at T = 200 in five simulations (five different initial random assignment of the (θ, α, r) parameter values) of a QRBOS(5,1)-CA with variable entanglement factor γ. Figure 8.16 shows in its right frame the mean parameter values across the lattice and in its left frame the mean payoffs across the lattice $(\overline{p})$ and the mean-field payoffs (p^*) of both player types. The results with

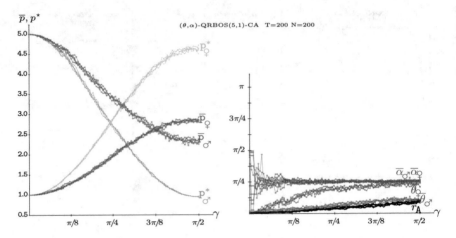

Fig. 8.16 The two-parameter QRBOS(5,1)-CA with variable entanglement factor γ. Five simulations at T = 200. Left: Mean payoffs ($\overline{p}$) and mean-field payoffs (p^*). Right: Mean quantum parameters and r_{σ} values

$\gamma = 0$, $\gamma = \pi/4$ and $\gamma = \pi/2$ in Fig. 8.16 are foreseeable from Fig. 8.14. Thus, regarding the payoffs, with $\gamma = 0$ (with no spatial effects): $\overline{p}_{\sigma} = p^*_{\sigma} = R = 5 > \overline{p}_{\female} = p^*_{\female} = \rho = 1$; with $\gamma = \pi/4$: $\overline{p}_{\sigma} > p^*_{\sigma} = p^*_{\female} > \overline{p}_{\female}$; and with $\gamma = \pi/2$: $p^*_{\female} > \overline{p}_{\female} > \overline{p}_{\sigma} > p^*_{\sigma}$. The main conclusion derived from the payoff frame (left) in Fig. 8.16 is that the mean payoff ($\overline{p}$) of the accelerated male overrates the inertial female if $\gamma < 3\pi/8$, whereas the inertial female overrates the accelerated male if $\gamma > 3\pi/8$. It seems that the bias towards the female player mentioned in Sect. 2.1 prevails in contests with very high entanglement. The mean-field payoff approaches (p^*) also react to the increase of γ in this way, but with payoff equalization achieved before, at $\gamma = \pi/4$. The right frame of Fig. 8.16 shows that both $\overline{r}_{\sigma}$ and $\overline{\theta}_{\sigma}$ slowly grow as γ increase, whereas $\overline{\theta}_{\female}$ grows much faster, reaching $\overline{\theta}_{\female} \simeq \pi/4$ at $\gamma = \pi/2$. Both $\overline{\alpha}_{\sigma}$ and $\overline{\alpha}_{\female}$ oscillate nearly $\pi/4$ after a noisy regime for low values of γ.

Unfair Contests

In *unfair* contexts, a type of players is restricted to classical strategies $\tilde{U}(\theta, 0)$, whereas the other type of players may use quantum $\hat{U}(\theta, \alpha)$ ones [11].

Figure 8.17 deals with five simulations of an accelerated (θ, α)-male player (red) versus an inertial θ-female player (blue) in a QRBOS(5,1)-CA with variable entanglement factor γ. Its left frame shows that the (θ, α)-male player fully overrates the θ-female player regardless of γ. This is so with no spatial effects as the mean-field payoff approaches (p^*) fully coincide with the actual mean payoffs ($\overline{p}$) for both player types. The right frame of Fig. 8.17 shows how every parameter, including $\overline{r}_{\sigma}$, plummets to zero in the long term, so that $\pi_{11} = 1$ and consequently $p_{\sigma} = R = 5$ and $p_{\female} = \rho = 1$.

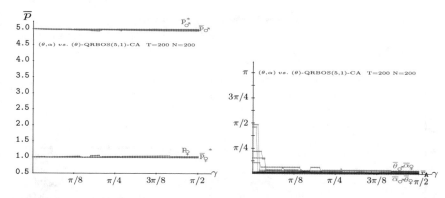

Fig. 8.17 The unfair quantum accelerated (θ, α)-male player (red) versus inertial θ-female player (blue) QRBOS(5,1)-CA game. Five simulations at $T = 2000$. Variable entanglement factor γ. Left: Mean payoffs $(\overline{p})$ and mean-field payoffs (p^*). Right: Mean quantum parameters and r_A values

Figure 8.18 deals with five simulations of an accelerated θ-male player (red) versus an inertial (θ, α)-female player (blue) in a QRBOS(5,1)-CA with variable γ. Thus, the scenario of Fig. 8.17 but with reversed unfairness, now favouring the female player. The left frame of Fig. 8.18 shows that the *favoured* player, i.e., the female, overrates the male beyond $\gamma = \pi/4$, but that, fairly contrary to expectation after Fig. 8.17, the player with only the θ parameter available, i.e., the male, overrates the female below $\gamma = \pi/4$. In the same vein as in the simulations of Fig. 8.17, no spatial effects emerge in the simulations of Fig. 8.18 as $p^* \simeq \overline{p}$ for both player types. The right frame of Fig. 8.18 shows that the $\alpha_{\female}$ parameter keeps $\overline{\alpha}_{\female} \simeq \pi/2$ regardless of γ in the long term, whereas the remaining parameters, including $\overline{r}_{\male}$, plummets to zero up to γ nearly the midpoint of the $[3\pi/8, \pi/2]$ interval, referred to as γ^* in what follows. We do not have any explanation to justify the odd behaviour detected in Fig. 8.18 for high gamma, particularly, regarding the *reappearance* of the r parameter. A similar phenomenon was found when dealing with the PD game (Fig. 8 in [1]). Any attempt to explain it is postponed to a further study dealing with the effect of quantum noise. With $\gamma > \gamma^*$ the parameters show a rather erratic behaviour that is reflected in the payoffs values in the left frame. But with $\gamma < \gamma^*$, where $\overline{r}_{\male} \simeq \overline{\theta}_{\male} \simeq \overline{\theta}_{\female} \simeq 0$, the mean-field estimation leads to a diagonal Π, i.e., non-factorizable, with $\pi_{22} = \sin^2(\alpha_{\male} + \alpha_{\female}) \sin^2 \gamma$, $\pi_{11} = 1 - \pi_{22}$; if additionally $(\alpha_{\male} + \alpha_{\female}) = \pi/2$ it turns out $\pi_{11} = \cos^2 \gamma$, $\pi_{22} = \sin^2 \gamma$, so that $p_{\male}(\gamma) = R \cos^2 \gamma + \rho \sin^2 \gamma$, $p_{\female}(\gamma) = \rho \cos^2 \gamma + R \sin^2 \gamma$. The latter formulas correspond to the *scissors*-like shape of the payoffs before γ^* in the left frame of Fig. 8.18. In particular, $p_{\male}(0) = R = 5$, $p_{\female}(0) = \rho = 1$; $p_{\male}(\pi/4) = p_{\female}(\pi/4) = (R + \rho)/2 = 3$.

Three-Parameter Strategies

This section deals with the full space of strategies SU(2), operating with three parameters as given in (2.3) in the QRBOS game.

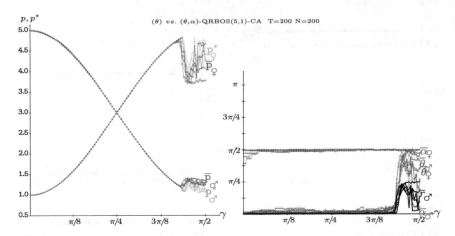

Fig. 8.18 The unfair accelerated θ-male player (red) versus inertial (θ, α)-female player (blue) QRBOS(5,1)-CA game. Variable entanglement factor γ. Five simulations at $T = 200$. Left: Mean payoffs $(\overline{p})$ and mean-field payoffs (p^*). Right: Mean quantum parameters values and r_A values

At variance with what happens in the two-parameter inertial frame described in Sect. 2.1, with middle-level election of the (θ, α, β) parameters, the probabilities in Π turns out equalized regardless of γ in the inertial frame [15, 16].

Figure 8.19 shows the results achieved in the scenario of Fig. 8.16, but in the three-parameter strategies model. The graph of the payoffs in Fig. 8.19 is reminiscent of that achieved in the scenario of Fig. 8.16 for low values of γ, so that the accelerated male overrates the inertial female in a decreasing manner as γ increases and the mean-field payoff estimates approach fairly well to the actual mean payoffs. But the structure of the payoff graphs for $\gamma > \pi/4$ notably differs when comparing Figs. 8.19 and 8.16. Thus, the actual payoffs tend to equalize rather than to diverge, and the mean-field estimates show an erratic behaviour instead of the crisp aspect shown not only in Fig. 8.16 but also in the unfair contests simulations. Unexpectedly, the form of the payoff graphs (both actual mean and mean-field) in the left frame of Fig. 8.19 very much resembles that in the 3P-QBOS with imperfect information of Fig. 11.4.

The graphs of the mean parameter values (right frame) in Fig. 8.19 indicate that the particularly high variability in the $\overline{\theta}$ parameters for $\gamma > \pi/4$ would explain that variability in the mean-field payoff estimates for high values of γ over approximately its midpoint. Remarkably, this dramatic variability in the mean-field payoff estimates p^* does not correspond with a similar variability in the actual mean payoff $\overline{p}$ which vary in a smooth way as γ grows, pointing out to notable spatial effects, that Fig. 8.20 exemplifies. The $\overline{r}_{\sigma^{\!*}}$ parameter varies in Fig. 8.19 much as in Fig. 8.16, slowly growing as γ increases.

The Nash Equilibrium (NE) strategies in the noninertial QBOS frame in the full space of strategies is studied in [14]. It is concluded in [14] that the NE in the classical inertial frame based in pure strategies are still in equilibrium in the noninertial frame, albeit the payoffs of both players are notably influenced by both γ and r as quantified in Eq. (8.7). Apparently, in the CA-simulations of Fig. 8.19 the accelerated-male

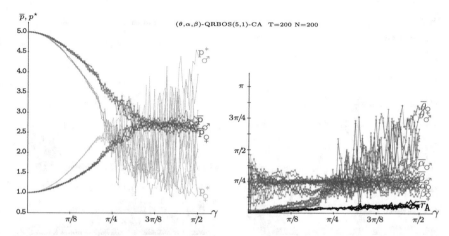

Fig. 8.19 The three-parameter QRBOS(5,1)-CA with variable entanglement factor γ. Five simulations at T = 200. Left: Mean payoffs ($\overline{p}$) and mean-field payoffs (p^*). Right: Mean quantum parameters and r_{σ} values

Fig. 8.20 Patterns at $T = 200$ in a simulation in the $\gamma = \pi/2$ scenario of Fig. 8.19. From left to right: $(\theta, \alpha, \beta, r)$ parameter patterns, and payoff pattern. Increasing grey levels indicate increasing values

player manages to impose strategies favouring him, i.e., with himself choosing F, for low values of γ; in the extreme case (F,F) with $\overline{r}_{\sigma} = 0$ if $\gamma = 0$. Let us mention here that this advantageous result achieved for the accelerated σ-player was not found in the 3P CA-simulations in the inertial frame reported in [17]. For high values of γ the equalization of payoffs is achieved above but near $\overline{p} = 2.5$, approximately at the same level as in the 3P CA-simulations for high γ in the inertial frame [17]. Recall that the maximum equal payoff attainable in the classical inertial frame is $p^+ = (R + \rho)/4 = 1.5$ in the uncorrelated model and $p^+ = (R + \rho)/2 = 3.0$ in the correlated model. Incidentally, the NE based on mixed strategies in the classical uncorrelated inertial frame gives the much lower equalized payoffs $p = Rr/(R + \rho) = 0.833$. This payoff, as well as the strategies that induce it, are retrieved making $\gamma = r = 0$ in the non-pure strategies in NE in the quantum non-inertial frame reported in [14].

Figure 8.21 deals with five simulations of an accelerated (θ, α, β)-male player (red) versus an inertial θ-female player (blue) in a QRBOS(5,1)-CA with variable entanglement factor γ. Thus, the 3P analogue to the 2P Fig. 8.17. Initially, for low γ, the general appearance of the payoff curves in Fig. 8.21 is the same as that of the Fig. 8.17, so that the quantum-accelerated male player fully overrates the classical inertial female player. But before $\gamma = \pi/4$, a kind of phase transition rockets the

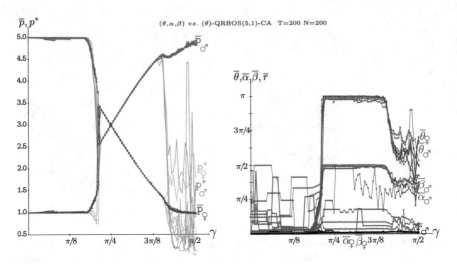

Fig. 8.21 The unfair quantum accelerated (θ, α, β)-male player (red) versus inertial θ-female player (blue) QRBOS(5,1)-CA game. Variable entanglement factor γ. Five simulations at $T = 200$. Left: Mean payoffs ($\overline{p}$) and mean-field payoffs (p^*). Right: Mean quantum parameters and r_A values

payoff of the female players and plummets that of the male players, so that contrary to expectation, the classical inertial female player overrates the quantum-accelerated male player (this is so up to $\gamma = \pi/8$). Immediately after this episode, both types of payoffs commence to approach, so that by $\gamma = \pi/4$ they equalize at the level $\overline{p} = (R + \rho)/2 = 3.0$. After this middle-level of γ, the quantum-accelerated male player increasingly overrates the classical female player. Spatial effects emerge only in one of the simulations with very high levels of γ in Fig. 8.17. The right frame of Fig. 8.21 indicates that $\overline{r}_{\sigma}$ is led to zero regardless of γ in the long term (as in Fig. 8.17), i.e., the asymptotic scenario is that of the inertial frame.

Figure 8.22 deals with five simulations of an accelerated θ-male player (red) versus an inertial (θ, α, β)-female player (blue) in a QRBOS(5,1)-CA with variable entanglement factor γ. Thus, the 3P analogue to the 2P Fig. 8.18, to whom Fig. 8.22 fully resembles. This so in particular regarding the *scissors*-like shape of the payoffs (left frame) for not too high γ.

The findings found regarding the CA the simulations of the quantum-relativistic BOS game can be summarized as follows: In the fair 2P model the evolving dynamics favours the accelerated players when the entanglement factor lies below its midpoint, and to the inertial ones above the midpoint entanglement. In unfair 2P contests, the quantum-accelerated players become fully privileged when faced with classical-inertial ones regardless of γ; whereas, in unfair classical-accelerated versus quantum-inertial games, the former players overrate the later ones with entanglement under its midpoint, whereas the payoffs order reverses with entanglement over its midpoint. In the three-parameter quantum model, the accelerated players overrate the inertial ones up to a high level of the entanglement factor, above which the payoffs of both player types tend to equalize. Heavy spatial effects emerge in CA simulations with no negligible entanglement, particularly in fair simulations, so that mean-field payoff estimates fail to be a good approach to the actual mean payoffs.

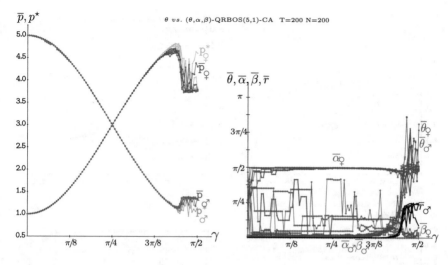

Fig. 8.22 The unfair accelerated θ-male player (red) versus inertial (θ, α, β)-female player (blue) QRBOS(5,1)-CA game. Five simulations at $T = 200$. Variable entanglement factor γ. Left: Actual mean payoffs $(\overline{p})$, and mean-field payoffs (p^*). Right: Mean quantum parameters and r_A values

8.4   Effect of Noise in a Quantum-Relativistic PD Game

As indicated in the Sect. 8.1 section of this chapter, in the Quantum-Relativistic PD (QRPD) we deal with, the player A moves with a uniform acceleration, and the player B remains stationary. In this scenario, the Unruh effect induces the initial density matrix given in Eqs. (8.4), and (8.7) gives Π for mutual cooperation coincident with Π^{QQ}. Permuting the columns, the rows, and the rows and columns of Π^{CC}, generate Π^{CD}, Π^{DC}, and Π^{DD}. As a result: $p_A < p_B$ if (C, C) or (C, D), and $p_A > p_B$ if (D, C) or (D, D).

Figure 8.23 shows under the label $\mu = 0$ the payoffs of the pure strategies in the PD with (5,3,2,1) parameters at $r = \pi/4$ for variable γ. These graph payoffs were already shown in Fig. 8.1, so that the considerations made when commenting that figure are also relevant here when referring to QRPD noiseless simulations.

The effect of noise in a QRPD is studied in this section by allowing the density matrix ρ to tackle the problem of the presence of noisy channels. Incidentally, the Unruh effect itself may be in turn interpreted as a quantum noise channel of amplitude-damping type [18].

We will consider in this section the effect of amplitude-damping noise type, characterized by the Kraus operators given in Eqs. 7.6, so that

$$\rho_{B,I}^* = \sum_{i=1}^{2} \sum_{j=1}^{2} (K_i \otimes K_j) \rho_{B,I} (K_i \otimes K_j)^\dagger. \tag{8.10}$$

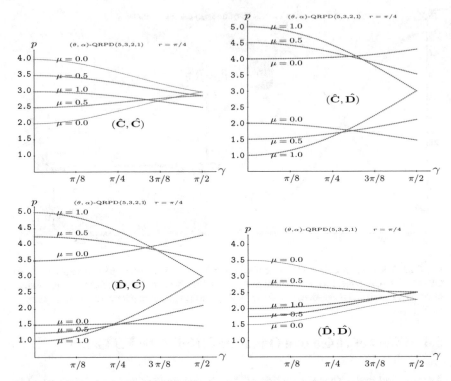

Fig. 8.23 Payoffs of the interactions of the pure strategies in a QRPD(5,3,2,1) at $r = \pi/4$ with variable entanglement factor γ. Top-Left: $(\hat{C}, \hat{C})$, Top-Right: $(\hat{C}, \hat{D})$, Bottom-Left: $(\hat{D}, \hat{C})$, Bottom-Right: $(\hat{D}, \hat{D})$. Three levels of noise are considered in each frame. Red: payoffs of player A, Blue: payoffs of player B

With maximum noise it is $\rho_{B,I}^{*}(\mu = 1) = \begin{pmatrix} 1 & 0 & 0 & 0 \\ 0 & 0 & 0 & 0 \\ 0 & 0 & 0 & 0 \\ 0 & 0 & 0 & 0 \end{pmatrix}$, regardless of r.[2] Therefore, as in

the inertial scenario it is $\Pi^{CC} = \begin{pmatrix} \cos^2 \frac{\gamma}{2} & 0 \\ 0 & \sin^2 \frac{\gamma}{2} \end{pmatrix}$. Permuting the columns, the rows,

---

[2] $(K_1 \otimes K_1)\rho_{B,I}(K_1 \otimes K_1)^{\dagger} = \begin{pmatrix} \cos^2 \frac{\gamma}{2} \cos^2 r & 0 & 0 & 0 \\ 0 & 0 & 0 & 0 \\ 0 & 0 & 0 & 0 \\ 0 & 0 & 0 & 0 \end{pmatrix}$, $(K_1 \otimes K_2)\rho_{B,I}(K_1 \otimes K_2)^{\dagger} =$

$\begin{pmatrix} \cos^2 \frac{\gamma}{2} \sin^2 r & 0 & 0 & 0 \\ 0 & 0 & 0 & 0 \\ 0 & 0 & 0 & 0 \\ 0 & 0 & 0 & 0 \end{pmatrix}$, $(K_2 \otimes K_1)\rho_{B,I}(K_2 \otimes K_1)^{\dagger} = \mathbf{0}$, $(K_2 \otimes K_2)\rho_{B,I}(K_2 \otimes K_2)^{\dagger} =$

$\begin{pmatrix} \sin^2 \frac{\gamma}{2} & 0 & 0 & 0 \\ 0 & 0 & 0 & 0 \\ 0 & 0 & 0 & 0 \\ 0 & 0 & 0 & 0 \end{pmatrix}$. $\cos^2 \frac{\gamma}{2} \cos^2 r + \cos^2 \frac{\gamma}{2} \sin^2 r + \sin^2 \frac{\gamma}{2} = 1$.

and the rows and columns of Π^{CC}, generate Π^{CD}, Π^{DC}, and Π^{DD}, leading to $p_A^{CC} = p_B^{CC} = R\cos^2\frac{\gamma}{2} + P\sin^2\frac{\gamma}{2}$, $p_A^{DD} = p_B^{DD} = R\sin^2\frac{\gamma}{2} + P\cos^2\frac{\gamma}{2}$, $p_A^{CD} = p_B^{CD} = \mathfrak{T}\sin^2\frac{\gamma}{2} + S\cos^2\frac{\gamma}{2}$, and $p_A^{DC} = p_B^{CD} = S\sin^2\frac{\gamma}{2} + \mathfrak{T}\cos^2\frac{\gamma}{2}$. These payoff dependences on γ of the pure strategies with full noise may be checked in Fig. 8.23 under the label $\mu = 1$ in the PD with (5,3,2,1) parameters. Note that the payoffs of both players coincide in the (C,C) and (D,D) games because in both cases Π is diagonal.

The expression of $\rho_{B,I}^*$ with middle level $\mu = 0.5$ noise is given in Eq. (8.11).[3]

$$\rho_{B,I}^* = \begin{pmatrix} \cos^2\frac{\gamma}{2}(\cos^2 r + \frac{1}{2}\sin^2 r) + \frac{1}{4}\sin^2\frac{\gamma}{2} & 0 & 0 & -i\frac{1}{4}\sin\gamma\cos r \\ 0 & \frac{1}{2}\cos^2\frac{\gamma}{2}\sin^2 r + \frac{1}{4}\sin^2\frac{\gamma}{2} & 0 & 0 \\ 0 & 0 & \frac{1}{4}\sin^2\frac{\gamma}{2} & 0 \\ \frac{1}{4}i\sin\gamma\cos r & 0 & 0 & \frac{1}{4}\sin^2\frac{\gamma}{2} \end{pmatrix}$$

(8.11)

The expression of $\rho_{B,I}^*(\mu = 0.5)$ given in Eq. (8.11) with maximum r reduces to

$$\rho_{B,I}^*[r = \pi/4] = \frac{1}{4}\begin{pmatrix} 3 - 2\sin^2\frac{\gamma}{2} & 0 & 0 & -i\frac{1}{\sqrt{2}}\sin\gamma \\ 0 & 1 & 0 & 0 \\ 0 & 0 & \sin^2\frac{\gamma}{2} & 0 \\ i\frac{1}{\sqrt{2}}\sin\gamma & 0 & 0 & \sin^2\frac{\gamma}{2} \end{pmatrix}.$$ [4] The payoffs dependence on γ of

---

[3] $(K_1 \otimes K_1)\rho_{BI}(K_1 \otimes K_1)^\dagger = \begin{pmatrix} 1 & 0 & 0 & 0 \\ 0 & \frac{1}{\sqrt{2}} & 0 & 0 \\ 0 & 0 & \frac{1}{\sqrt{2}} & 0 \\ 0 & 0 & 0 & \frac{1}{2} \end{pmatrix}\begin{pmatrix} \cos^2\frac{\gamma}{2}\cos^2 r & 0 & 0 & -i\frac{1}{2}\sin\gamma\cos r \\ 0 & \frac{1}{\sqrt{2}}\cos^2\frac{\gamma}{2}\sin^2 r & 0 & 0 \\ 0 & 0 & 0 & 0 \\ i\frac{1}{2}\sin\gamma\cos r & 0 & 0 & \frac{1}{2}\sin^2\frac{\gamma}{2} \end{pmatrix} =$

$\begin{pmatrix} \cos^2\frac{\gamma}{2}\cos^2 r & 0 & 0 & -i\frac{1}{2}\sin\gamma\cos r \\ 0 & \frac{1}{2}\cos^2\frac{\gamma}{2}\sin^2 r & 0 & 0 \\ 0 & 0 & 0 & 0 \\ i\frac{1}{2}\sin\gamma\cos r & 0 & 0 & \frac{1}{4}\sin^2\frac{\gamma}{2} \end{pmatrix}$, $(K_1 \otimes K_2)\rho_{B,I}(K_1 \otimes K_2)^\dagger = \begin{pmatrix} 0 & \frac{1}{\sqrt{2}} & 0 & 0 \\ 0 & 0 & 0 & 0 \\ 0 & 0 & 0 & \frac{1}{2} \\ 0 & 0 & 0 & 0 \end{pmatrix}$

$\begin{pmatrix} 0 & 0 & -i\frac{1}{2}\sin\gamma\cos r & 0 \\ \frac{1}{\sqrt{2}}\cos^2\frac{\gamma}{2}\sin^2 r & 0 & 0 & 0 \\ 0 & 0 & 0 & 0 \\ 0 & 0 & \frac{1}{2}\sin^2\frac{\gamma}{2} & 0 \end{pmatrix} = \begin{pmatrix} \frac{1}{2}\cos^2\frac{\gamma}{2}\sin^2 r & 0 & 0 & 0 \\ 0 & 0 & 0 & 0 \\ 0 & 0 & \frac{1}{4}\sin^2\frac{\gamma}{2} & 0 \\ 0 & 0 & 0 & 0 \end{pmatrix}$, $(K_2 \otimes K_1)\rho_{B,I}(K_2 \otimes K_1)^\dagger =$

$\begin{pmatrix} 0 & 0 & \frac{1}{\sqrt{2}} & 0 \\ 0 & 0 & 0 & \frac{1}{2} \\ 0 & 0 & 0 & 0 \\ 0 & 0 & 0 & 0 \end{pmatrix}\begin{pmatrix} 0 & -\frac{1}{\sqrt{2}}i\cos\frac{\gamma}{2}\sin\frac{\gamma}{2}\cos r & 0 & 0 \\ 0 & 0 & 0 & 0 \\ 0 & 0 & 0 & 0 \\ 0 & \frac{1}{2}\sin^2\frac{\gamma}{2} & 0 & 0 \end{pmatrix} = \begin{pmatrix} 0 & 0 & 0 & 0 \\ 0 & \frac{1}{4}\sin^2\frac{\gamma}{2} & 0 & 0 \\ 0 & 0 & 0 & 0 \\ 0 & 0 & 0 & 0 \end{pmatrix}$, $(K_2 \otimes K_2)\rho_{B,I}(K_2 \otimes K_2)^\dagger =$

$\begin{pmatrix} 0 & 0 & 0 & \frac{1}{2} \\ 0 & 0 & 0 & 0 \\ 0 & 0 & 0 & 0 \\ 0 & 0 & 0 & 0 \end{pmatrix}\begin{pmatrix} -\frac{1}{2}i\cos\frac{\gamma}{2}\sin\frac{\gamma}{2}\cos r & 0 & 0 & 0 \\ 0 & 0 & 0 & 0 \\ 0 & 0 & 0 & 0 \\ \frac{1}{2}\sin^2\frac{\gamma}{2} & 0 & 0 & 0 \end{pmatrix} = \begin{pmatrix} \frac{1}{4}\sin^2\frac{\gamma}{2} & 0 & 0 & 0 \\ 0 & 0 & 0 & 0 \\ 0 & 0 & 0 & 0 \\ 0 & 0 & 0 & 0 \end{pmatrix}$

[4] $\rho P = \frac{1}{4}\begin{pmatrix} 3 - 2\sin^2\frac{\gamma}{2} & 0 & 0 & -i\frac{1}{\sqrt{2}}\sin\gamma \\ 0 & 1 & 0 & 0 \\ 0 & 0 & \sin^2\frac{\gamma}{2} & 0 \\ i\frac{1}{\sqrt{2}}\sin\gamma & 0 & 0 & \sin^2\frac{\gamma}{2} \end{pmatrix}\begin{pmatrix} 0 & 0 & 0 & 1 \\ 0 & 0 & -1 & 0 \\ 0 & -1 & 0 & 0 \\ 1 & 0 & 0 & 0 \end{pmatrix} = \frac{1}{4}\begin{pmatrix} -i\frac{1}{\sqrt{2}}\sin\gamma & 0 & 0 & 3 - 2\sin^2\frac{\gamma}{2} \\ 0 & 0 & -1 & 0 \\ 0 & -\sin^2\frac{\gamma}{2} & 0 & 0 \\ \sin^2\frac{\gamma}{2} & 0 & 0 & i\frac{1}{\sqrt{2}}\sin\gamma \end{pmatrix}$,

$\rho P - P\rho = \frac{1}{4}\begin{pmatrix} -i\frac{2}{\sqrt{2}}\sin\gamma & 0 & 0 & 3 - 3\sin^2\frac{\gamma}{2} \\ 0 & 0 & -1 + \sin^2\frac{\gamma}{2} & 0 \\ 0 & -\sin^2\frac{\gamma}{2} + 1 & 0 & 0 \\ -3 + 3\sin^2\frac{\gamma}{2} & 0 & 0 & i\frac{2}{\sqrt{2}}\sin\gamma \end{pmatrix}$,

$P\rho = \frac{1}{4}\begin{pmatrix} i\frac{1}{\sqrt{2}}\sin\gamma & 0 & 0 & \sin^2\frac{\gamma}{2} \\ 0 & 0 & -\sin^2\frac{\gamma}{2} & 0 \\ 0 & -1 & 0 & 0 \\ 3 - 2\sin^2\frac{\gamma}{2} & 0 & 0 & -i\frac{1}{\sqrt{2}}\sin\gamma \end{pmatrix}$,

the pure strategies in the QRPD(5,3,2,1) with $\mu = 1/2$ and $r = \pi/4$ may be checked in Fig. 8.23 under the label $\mu = 1/2$. In particular, the case of the extremes $\gamma = 0$ with $\Pi^{CC} = \begin{pmatrix} 3/4 & 1/4 \\ 0 & 0 \end{pmatrix}$, and $\gamma = \pi/2$ with $\Pi^{CC} = \begin{pmatrix} 4/8 & 2/8 \\ 1/8 & 1/8 \end{pmatrix}$.

Figure 8.23 enables to observe the effect of noise given γ in the payoffs of the pure strategies at $r = \pi/4$. Thus for example, with no entanglement it is $\Pi^{CC}_{\gamma=0} = \frac{1}{2}\begin{pmatrix} 1+\mu & 1-\mu \\ 0 & 0 \end{pmatrix}$, and consequently $p^{CC}_A(\gamma = 0) = \frac{1}{2}(R + S + (R - S)\mu)$, so that in the top-left frame of Fig. 8.23 it is: $p^{CC}_A(\gamma = 0, \mu = 0.0) = \frac{1}{2}(R + S) = 2.0$, $p^{CC}_A(\gamma = 0, \mu = 0.5) = \frac{1}{4}(3R + S) = 2.5$, and $p^{CC}_A(\gamma = 0, \mu = 1.0) = R = 3.0$.

The Spatialized QRPD with Noise

As customary in the spatial simulations of this book, in the QRPD-CA with noise simulations that follow, the A and B players alternate in the site occupation in a 2D lattice, interact in the cellular automata manner and the evolution is ruled by the deterministic imitation of the best paid mate neighbour. Also as usual, the simulations in this section are run up to T= 200 in a $N = 200$ lattice with periodic boundary conditions, and five different initial random assignment of the (θ, α, r) parameter values are implemented in every studied scenario [19].

The Figs. 8.24 and 8.27 deal with the results obtained in QRPD(5,3,2,1)-CA simulations with variable entanglement factor γ and $\mu = 0.5$ and full noise respectively. These figures show in their left frame the mean payoffs across the lattice $(\overline{p})$ of both player types, and in their right frame the mean parameter values across the lattice. Only the mean payoffs are shown in the $\mu = 0.5$ simulations in Fig. 8.26. Recall that in the simulations free of noise of the QRPD(5,3,2,1)-CA game reported in Sect. 8.2, it was shown that the CA imitation dynamics enables the emergence of NE in a manner that is not far from the expected in the inertial model: $\{\hat{D}, \hat{D}\} \rightarrow \{\hat{D}, \hat{Q}\} \rightarrow \{\hat{Q}, \hat{Q}\}$ as γ increases, with thresholds: $\gamma^* = \arcsin(1/2) = \pi/6$, $\gamma^\bullet = \arcsin(1/\sqrt{2}) = \dfrac{\pi}{4}$.

Figure 8.24 deals with the results obtained in a QRPD(5,3,2,1)-CA simulation with $\mu = 0.5$ noise. Although the general form of the graphs of $(\overline{p}_A, \overline{p}_B)$ shown in the simulations free of noise reported in Sect. 8.2 it is still recognizable with $\mu = 0.5$ noise in Fig. 8.24, it is notably altered. This is so in two main features of the graphs: (i) the payoffs associated with mutual defection at maximum r persist beyond γ^*, up to approximately the midpoint of the $(\gamma^*, \gamma^\bullet)$ interval, and (ii) the

---

$P\rho P = \frac{1}{4}\begin{pmatrix} \sin^2 \frac{\gamma}{2} & 0 & 0 & i\frac{1}{\sqrt{2}}\sin\gamma \\ 0 & \sin^2 \frac{\gamma}{2} & 0 & 0 \\ 0 & 0 & 1 & 0 \\ -i\frac{1}{\sqrt{2}}\sin\gamma & 0 & 0 & 3 - 2\sin^2 \frac{\gamma}{2} \end{pmatrix}$. $\rho(2,2) = \frac{1}{4}((1 - \sin^2 \frac{\gamma}{2}) + \sin^2 \frac{\gamma}{2}\sin^2 \frac{\gamma}{2}) = 1 - \frac{1}{4}\sin^2 \gamma)$,

$\rho(3,3) = \frac{1}{4}((1 - \sin^2 \frac{\gamma}{2})\sin^2 \frac{\gamma}{2} + \sin^2 \frac{\gamma}{2}) = 1 + \frac{1}{4}\sin^2 \gamma)$, $\rho(1,1) = \frac{1}{4}((1 - \sin^2 \frac{\gamma}{2})(3 - 2\sin^2 \frac{\gamma}{2}) + \sin^2 \frac{\gamma}{2}\sin^2 \frac{\gamma}{2} + i\frac{1}{2}\sin\gamma(-i\frac{2}{\sqrt{2}}\sin\gamma) = \frac{1}{4}(3 - 2\sin^2 \frac{\gamma}{2} - 3\sin^2 \frac{\gamma}{2} + 2\sin^4 \frac{\gamma}{2} + \sin^4 \frac{\gamma}{2} + \frac{1}{\sqrt{2}}\sin^2 \gamma)$,

$\rho(4,4) = \frac{1}{4}((1 - \sin^2 \frac{\gamma}{2})\sin^2 \frac{\gamma}{2} + \sin^2 \frac{\gamma}{2}(3 - 2\sin^2 \frac{\gamma}{2}) + i\frac{1}{2}\sin\gamma(i\frac{2}{\sqrt{2}}\sin\gamma) = \frac{1}{4}(\sin^2 \frac{\gamma}{2} - \sin^4 \frac{\gamma}{2} + 3\sin^2 \frac{\gamma}{2} - 2\sin^4 \frac{\gamma}{2} - \frac{1}{\sqrt{2}}\sin^2 \gamma)$.

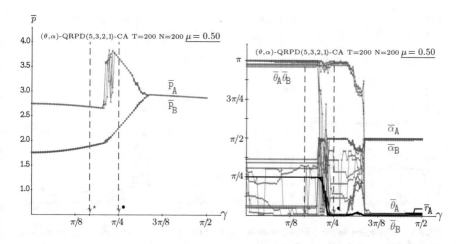

Fig. 8.24 The QRPD(5,3,2,1)-CA with $\mu = 0.5$ noise factor. Variable entanglement factor γ. Five simulations at T = 200. Left: Mean payoffs. Right: Mean quantum parameters and mean value of r_A

payoffs equalization is reached beyond $\gamma^\bullet$ and shows a small negative slope. The latter occurs because the final density matrix for $\mu = 1/2$ and $r = 0$ turns out to be $\rho_{B,I}^* = \begin{pmatrix} \cos^2 \frac{\gamma}{2} + \frac{1}{4}\sin^2 \frac{\gamma}{2} & 0 & 0 & -i\frac{1}{4}\sin\gamma \\ 0 & \frac{1}{4}\sin^2 \frac{\gamma}{2} & 0 & 0 \\ 0 & 0 & \frac{1}{4}\sin^2 \frac{\gamma}{2} & 0 \\ i\frac{1}{4}\sin\gamma & 0 & 0 & \frac{1}{4}\sin^2 \frac{\gamma}{2} \end{pmatrix}$, so that the equalized payoffs

become: $p_A^{QQ} = p_B^{QQ} = R\cos^2 \frac{\gamma}{2} + \frac{1}{4}(R + P + S + \mathfrak{T})\sin^2 \frac{\gamma}{2}$, which vary from $p = R = 3.0$ at $\gamma = 0.0$ to $p = \frac{1}{2}R + \frac{1}{8}(R + P + S + \mathfrak{T}) = 2.875$ at $\gamma = \pi/2$.

Figure 8.25 shows the dynamics in simulations in the QRPD(5,3,2,1)-CA scenario of Fig. 8.24 with $\gamma = 0$ (left), $\gamma = \pi/4$ (center), and $\gamma = \pi/2$ (right). As a result of the initial random assignment of the parameter values, it is initially in both frames: $\bar{\theta} \simeq \pi/2 = 1.570$, $\bar{\alpha} \simeq \pi/8 = 0.785$, and $\bar{r} \simeq \pi/8 = 0.196$. With $\gamma = 0$, both $\bar{\theta}$'s rocket to π (which makes $\bar{\alpha}$'s irrelevant), and $\bar{r}$ rockets to $\pi/4$. In the central frame with $\gamma = \pi/4$, the parameter stabilization is achieved also in fast manner ($\bar{\alpha}_B \rightarrow \pi/2\, \bar{\theta}_A \rightarrow \pi, \bar{\theta}_B \rightarrow 0, \bar{r} \rightarrow 0$), with the exception of $\bar{\alpha}_A$, whose value becomes in fact irrelevant as $\bar{\theta}_A \rightarrow \pi$. With $\gamma = \pi/2$, both $\bar{\alpha}$'s rocket towards $\pi/2$, both $\bar{\theta}$'s plummet to zero, i.e., the parameters of the Q strategy, and $\bar{r}$ decreases towards zero. Remarkably, the parameter tendencies heavily emerge, as a rule, from the very beginning, despite the full range of parameters initially accessible in the CA local interactions.

The general form of the plots in Fig. 8.24 remains in CA-simulations with choices of the PD parameters different from that adopted here, i.e., (5,3,2,1) so that $\gamma^* < \gamma^\bullet$. But the details of the transition from $\{\hat{D}, \hat{D}\}$ to $\{\hat{Q}, \hat{Q}\}$ vary notably accordingly to the hierarchy of the γ^* and $\gamma^\bullet$ values. Figure 8.26 shows one example of PD-parameters

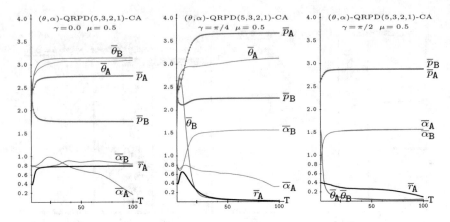

Fig. 8.25 Dynamics in three simulations in the QRPD(5,3,2,1)-CA scenario of Fig. 8.24. Left: $\gamma = 0$, Center: $\gamma = \pi/4$, Right: $\gamma = \pi/2$

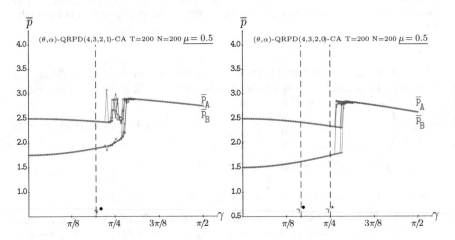

Fig. 8.26 Mean payoffs in the QRPD-CA with $\mu = 0.5$ noise. Variable entanglement factor γ. Five simulations at T = 200. Left: (4,3,2,1)-PD payoffs. Right: (4,3,2,0)-PD payoffs

with $\gamma^* = \gamma^\bullet$, that of (4,3,2,1), and one example of PD-parameters with $\gamma^\bullet < \gamma^*$, that of (4,3,2,0). In the latter case, both $\{\hat{Q}, \hat{Q}\}$ and $\{\hat{D}, \hat{D}\}$ are in NE in the $(\gamma^\bullet, \gamma^*)$ interval in conventional inertial quantum games free of noise. The accelerated player seems to take advantage of this by *imposing* mutual defection beyond $\gamma^\bullet$. This is so in the (4,3,2,0)-PD simulations free of noise shown in Sect. 8.2, and a greater degree in the $\mu = 0.5$ simulations in the right frame of Fig. 8.26 where mutual defection keeps in the full $(\gamma^\bullet, \gamma^*)$ interval.

Mutual defection arising with low entanglement in Figs. 8.24 and 8.26 makes irrelevant the α parameters (as $\cos \pi/2 = 0$ annihilates the influence of α in (2.3)), before the emergence of the $\{\hat{D}, \hat{Q}\}$ pair in these figures. Still in these figures, with low entanglement it is $\bar{r} \simeq \pi/4$, which in the particular case of no entanglement

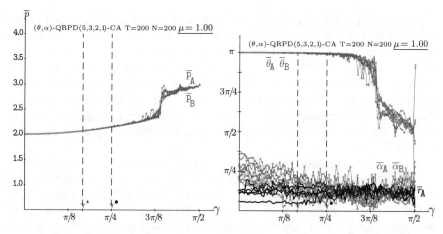

Fig. 8.27 The QRPD(5,3,2,1)-CA with full noise. Variable entanglement factor γ. Five simulations at T = 200. Left: Mean payoffs. Right: Mean quantum parameters and mean value of r_A

produces $\Pi_{\gamma=0}^{DD} = \frac{1}{2}\begin{pmatrix} 0 & 0 \\ 1-\mu & 1+\mu \end{pmatrix}$, leading to $p_A^{DD} = \frac{1}{2}(\mathfrak{T} + P + (P - \mathfrak{T})\mu)$, $p_B^{DD} = \frac{1}{2}(P + S + (P - S)\mu)$. Consequently, in the simulations free of noise reported in Sect. 8.2, with no entanglement it is $p_A^{DD} = \frac{1}{2}(\mathfrak{T} + P) = 3.5$, $p_B^{DD} = \frac{1}{2}(P + S) = 1.5$. In $\mu = 0.5$ simulations with no entanglement it is $p_A^{DD} = \frac{1}{4}(3P + \mathfrak{T})$, $p_B^{DD} = \frac{1}{4}(3P + S)$, giving $p_A^{DD} = 2.75$, $p_B^{DD} = 1.75$ in Fig. 8.24; in the two scenarios of Fig. 8.26, it is $p_A^{DD} = 2.50$, whereas it is $p_B^{DD} = 1.75$ in its left frame and $p_B^{DD} = 1.50$ in its right frame. With $\mu = 1.0$, as in Fig. 8.27, it is $p_B^{DD}(\gamma = 0, \mu = 1.0) = P = 2.0$ regardless of r. Thus, the payoffs at $\gamma = 0$ emerging in the CA simulations reported in this section correspond to the $\{\hat{D}, \hat{D}\}$ theoretical ones.

Figure 8.27 deals with the results obtained in the QRPD(5,3,2,1)-CA scenario of Fig. 8.24 but with full $\mu = 1.0$ noise. Full noise eliminates the role of the $\gamma^\star, \gamma^\bullet$ thresholds, so that the graphs of both mean payoffs tend to fit that of mutual defection at $\mu = 1.0$, $p_A^{DD} = p_B^{DD} = R \sin^2 \frac{\gamma}{2} + P \cos^2 \frac{\gamma}{2}$ regardless of r, up to approximately $\gamma = 3\pi/8$. Beyond this value of γ, as a result of the notable decrease of both $\overline{\theta}_A$ and $\overline{\theta}_B$ parameters shown in the right frame of Fig. 8.27, both mean payoffs increase their value compared to that of mutual defection, so that at $\gamma = \pi/2$ it is $\overline{p}_A \simeq \overline{p}_B \simeq 3.0$, i.e., the payoff of mutual cooperation.

Unfair Contests

Let us assume the unfair situation: A type of players is restricted to classical strategies $\tilde{U}(\theta, 0)$, whereas the other type of players may use quantum $\hat{U}(\theta, \alpha)$ ones [11, 20].

Figures 8.28 and 8.31 deal with five simulations of an accelerated (θ, α)-player A (red) versus an inertial θ-player B (blue) in a QRPD(5,3,2,1)-CA with variable entanglement factor γ. Its left frame shows the asymptotic mean payoffs across the lattice $(\overline{p})$ of both player types, and in its right frame the mean parameter values where $\overline{\alpha}_B = 0.0$. The left frames of these figures also show the mean-field payoffs (p^*) achieved in a single hypothetical two-person game with players adopting the mean parameters appearing in the spatial dynamic simulation, those given in the right panels of Figs. 8.28 and 8.31. Namely, with strategies of the form given in Eqs. (3.1) and $r = \overline{r}$.

In the unfair simulations free of noise reported in Sect. 8.2, mutual defection at $r = \pi/4$ keeps up to close $\gamma^\bullet$, where a kind of discontinuity emerges so that the trend to $\overline{p}_A$ and $\overline{p}_B$ to approach ceases, and it is replaced by a trend to the separation of both mean payoffs, so that at $\gamma = \pi/2$ both payoffs are separated nearly as at $\gamma = 0$. As a result, the quantum player overrates that classical player regardless of γ. In a conventional (non CA) unfair inertial QPD game, the quantum player is well advised to play the *miracle* strategy $\hat{M} = \hat{U}(\pi/2, \pi/2)$, that if $\gamma > \pi/4$ ensures him to over-rate the classical player, even if the latter chooses $\hat{D}$. In the case of full entangling, $\hat{M}$ versus $\hat{D}$ will provide the payoffs $\dfrac{\mathfrak{T} + P}{2}$=3.5 and $\dfrac{S + P}{2}$=1.5 to both the quantum and classical player respectively, payoffs close to those arising in the noiseless unfair simulations reported in Sect. 8.2.

Figure 8.28 shows the results in the above described unfair scenario with $\mu = 0.5$ noise. Mutual cooperation at $\overline{r} = \pi/4$ keeps in this figure over, but not far, $\gamma^\bullet$. Then, during a short γ interval, the pair $\{\hat{Q}, \hat{D}\}$ at $\overline{r} = 0$ emerges, producing payoffs that behave in a fairly linear way, with the payoff of the classical player overrating that of the classical player before the intersection of both player payoffs. After this short γ-interval, over, but not far, $3\pi/8$, the payoff of the quantum player overrates that of the classical player. Much as expected. The pair $\{\hat{Q}, \hat{D}\}$ at $r = 0$ generates the joint probabilities: $\pi_{11} = \pi_{44} = \frac{1}{4}\sin^2\frac{\gamma}{2}$ $\pi_{12} = \cos^2\frac{\gamma}{2} - \frac{1}{4}\sin^2\frac{\gamma}{2}(7 - 8\sin^2\frac{\gamma}{2})$, $\pi_{21} = \frac{1}{4}\sin^2\frac{\gamma}{2}(9 - 8\sin^2\frac{\gamma}{2})$. So that the payoffs intersection occurs at $\gamma^+ = 2\arcsin\frac{1}{2} = 2\gamma^\star = \pi/3 = 1.047$, giving the equalized payoff $p^= = \frac{1}{16}(R + P + 7(S + \mathfrak{T})) = 2.937$ in Fig. 8.28. Please, note that γ^+ does not depend on the PD-parameters, so that, for example, it applies in simulations with (4,3,2,1) and (4,3,2,0) PD-parameters, giving $p^= = 2.500$ and $p^= = 2.062$ respectively.

The mean-field payoffs estimations fit fairly perfectly to the actual ones up to γ near to $\gamma^\bullet$ in simulations free of noise, and up to $\gamma \simeq 3\pi/8$ in Fig. 8.28 But for higher γ values the mean-field p^* values notably differ from the actual $\overline{p}$ ones. This is so even qualitatively, in the sense that $\overline{p}_A > \overline{p}_B$ whereas $\overline{p}_A^* < \overline{p}_B^*$. The explanation of this divergence relies in spatial effects (fairly absent in fair contests), as commented below when dealing with Fig. 8.29.

Figure 8.29 deals with a simulation in the $\alpha_B = 0$ unfair scenario of Fig. 8.28 with $\gamma = \pi/2$ up to $T = 200$. The far left panel of the figure shows the evolution up to $T = 200$ of the mean values across the lattice of θ, α and r as well of the actual and mean-field payoff estimations. At variance with what happens in the right frame of Fig. 8.25, the $\overline{\theta}$ parameters do not plummet to zero, and $\overline{r}$ remains at levels not close

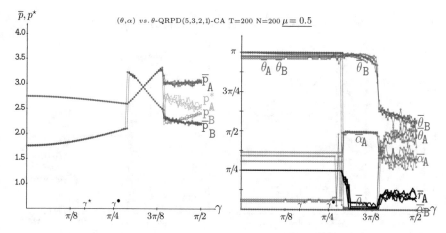

Fig. 8.28 The QRPD(5,3,2,1)-CA unfair quantum accelerated (θ, α)-player A (red) versus inertial θ-player B (blue) with $\mu = 0.5$ noise. Variable entanglement γ. Five simulations at T = 200. Left: Mean payoffs ($\overline{p}$) and mean-field payoffs (p^*). Right: Mean quantum parameters and mean values of r_A

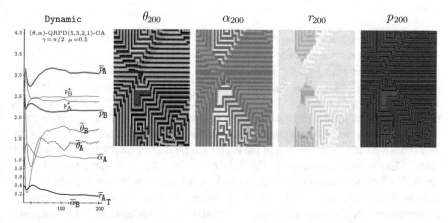

Fig. 8.29 A simulation in the $\gamma = \pi/2$ unfair scenario of Fig. 8.28. Far Left: Evolving mean parameters and payoffs. Center: parameter patterns. Far Right: Payoff pattern. Increasing grey levels indicate increasing parameter values

to zero. The parameter patterns at $T = 200$ in Fig. 8.29 show a kind of maze-like aspect, particularly crisp in the θ pattern, a kind of spatial heterogeneity that explains why the mean-field estimations of the payoffs p^* differ from the actual ones $\overline{p}$. This divergence becomes apparent from the initial iterations in the left frame of Fig. 8.29, so that after a short transition time, it is $\overline{p}_A > \overline{p}_B^* > p_A^* > \overline{p}_B$.

Figure 8.30 deals with simulations in the scenario of Fig. 8.28 but with reversed unfairness: Accelerated θ-player A versus inertial (θ, α)-player B. Unexpectedly, the accelerated θ-player overrates the inertial (θ, α)-player for not high values of

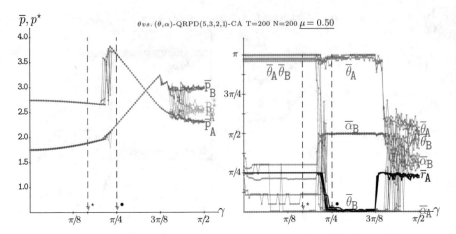

Fig. 8.30 The QRPD(5,3,2,1)-CA unfair accelerated θ-player A (red) versus an inertial (θ, α)-player B (blue) with $\mu = 0.5$ noise. Variable entanglement γ. Five simulations at T = 200. Left: Mean payoffs ($\overline{p}$) and mean-field payoffs (p^*). Right: Mean quantum parameters and mean value of r_A

γ; via mutual defection at $r = \pi/4$ (as in Fig. 8.28) up to close γ^*, and via $\{\hat{D}, \hat{Q}\}$ at $\overline{r} = 0$ for higher values of γ up to the payoff equalization at $\gamma = \gamma^= = 1.047$. For $\gamma > \gamma^=$, the (θ, α)-player B overrates the θ-player B in Fig. 8.30, reversing, as somehow expected, what happens in Fig. 8.28. At approximately $\gamma = 3\pi/8$, $\overline{r}$ rockets up $\gamma = \pi/4$ and close beyond this γ $\{\hat{D}, \hat{Q}\}$ ceases to operate. In parallel, the mean-field payoffs estimations p^* that fit perfectly the actual ones $\overline{p}$ up to some value just over $\gamma = 3\pi/8$, appreciably differ, much as already observed in Fig. 8.28.

Figure 8.31 shows the results in the unfair scenario of Fig. 8.28, but with $\mu = 1.0$ noise. Mutual defection keeps in Fig. 8.31 as in Fig. 8.27 up to approximately $\gamma = 3\pi/8$. With higher entanglement the payoff of the quantum player overrates that of the classical player in a short range of γ before their equalization for the highest entanglement. In the latter scenario, soft spatial effects arise, so that the actual and mean-field payoffs are not coincident.

Three-Parameter Strategies

This section deals with the full space of strategies SU(2), operating with three parameters (3P) as given in Eq. (2.3).

In the 3P strategies QRPD(5,3,2,1)-CA simulations free of noise reported in Sect. 8.2 the graph of the actual mean payoffs is reminiscent of that of the $(\hat{D}, \hat{D})$ game at $r = \pi/4$ all along the γ variation. It was also reported in Sect. 8.2 that for low γ the mean-field payoff estimations fit perfectly the actual mean payoffs, but that beyond $\gamma = \pi/8$ heavy spatial effects emerge, so that p^* and $\overline{p}$ dramatically diverge as γ grows.

Figure 8.32 deals with three-parameter strategies QRPD(5,3,2,1)-CA simulations with $\mu = 0.5$ noise. The form of the graphs of the actual mean payoffs (left frame)

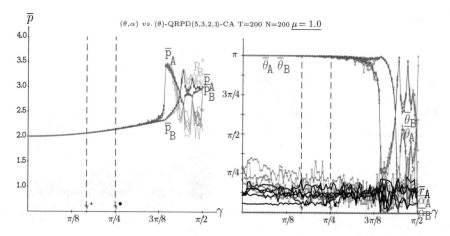

Fig. 8.31 The QRPD(5,3,2,1)-CA unfair quantum accelerated (θ, α)-player A (red) versus inertial θ-player B (blue) with full noise. Variable entanglement γ. Five simulations at T = 200. Left: Mean payoffs $(\overline{p})$ and mean-field payoffs (p^*). Right: Mean quantum parameters and mean values of r_A

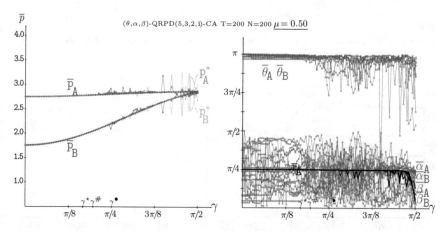

Fig. 8.32 The three-parameter QRPD(5,3,2,1)-CA with $\mu = 0.5$ noise. Variable entanglement factor γ. Five simulations at T = 200. Left: Mean payoffs $(\overline{p})$ and mean-field payoffs (p^*). Right: Mean quantum parameters and mean value of r_A

of Fig. 8.32 notably corresponds with those shown in the (D, D)-μ=0.5 bottom right frame of Fig. 8.23. The right frame of Fig. 8.32 clearly informs about the origin of this payoffs structure as, $\overline{\theta}_A \simeq \overline{\theta}_B \simeq \pi$ regardless of γ (which makes irrelevant the α parameters), and $\overline{r}$ keeps close to $\pi/4$, except when γ approaches its maximum $\pi/2$. Such a fairly crisp definition of the mean parameters indicates the absence of relevant spatial effects, and in consequence a good agreement between actual and mean-field payoffs as shown in the left frame of Fig. 8.32.

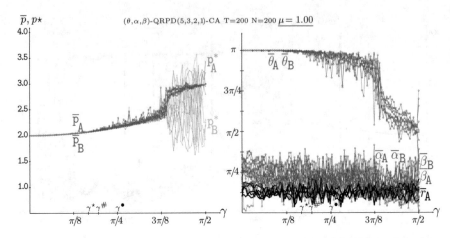

Fig. 8.33 The three-parameter QRPD(5,3,2,1)-CA $\mu = 1.0$ noise. Variable entanglement γ. Five simulations at T = 200. Left: Mean payoffs ($\overline{p}$) and mean-field payoffs (p^*). Right: Mean quantum parameters and mean value of r_A

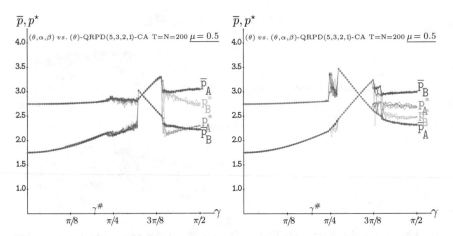

Fig. 8.34 Mean payoffs ($\overline{p}$) and mean-field (p^*) payoffs in an unfair three-parameter QRPD(5,3,2,1)-CA with $\mu = 0.5$ noise. Variable entanglement γ. Five simulations at T = 200. Left: Quantum accelerated (θ, α, β)-player A versus inertial θ-player B. Right: θ-player A versus inertial (θ, α, β)-player B

Figure 8.33 deals with the results achieved in the scenario of Fig. 8.32, but with full noise. Mutual defection prevails in Fig. 8.33, much as in the full noise simulations in Figs. 8.27 and 8.31. Spatial effects arise over approximately $\gamma = 3\pi/8$ so that the actual and mean-field payoffs are not coincident with very high entanglement.

The left and right frames of the unfair 3P-QRPD(5,3,2,1) simulations with $\mu = 0.5$ noise shown in Fig. 8.34 very much resemble those of the left frames of Figs. 8.28 and 8.30 respectively. More importantly, the advantage that a quantum player achieves over a player restricted to classical strategies in the scenarios of Fig. 8.34 is diminished, as in the two-parameter model, in the presence of quantum noise as it is demonstrated by the comparison of the payoff graphs in Fig. 8.34 to those corresponding ones given free of noise in Sect. 8.2. Note that in these 3P-QRPD unfair scenarios, the $\gamma^{\#} = \arcsin(\sqrt{\frac{P-S}{\mathfrak{T}+P-R-S}} = \frac{1}{3})$ threshold does not play any role.

Let us summarize the findings reported in this section regarding the effect of noise in the QRPD game by concluding that noise reduces the unbalance in payoffs induced by the acceleration of one player, and delays the emergence of the Q strategy. Noise at middle level ($\mu = 0.5$) clearly illustrates these effects, which are led to its maximum with full noise, in which case mutual defection prevails, even with high entanglement.

References

1. Alonso-Sanz, R., Carvalho, C., Situ, H.: A quantum relativistic prisoner's dilemma cellular automaton. Int. J. Theor. Phys. **55**(10), 4310–4323 (2016)
2. Alonso-Sanz, R., Situ, H.Z.: A quantum relativistic battle of the sexes cellular automaton. Phys. A **468**, 267–277 (2017)
3. Khan, S., Khan, M.K.: Noisy relativistic quantum games in noninertial frames. Quantum Inf. Process. **12**(2), 1351–1363 (2013)
4. Khan, S., Khan, M.K.: Relativistic quantum games in noninertial frames. J. Phys. A Math. Theor. **44**, 355302 (2011)
5. Alsing, P.M., Fuentes-Schuller, I., Mann, R.B., Tessier, T.E.: Entanglement of Dirac fields in noninertial frames. Phys. Rev. A **74**, 032326 (2006)
6. Takagi, S.: Vacuum noise and stress induced by uniform acceleration Hawking-Unruh effect in rindler manifold of arbitrary dimension. Prog. Theor. Phys. Suppl. **88**, 1–142 (1986)
7. Du, J.F., Xu, X.D., Li, H., Zhou, X., Han, R.: Entanglement playing a dominating role in quantum games. Phys. Lett. A **89**(1–2), 9–15 (2001)
8. Du, J.F., Li, H., Xu, X.D., Zhou, X., Han, R.: Phase-transition-like behaviour of quantum games. J. Phys. A Math. Gen. **36**(23), 6551–6562 (2003)
9. Alonso-Sanz, R.: Variable entangling in a quantum prisoner's dilemma cellular automaton. Quantum Inf. Process. **14**, 147–164 (2015)
10. Alonso-Sanz, R.: A quantum prisoner's dilemma cellular automaton. Proc. R. Soc. A **470**, 20130793 (2014)
11. Flitney, A.P., Abbott, D.: Advantage of a quantum player over a classical one in 2x2 quantum games. Proc. R. Soc. Lond. A **459**(2038), 2463–2474 (2003)
12. Eisert, J., Wilkens, M.: Quantum games. J. Mod. Opt. **47**(14–15), 2543–2556 (2000)
13. Situ, H.Z., Huang, Z.M.: Relativistic quantum Bayesian game under decoherence. Int. J. Theor. Phys. **55**, 2354–2363 (2016)
14. Weng, G., Yu, Y.: A Quantum battle of the sexes in noninertial frame. J. Mod. Phys. **5**(9) (2014). https://doi.org/10.4236/jmp.2014.59094
15. Alonso-Sanz, R.: A quantum battle of the sexes cellular automaton. Proc. R. Soc. A **468**, 3370–3383 (2012)
16. Alonso-Sanz, R.: On a three-parameter quantum battle of the sexes cellular automaton. Quantum Inf. Process. **12**(5), 1835–1850 (2013)
17. Alonso-Sanz, R.: Variable entangling in a quantum battle of the sexes cellular automaton. In: ACRI-2014. LNCS, vol. 8751, pp. 125–135 (2014)

18. Omkar, S., Srikanth, R., Banerjee, S., Alok, A.K.: The Unruh effect interpreted as a quantum noise channel. Quantum Inf. Comput. **9–10**, 0757–0770 (2016)
19. Alonso-Sanz, R., Situ, H.: On the effect of quantum noise in a quantum relativistic prisoner's dilemma cellular automaton. Int. J. Theor. Phys. **55**(12), 5265–5279 (2016)
20. Flitney, A.P., Abbott, D.: Quantum games with decoherence. J. Phys. A Math. Gen. **38**(2), 449 (2004)

Chapter 9
Quantum Memory

The disrupting effect of quantum memory on the dynamics of a spatial quantum formulation of the iterated Prisoner's Dilemma game with variable entangling is studied in this chapter. The main findings of this chapter refer to the shrinking effect of memory on the disruption induced by correlated noise on quantum communication in a complex system formed by several interacting players. The importance of quantum noise lies on the fact that it yields decoherence, and then dramatically affects the information transmitted through a quantum channel.

To play a quantum game, an arbiter must prepare a usually entangled initial state, and transfer it to the players. The correlations in the initial state are extremely fragile, and then any interaction with the environment can lead to decoherence, i.e., a loss of information. Consequently, a more accurate description of the communication process through a quantum channel always requires accounting for the random noise exerted by the environment, which has a dramatic effect on entangled states [1]. Furthermore, when the time scales for successive uses of the quantum channels are not sufficiently large, the noise is correlated, and then it additionally accounts for some memory effects [2].

9.1 Quantum Uncorrelated Noise

Let us recall here from Chap. 7 that within the EWL quantization scheme [3], the initial density matrix (ρ_i) turns out to be that given in Eq. 7.2, and that after the application of the quantum strategies, the final density matrix becomes that given in Eq. 7.3. Also from Chap. 7 it is known that if the interaction of a system with the environment can be described through noisy operators, and it is assumed that the environment thermalizes sufficiently fast, this interaction is adequately described by uncorrelated noise, in which case the noise distorts the initial density matrix to as,

© Springer Nature Switzerland AG 2019
R. Alonso-Sanz, *Quantum Game Simulation*, Emergence, Complexity
and Computation 36, https://doi.org/10.1007/978-3-030-19634-9_9

$$\rho_i^{(u)}(\mu) = \sum_{i=1}^{2}\sum_{j=1}^{2}(K_i^u \otimes K_j^u)\rho_i(K_i^u \otimes K_j^u)^\dagger \tag{9.1}$$

where the K_i^u denote the Kraus operators that describe the effect of amplitude-damping given in Eq. 7.11.

For convenience, in this chapter the superscript ^{u} will be added when dealing with uncorrelated noise. Thus, for example, the distorted initial density matrix with general noise strength μ that was given in Eq. (7.7) will be here referred in this chapter to as $\rho_i^{(u)}(\mu)$, in particular, the density matrix with $\mu = 0.5$ that was given given in Eq. (7.8) will be here referred to as $\rho_i^{(u)}(0.5)$. Analogously, the joint probability matrices of mutual $\hat{Q}$ and $(\hat{Q}, \hat{D})$ with $\mu = 0.5$ in the QPD($\mathfrak{T}$,R,P,S) given in Eqs. (7.9)–(7.10) will be here referred to as $\Pi^u(0.5)$, and the payoffs of the $\hat{Q}$, $\hat{D}$ and $\hat{C}$ strategy combinations for $\mu = 0.5$ given in Eq. (7.11) will be referred here to as $p^{(u)}$.

Let us recall here that the payoffs given in Eq. (7.11) for the QPD ($\mu = 0.5$) with general payoffs, with (5, 3, 2, 1) payoffs reduce to,

$$p_{A,B}^{\hat{C},\hat{C}(u)}(0.5) = p_{A,B}^{\hat{Q},\hat{Q}(u)}(0.5) = 3 - \frac{1}{4}\sin^2\frac{\gamma}{2} \tag{9.2a}$$

$$p_{A,B}^{\hat{D},\hat{D}(u)}(0.5) = 2 + \frac{3}{4}\sin^2\frac{\gamma}{2} \tag{9.2b}$$

$$p_A^{\hat{Q},\hat{D}(u)}(0.5) = p_B^{\hat{D},\hat{Q}(u)}(0.5) = 1 + \frac{39}{4}\sin^2\frac{\gamma}{2} - 8\sin^4\frac{\gamma}{2} \tag{9.2c}$$

$$p_A^{\hat{D},\hat{Q}(u)}(0.5) = p_B^{\hat{Q},\hat{D}(u)}(0.5) = 5 - \frac{41}{4}\sin^2\frac{\gamma}{2} + 8\sin^4\frac{\gamma}{2} \tag{9.2d}$$

$$p_{A,B}^{\hat{Q},\hat{C}(u)}(0.5) = p_{A,B}^{\hat{C},\hat{Q}(u)}(0.5) = 3 - \frac{9}{4}\sin^2\frac{\gamma}{2} + 2\sin^4\frac{\gamma}{2} \tag{9.2e}$$

The payoffs of Eq. (9.2) were already plotted in Fig. 7.1 and are also shown here in Fig. 9.1, labeled with $\eta = 0.0$. As can be there seen, while the payoffs $p_A^{\hat{Q},\hat{D}(u)}(0.5) = p_B^{\hat{D},\hat{Q}(u)}(0.5)$ strongly increase with the entanglement strength, γ, being equal to 1 for $\gamma = 0$ and to 3.875 for $\gamma = \pi/2$, the payoffs $p_{A,B}^{\hat{D},\hat{D}(u)}(0.5)$ have a much moderate increment in the same interval, changing from 2 up to 2.375. On the contrary, the payoffs $p_A^{\hat{D},\hat{Q}(u)}(0.5) = p_B^{\hat{Q},\hat{D}(u)}(0.5)$ strongly decrease with the entanglement factor, their value being reduced from 5.0 to 1.875. The payoffs $p_{A,B}^{\hat{Q},\hat{C}(u)}(0.5) = p_{A,B}^{\hat{C},\hat{Q}(u)}(0.5)$ also dismiss with the entanglement factor, but much less: the initial and final values are respectively equal to 3 and 2.375. Finally, notice that the payoffs $p_{A,B}^{\hat{Q},\hat{Q}(u)}(0.5)$ are almost independent of the entanglement factor, being the difference between the initial and final values equal to solely 0.125.

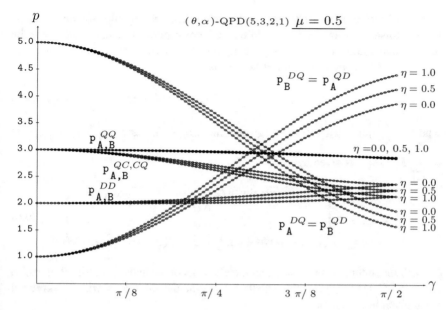

Fig. 9.1 Payoffs of the interactions of the $\hat{Q}$, $\hat{D}$ and $\hat{C}$ strategies in a QPD(5, 3, 2, 1) with variable entanglement factor γ, noise strength $\mu = 0.5$ and memory factor $\eta = 0$ (uncorrelated noise), $\eta = 0.5$, and $\eta = 1.0$ (maximally correlated noise)

9.2 Quantum Correlated Memory Noise

When also accounting for the dynamics of the environment, the successive uses of a quantum channel can render some memory effects that lead to correlated noise. The effect of this noise on the initial density matrix (7.2) reads

$$\rho_i(\mu, \eta) = (1 - \eta) \sum_{i=1}^{2} \sum_{j=1}^{2} (K_i^u \otimes K_j^u) \rho_i (K_i^u \otimes K_j^u)^\dagger + \eta \sum_{i=1}^{2} K_i^c \rho_i K_i^{c\ \dagger}, \quad (9.3)$$

where the parameter $0 \leq \eta \leq 1$ accounts for the memory strength, and the Kraus operators K_i^u are given by Eq. (7.6). The operators K_1^c and K_2^c that describe the correlated part of the noise are given by [2]

$$K_1^c = \begin{pmatrix} 1 & 0 & 0 & 0 \\ 0 & 1 & 0 & 0 \\ 0 & 0 & 1 & 0 \\ 0 & 0 & 0 & \sqrt{1-\mu} \end{pmatrix}, \quad \text{and} \quad K_2^c = \begin{pmatrix} 0 & 0 & 0 & \sqrt{\mu} \\ 0 & 0 & 0 & 0 \\ 0 & 0 & 0 & 0 \\ 0 & 0 & 0 & 0 \end{pmatrix}. \quad (9.4)$$

According to Eq. (9.3), for $\eta = 0$ one recovers the transformation for totally uncorrelated noise given by Eq. (9.1), while $\eta = 1$ corresponds to a situation where the noise is maximally correlated. In general, the transformed matrix $\rho_i(\mu, \eta)$ given by Eq. (9.3) can be split as the sum of two terms,

$$\rho_i(\mu, \eta) = (1 - \eta)\rho_i^{(u)}(\mu) + \eta\, \rho_i^{(c)}(\mu), \qquad (9.5)$$

where $\rho_i^{(u)}(\mu) \equiv \rho_i(\mu, 0)$ is given in Eq. (9.1), and $\rho_i^{(c)}(\mu) \equiv \rho_i(\mu, 1)$
$= \sum_{i=1}^{2} K_i^c \rho_i K_i^{c\,\dagger}$ is given by

$$\rho_i^{(c)}(\mu) = \begin{pmatrix} 1 - (1-\mu)\sin^2 \gamma/2 & 0 & 0 & -i\sqrt{1-\mu}\cos\frac{\gamma}{2}\sin\frac{\gamma}{2} \\ 0 & 0 & 0 & 0 \\ 0 & 0 & 0 & 0 \\ i\sqrt{1-\mu}\cos\frac{\gamma}{2}\sin\frac{\gamma}{2} & 0 & 0 & (1-\mu)\sin^2\frac{\gamma}{2} \end{pmatrix} \qquad (9.6)$$

With maximum noise ($\mu = 1$), all the elements of the density matrix (9.6) vanish except its (1,1) element, which equals 1, as in the density matrix with uncorrelated noise given in Eq. (7.7).

From Eq. (9.6), the joint probability can be split as the sum of an uncorrelated term, Π_u, plus a totally correlated one, Π_c, as

$$\Pi(\mu, \eta) = (1 - \eta)\,\Pi_u(\mu) + \eta\,\Pi_c(\mu). \qquad (9.7)$$

Consequently, the total payoffs can be similarly split as

$$p_j \equiv p_j(\mu, \eta) = (1 - \eta)p_j^{(u)}(\mu) + \eta\, p_j^{(c)}(\mu), \qquad j = A, B. \qquad (9.8)$$

The joint probabilities of the $\hat{Q}$, $\hat{D}$ and $\hat{C}$ strategy combinations for maximally correlated noise ($\eta = 1$) are equal to

$$\Pi_c^{\hat{C},\hat{C}}(\mu) = \Pi_c^{\hat{Q},\hat{Q}}(\mu) = \begin{pmatrix} 1 - \pi_{c(2,2)}^{\hat{Q},\hat{Q}} & 0 \\ 0 & \pi_{c(2,2)}^{\hat{Q},\hat{Q}} \end{pmatrix}, \; \Pi_c^{\hat{D},\hat{D}}(\mu) = \begin{pmatrix} \pi_{c(2,2)}^{\hat{Q},\hat{Q}} & 0 \\ 0 & 1 - \pi_{c(2,2)}^{\hat{Q},\hat{Q}} \end{pmatrix}$$
$$(9.9a)$$

$$\Pi_c^{\hat{Q},\hat{D}}(\mu) = \begin{pmatrix} 0 & 1 - \pi_{c(2,1)}^{\hat{Q},\hat{D}} \\ \pi_{c(2,1)}^{\hat{Q},\hat{D}} & 0 \end{pmatrix}, \; \Pi_c^{\hat{D},\hat{Q}}(\mu) = \begin{pmatrix} 0 & \pi_{c(2,1)}^{\hat{Q},\hat{D}} \\ 1 - \pi_{c(2,1)}^{\hat{Q},\hat{D}} & 0 \end{pmatrix} \qquad (9.9b)$$

$$\Pi_c^{\hat{Q},\hat{C}}(\mu) = \Pi_c^{\hat{C},\hat{Q}}(\mu) = \begin{pmatrix} 1 - \pi_{c(2,1)}^{\hat{Q},\hat{D}} & 0 \\ 0 & \pi_{c(2,1)}^{\hat{Q},\hat{D}} \end{pmatrix} \qquad (9.9c)$$

where $\pi_{c(2,2)}^{\hat{Q},\hat{Q}} \equiv \pi_{c(2,2)}^{\hat{Q},\hat{Q}}(\mu) = (2 - \mu - 2\sqrt{1-\mu})\sin^2\frac{\gamma}{2} - 2(1 - \mu - \sqrt{1-\mu})$ $\sin^4\frac{\gamma}{2}$, and $\pi_{c(2,1)}^{\hat{Q},\hat{D}} \equiv \pi_{c(2,1)}^{\hat{Q},\hat{D}}(\mu) = (2 - \mu + 2\sqrt{1-\mu})\sin^2\frac{\gamma}{2} - 2(1 - \mu + \sqrt{1-\mu})$

$\sin^4 \frac{\gamma}{2}$. Notice that in the noiseless case ($\mu = 0$), we have that $\pi_{c(1,1)}^{\hat{Q},\hat{Q}}(0)=1$, and $\pi_{c(2,1)}^{\hat{Q},\hat{D}}(0) = \sin^2 \gamma$. Equation (9.9) yield the following *correlated* contributions to the total payoffs [see Eq. (9.8)],

$$p_{A,B}^{\hat{C},\hat{C}(c)}(\mu) = p_{A,B}^{\hat{Q},\hat{Q}(c)}(\mu) = P\pi_{c(2,2)}^{\hat{Q},\hat{Q}} + R\left(1 - \pi_{c(2,2)}^{\hat{Q},\hat{Q}}\right) = R - (R - P)\pi_{c(2,2)}^{\hat{Q},\hat{Q}}$$

(9.10a)

$$p_{A,B}^{\hat{D},\hat{D}(c)}(\mu) = P\left(1 - \pi_{c(2,2)}^{\hat{Q},\hat{Q}}\right) + R\pi_{c(2,2)}^{\hat{Q},\hat{Q}} = P + (R - P)\pi_{c(2,2)}^{\hat{Q},\hat{Q}}$$

(9.10b)

$$p_A^{\hat{Q},\hat{D}(c)}(\mu) = p_B^{\hat{D},\hat{Q}(c)}(\mu) = S\left(1 - \pi_{c(2,1)}^{\hat{Q},\hat{D}}\right) + \mathfrak{T}\pi_{c(2,1)}^{\hat{Q},\hat{D}} = S + (\mathfrak{T} - S)\pi_{c(2,1)}^{\hat{Q},\hat{D}}$$

(9.10c)

$$p_A^{\hat{D},\hat{Q}(c)}(\mu) = p_B^{\hat{Q},\hat{D}(c)}(\mu) = \mathfrak{T}\left(1 - \pi_{c(2,1)}^{\hat{Q},\hat{D}}\right) + S\pi_{c(2,1)}^{\hat{Q},\hat{D}} = \mathfrak{T} - (\mathfrak{T} - S)\pi_{c(2,1)}^{\hat{Q},\hat{D}}$$

(9.10d)

$$p_{A,B}^{\hat{Q},\hat{C}(c)}(\mu) = p_{A,B}^{\hat{C},\hat{Q}(c)}(\mu) = R\left(1 - \pi_{c(2,1)}^{\hat{Q},\hat{D}}\right) + P\pi_{c(2,1)}^{\hat{Q},\hat{D}} = R - (R - P)\pi_{c(2,1)}^{\hat{Q},\hat{D}}$$

(9.10e)

which in the PD(5, 3, 2, 1) scenario further simplify to

$$p_{A,B}^{\hat{C},\hat{C}(c)}(0.5) = p_{A,B}^{\hat{Q},\hat{Q}(c)}(0.5) = 3 - \left[\left(\frac{3}{2} - \sqrt{2}\right)\sin^2 \frac{\gamma}{2} - (1 - \sqrt{2})\sin^4 \frac{\gamma}{2}\right]$$

(9.11a)

$$p_{A,B}^{\hat{D},\hat{D}(c)}(0.5) = 2 + \left[\left(\frac{3}{2} - \sqrt{2}\right)\sin^2 \frac{\gamma}{2} - (1 - \sqrt{2})\sin^4 \frac{\gamma}{2}\right]$$

(9.11b)

$$p_A^{\hat{Q},\hat{D}(c)}(0.5) = p_B^{\hat{D},\hat{Q}(c)}(0.5) = 1 + 4\left[\left(\frac{3}{2} + \sqrt{2}\right)\sin^2 \frac{\gamma}{2} - (1 + \sqrt{2})\sin^4 \frac{\gamma}{2}\right]$$

(9.11c)

$$p_A^{\hat{D},\hat{Q}(c)}(0.5) = p_B^{\hat{Q},\hat{D}(c)}(0.5) = 5 - 4\left[\left(\frac{3}{2} + \sqrt{2}\right)\sin^2 \frac{\gamma}{2} - (1 + \sqrt{2})\sin^4 \frac{\gamma}{2}\right]$$

(9.11d)

$$p_{A,B}^{\hat{Q},\hat{C}(c)}(0.5) = p_{A,B}^{\hat{C},\hat{Q}(c)}(0.5) = 3 - \left[\left(\frac{3}{2} + \sqrt{2}\right)\sin^2 \frac{\gamma}{2} - (1 + \sqrt{2})\sin^4 \frac{\gamma}{2}\right]$$

(9.11e)

Equation (9.11) are represented in Fig. 9.1 under the labels $\eta = 1.0$. As can be seen, correlations increase the payoffs for $p_A^{\hat{Q},\hat{D}(c)}(0.5) = p_B^{\hat{D},\hat{Q}(c)}(0.5)$ respect to the uncorrelated results (their final values for $\gamma = \pi/2$ and $\eta = 1$ equal 4.4142 instead of 3.875). However, memory reduces the value of the remaining payoffs shown in the figure. In particular, for $\gamma = \pi/2$, we have that $p_{A,B}^{\hat{D},\hat{D}(c)}(0.5) = p_{A,B}^{\hat{Q},\hat{C}(c)}(0.5) = p_{A,B}^{\hat{C},\hat{Q}(c)}(0.5)$ are reduced by 0.2286, while the payoffs $p_A^{\hat{Q},\hat{D}(c)}(0.5) = p_B^{\hat{Q},\hat{D}(c)}(0.5)$

are reduced by 0.2892 with respect to the uncorrelated results. To conclude, notice that the $\{\hat{Q}, \hat{Q}\}$ payoffs for memoryless and maximally correlated noise are indistinguishable to the naked eye. Actually, by combination of Eqs. (9.2a), (9.8), and (9.11a) one can demonstrate that the difference between them is $p_{A,B}^{\hat{Q},\hat{Q}\,(u)}(0.5) - p_{A,B}^{\hat{Q},\hat{Q}\,(c)}(0.5) = (5/4 - \sqrt{2})\sin^2\frac{\gamma}{2} - (1 - \sqrt{2})\sin^4\frac{\gamma}{2}$, which has a negligible value of only 0.0214 for $\gamma = \pi/2$.

Figure 9.1 also shows the payoffs for an intermediate value of memory equal to $\eta = 0.5$. All payoffs for this memory factor correspond to the average value between $\eta = 0$ and $\eta = 1$, as expected from Eq. (9.8).

9.3   The Spatialized Quantum Prisoner's Dilemma with Memory

The spatial simulations in this section have been performed regarding the Prisoner's Dilemma game with parameters $(\mathfrak{T}, R, P, S) = (5, 3, 2, 1)$ [4].

9.3.1   Two Parameter Strategy Simulations

Fair Quantum Games

The left and right columns in Fig. 9.2 show, respectively, the mean payoffs and the mean parameter values of A and B, associated with a $(5, 3, 2, 1)$-QPD-CA evolving in the presence of a quantum noise channel with noise strength equal to $\mu = 0.5$, and memory factor $\eta = 0.0$ (top), $\eta = 0.5$ (middle), and $\eta = 1.0$ (bottom). In all these three scenarios, the QPD-CA exhibits three different behaviours depending on the value of the entanglement factor, which is pointed out by two landmark values, $\gamma^\star \equiv \gamma_{\mu,\eta}^\star$ and $\gamma^\bullet \equiv \gamma_{\mu,\eta}^\bullet$. These regions have the imprint of the NE that appear in two-player quantum PD games with $P - S \leq \mathfrak{T} - R$: $\{\hat{D}, \hat{D}\}$ for $\gamma < \gamma^\star$, $\{\hat{Q}, \hat{Q}\}$ for $\gamma > \gamma^\star$, and both $\{\hat{Q}, \hat{D}\}$ and $\{\hat{D}, \hat{Q}\}$ in the $(\gamma^\star, \gamma^\bullet)$ transition interval.

The CA simulations in Fig. 9.2 quickly evolve to detect the single NE outside the $(\gamma^\star, \gamma^\bullet)$ transition interval. In this way, the panels in the right columns of Fig. 9.2 show that the average value of both $\bar{\theta}$'s well approximates π for $\gamma < \gamma^\star$, making as a consequence $\bar{\alpha}$'s irrelevant, as $\cos(\pi/2) = 0$ annihilates the influence of α's in the two-parameter subset of SU(2), whereas for $\gamma > \gamma^\bullet$, both $\bar{\theta}$'s well approximate 0 and both $\bar{\alpha}$'s well approximate $\pi/2$. In other words, the system evolves to a $\{\hat{D}, \hat{D}\}$-NE for $\gamma < \gamma^\star$ and to a $\{\hat{Q}, \hat{Q}\}$-NE for $\gamma > \gamma^\bullet$. Consequently, the payoffs in the left columns of Fig. 9.2 out of the transition interval correspond to those shown in Fig. 9.1, namely $p_{A,B}^{\hat{D},\hat{D}}$ for $\gamma < \gamma^\star$ and $p_{A,B}^{\hat{Q},\hat{Q}}$ for $\gamma > \gamma^\bullet$. Let us point out here again that for $\gamma < \gamma^\star$, $p_{A,B}^{\hat{D},\hat{D}}$ keeps close to $P = 2$, while for $\gamma > \gamma^\bullet$, $p_{A,B}^{\hat{D},\hat{D}}$ keeps close to $R = 3$ regardless of η.

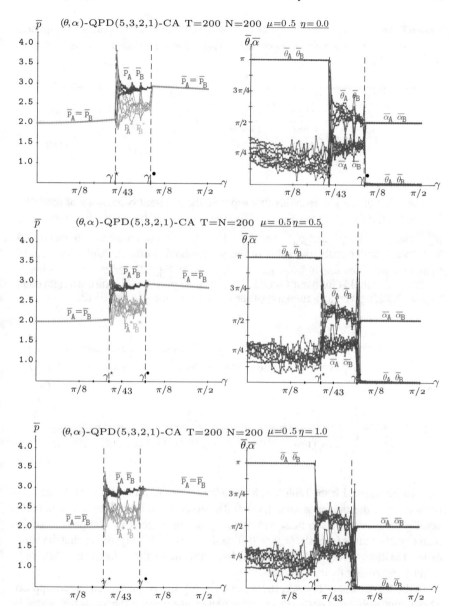

Fig. 9.2 The QPD(5, 3, 2, 1)-CA with noise strength $\mu = 0.5$ and memory factor $\eta = 0$ (top), $\eta = 0.5$ (center), and $\eta = 1.0$ (bottom). Variable entanglement factor γ. Five simulations at T = 200. Left: Mean payoffs and mean-field payoffs. Right: Mean quantum parameter values

Table 9.1 Value of the thresholds $\gamma^*_{\mu,\eta}$ and $\gamma^\bullet_{\mu,\eta}$ in the QPD(5, 3, 2, 1) played with two quantum players (Fair), and with a quantum player facing a classical player with $\hat{U}(\theta,0)$ strategies (Unfair)

μ	η	$\gamma^*_{\mu,\eta}$	Fair			Unfair	
			$\gamma^\bullet_{\mu,\eta}$	$\gamma^\bullet_{\mu,\eta}-\gamma^*_{\mu,\eta}$	$\gamma^\bullet_{\mu,\eta}$	$\gamma^\bullet_{\mu,\eta}-\gamma^*_{\mu,\eta}$
0.0	0.0	0.524	0.785	0.261	0.955	0.432
0.5	0.0	0.723	1.047	0.324	1.231	0.508
	0.5	0.668	0.999	0.332	1.202	0.534
	1.0	0.624	0.957	0.333	1.174	0.550

The value of the γ-thresholds that separate the different behaviours of the QPD-CA, due to the change in the NE as γ increases emerge at the payoffs intersections of $p_A^{\hat{D},\hat{D}}$ and $p_A^{\hat{Q},\hat{D}}$ for γ^*, and $p_A^{\hat{D},\hat{Q}}$ and $p_A^{\hat{Q},\hat{Q}}$ for $\gamma^\bullet$. In the particular case of the QPD(5, 3, 2, 1)-CA under study, the value of the γ-thresholds in the noiseless scenario [5, 6] and for the three sets of noise parameters used in Fig. 9.2, are given in Table 9.1.

The γ-thresholds for a fair QPD(5, 3, 2, 1) game played with noise strength equal to $\mu = 0.5$ are given as a function of the memory factor η in Eq. (9.12).

$$\gamma^*_{0.5,\eta} = 2\arcsin\left(\frac{2}{\sqrt{18 + (10\sqrt{2}-9)\eta + \sqrt{196 + (280\sqrt{2}-244)\eta + (281 - 180\sqrt{2})\eta^2}}}\right),$$

(9.12a)

$$\gamma^\bullet_{0.5,\eta} = 2\arcsin\left(\frac{2\sqrt{2}}{\sqrt{20 + (10\sqrt{2}-11)\eta + \sqrt{144 + 40(6\sqrt{2}-7)\eta + (321 - 220\sqrt{2})\eta^2}}}\right).$$

(9.12b)

As can be inferred from Table 9.1, both thresholds have their minimum values and separation in the noiseless case ($\mu = 0.0$). Moreover, their values and separation notably increase with the noise strength, μ, as can be seen by comparison of the memoryless results ($\eta = 0$) for $\mu = 0.0$[1] and $\mu = 0.5$.[2] Finally, notice that the value of both critical values ($\gamma^*, \gamma^\bullet$) diminishes with the memory factor η, while only slightly increasing their separation.

In the left panels of Fig. 9.2, the mean-field payoffs p^* have been superimposed. These approaches are obtained in a two-player game using the strategies given in Eq. (3.1) with the actual mean parameter values given in the right panels.

As can be seen in Fig. 9.2, the mean-field payoffs coincide with the actual mean payoffs in the three scenarios for $\gamma < \gamma^*$ and $\gamma > \gamma^\bullet$. However, the mean-field approximation notably differs from the actual mean payoffs in the ($\gamma^*,\gamma^\bullet$) interval.

---

[1] $\gamma^*_{0.0,0.0}=\arcsin\left(\sqrt{\frac{P-S}{\mathfrak{T}-S}}=\frac{1}{4}\right)=\frac{\pi}{6}\simeq 0.524$, $\gamma^\bullet_{0.0,0.0}=\arcsin\left(\sqrt{\frac{\mathfrak{T}-R}{\mathfrak{T}-S}}=\frac{2}{4}\right)=\frac{\pi}{4}\simeq 0.785$.

[2] $\gamma^*_{0.5,0.0}=2\arcsin\left(\sqrt{\frac{P-S}{2(\mathfrak{T}-S)}}=\frac{1}{8}\right)\simeq 0.723$, $\gamma^\bullet_{0.5,0.0}=2\arcsin\left(\sqrt{\frac{\mathfrak{T}-R}{2(\mathfrak{T}-S)}}=\frac{2}{8}\right)=\frac{\pi}{3}\simeq 1.047$.

Both the mean-field estimation of player A (brown) and that of the player B (green), as a rule, underestimate the actual mean payoffs of player A (red) and B (blue) in the γ-transition interval. Notice also that in the γ-transition interval both payoffs and mean parameter values notably oscillate, and have a strong dependence on the initial conditions. The ultimate reason for these disturbances relies on the existence of two NE in the γ-transition interval ($\{\hat{Q}, \hat{D}\}$ and $\{\hat{D}, \hat{Q}\}$), inducing the spatial effects that have been commented in the Chap. 3, after being reported in the articles [7–9]

Unfair Quantum Games

In this subsection, we consider an unfair situation, where the A players use full quantum $\hat{U}(\theta, \alpha)$ strategies whereas the B players are restricted to either classical strategies $\hat{U}(\theta, 0)$, or to purely *imaginary* strategies $\hat{U}(0, \alpha)$ [10, 11].

Figure 9.3 shows the results for QPD(5, 3, 2, 1)-CA contests involving a mixed system that combines quantum (θ, α)-players A (red) and classical θ-players B (blue), for a noise strength equal to $\mu = 0.5$, and a memory factor equal to $\eta = 0$ (top), $\eta = 0.5$ (middle), and $\eta = 1$ (bottom). In this case, as in the fair situation reported in Sect. 9.3.1, two thresholds, γ^* and $\gamma^\bullet$, separate the different behaviours in the studied game. While the lower threshold, γ^*, is the same as for fair games, the upper one, $\gamma^\bullet$, changes as it arises now at the intersection of the $p_{A,B}^{\hat{Q},\hat{C}}$ and $p_B^{\hat{Q},\hat{D}}$ payoffs. The $\gamma^\bullet$ threshold in the noiseless scenario is shown in the left frame of Fig. 9.4.[3]

The value of the γ-thresholds for the particular case of the QPD(5, 3, 2, 1)-CA under study, in the noiseless scenario and for the three sets of noise parameters used in Fig. 9.3, are listed in Table 9.1, and in general in the Eq. (9.13a). By comparison of the results obtained for the fair and unfair settings, it is clear that $\gamma^\bullet$ is always greater in the unfair scenario than in the fair one. As in this latter case, the $\gamma^\bullet$ threshold has its minimum value in the noiseless scenario, and its value similarly increases with the noise strength μ, and reduces with the memory factor η, albeit at a low extent. Contrary to the fair setting, the separation of both thresholds moderately decreases with the memory factor η in unfair games. For the unfair quantum versus classical case, the lowest threshold is given by Eq. (9.12a), while the largest one equals,

$$\gamma_{0.5,\eta}^\bullet = 2 \arcsin \left(\frac{2\sqrt{2}}{\sqrt{16 + (6\sqrt{2} - 7)\eta + \sqrt{64 + (96\sqrt{2} - 128)\eta + (121 - 84\sqrt{2})\eta^2}}} \right)$$

$$(9.13a)$$

As in the fair games, both $\bar{\theta}_A$ and $\bar{\theta}_B$ in Fig. 9.3 are equal to π for $\gamma < \gamma^*$, which corresponds to mutual defection, $\{\hat{D}, \hat{D}\}$, and then the mean payoffs equal those

---

[3]From Eq. (2.7) it is $p_B^{\hat{Q},\hat{D}(u)}(0) = \mathfrak{T} \cos^2 \gamma + S \sin^2 \gamma$ and $p_{A,B}^{\hat{Q},\hat{C}(u)}(0) = R \cos^2 \gamma + P \sin^2 \gamma$, whose equalization leads to $\gamma_{0.0,0.0}^\bullet = \arcsin\left(\sqrt{\frac{\mathfrak{T}-R}{\mathfrak{T}+P-R-S}}\right) = \arcsin\left(\sqrt{\frac{2}{3}}\right) \simeq 0.955$. In this scenario, it is $\gamma_{0.5,0.0}^\bullet = 2\arcsin\left(\sqrt{\frac{P-S}{\mathfrak{T}+P-R-S}} = \frac{1}{3}\right) \simeq 1.231$.

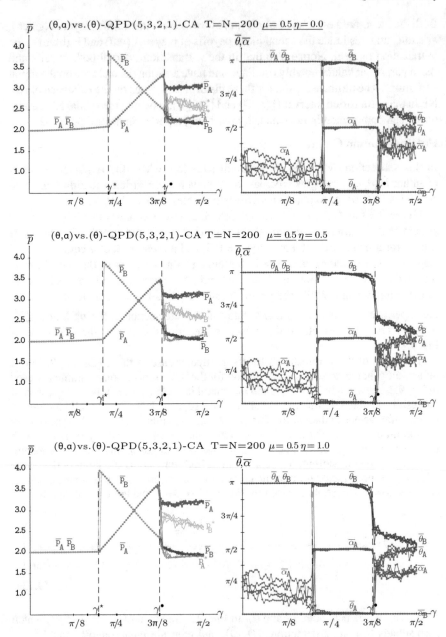

Fig. 9.3 The PD(5, 3, 2, 1)-CA with a quantum (θ, α)-player A (red) versus a classical θ-player B (blue). Noise strength $\mu = 0.5$, and memory factor $\eta = 0$ (top), $\eta = 0.5$ (middle), and $\eta = 1$ (bottom). Variable entanglement factor, γ. Five unfair simulations at T = 200. Left: Actual mean payoffs $)\overline{p}$) and mean-field payoffs (p^*). Right: mean quantum parameter values

in the fair case. A completely different behaviour respect to the fair setting arises in the transition interval, $(\gamma^*, \gamma^\bullet)$. In this case, the $\{\hat{Q}, \hat{D}\}$ strategy clearly emerges isolated, because the $\{\hat{D}, \hat{Q}\}$ pair, also in NE in the transition interval in fair contests, is now unfeasible as the player B may not resort to quantum strategies such as the $\hat{Q}$ one.

The CA results for $\gamma > \gamma^\bullet$ shown in Fig. 9.3 are strongly influenced by the so-called *Miracle* strategy $\hat{M} = \hat{U}(\pi/2, \pi/2)$. From the seminal EWL paper [3], it is known that in the two-player noiseless scenario, the quantum player may outperform the classical one by means of $\hat{M}$ provided that γ is high enough. With full entanglement in particular, the $\hat{M}$-player A outperforms the $\hat{U}(\theta_B, 0)$-player B for all θ_B, except for $\theta_B = \pi/2$, where the $\{\hat{M}, \hat{U}(\theta_B, 0)\}$ strategy pair produces the same payoff for both players (see Fig. 4.2).

The miracle strategy seems to play some role in the spatial simulations of the unfair contests of Fig. 9.3 for $\gamma > \gamma^\bullet$, because $\bar{\theta}_A$ is close to $\pi/2$, and player B, in turn, *replies* with $\bar{\theta}_B$ not far from $\pi/2$, particularly near $\gamma = \pi/2$. Nevertheless, $\bar{\alpha}_A$ in games with $\gamma > \gamma^\bullet$ does not approach to $\pi/2$ (the α parameter of $\hat{M}$), but to $3\pi/8$. Thus, the mean field payoff approaches in CA noiseless simulations for $\gamma > \gamma^\bullet$ [7] are similar to those generated by the variant of $\hat{M}$ that we call here $\hat{W} = \hat{U}(\pi/2, 3\pi/8)$. The payoffs induced by the $\{\hat{W}, \hat{U}(\pi/2, 0)\}^4$ together with the payoffs induced by the $\{\hat{M}, \hat{D}\}^5$ in a QPD(5, 3, 2, 1) game with two players and no noise are shown in the right panel of Fig. 9.4. Notice that, with the $\{\hat{W}, \hat{U}(\pi/2, 0)\}$ strategy pair, player B outperforms player A for all γ, whereas in the $\{\hat{M}, \hat{D}\}$ game it is $p_A > p_B$ from $\gamma = \pi/4$.

The structure of the mean-field payoffs estimations p^* in the CA simulations of the three scenarios of Fig. 9.3 beyond $\gamma^\bullet$ is reminiscent of that of the payoffs generated by the pair $\{\hat{W}, \hat{U}(\pi/2, 0)\}$, shown in the right panel of Fig. 9.4 for the noiseless scenario. These mean-field payoff estimations notably differ from the actual $\bar{p}$ payoffs, whose structure in turn is reminiscent of that generated by the $\{\hat{M}, \hat{D}\}$ pair, also shown in the right panel of Fig. 9.4 for the noiseless scenario. The discrepancy between both actual and mean-field payoffs is not only quantitative but qualitative, in the sense that

---

[4] $p_A^{\hat{W}, \hat{U}(\pi/2, 0)}(0, 0) = \frac{1}{4}\left[11 + \frac{3}{4}(2 + \sqrt{2})\sin^2\gamma - 3\sqrt{2 + \sqrt{2}}\sin\gamma\right]$, and $p_B^{\hat{W}, \hat{U}(\pi/2, 0)}(0, 0) = \frac{1}{4}\left[11 - \frac{5}{4}(2 + \sqrt{2})\sin^2\gamma + \sqrt{2 + \sqrt{2}}\sin\gamma\right]$. These payoffs are structurally similar to those generated by the $\{\hat{M}, \hat{U}(\pi/2, 0)\}$ pair from $\Pi_u^{\hat{M}, \hat{U}(\pi/2, 0)}(0, 0) = \frac{1}{4}\begin{pmatrix} \cos^2\gamma & \cos^2\gamma \\ (\cos\frac{\gamma}{2} - \sin\frac{\gamma}{2})^4 & (\cos\frac{\gamma}{2} + \sin\frac{\gamma}{2})^4 \end{pmatrix}$, which are equal to $p_A^{\hat{M}, \hat{U}(\pi/2, 0)}(0, 0) = \frac{1}{4}(11 + 3\sin^2\gamma - 6\sin\gamma)$, and $p_B^{\hat{M}, \hat{U}(\pi/2, 0)}(0, 0) = \frac{1}{4}(11 - 5\sin^2\gamma + 2\sin\gamma)$. In both games it is $p_A^{\hat{M}, \hat{U}(\pi/2, 0)}(0, 0) = p_B^{\hat{M}, \hat{U}(\pi/2, 0)}(0, 0) = 11/4$ at $\gamma = 0$, and $p_B^{\hat{M}, \hat{U}(\pi/2, 0)}(0, 0) > p_A^{\hat{M}, \hat{U}(\pi/2, 0)}(0, 0)$, even for $\gamma = \pi/2$ in the $\{\hat{W}, \hat{U}(\pi/2, 0)\}$ game, whereas $p_A^{\hat{M}, \hat{U}(\pi/2, 0)}(0, 0) = p_B^{\hat{M}, \hat{U}(\pi/2, 0)}(0, 0) = 2$ in the $\{\hat{M}, \hat{U}(\pi/2, 0)\}$ game.

[5] It is $\Pi_u^{\hat{M}, \hat{D}}(0, 0) = \frac{1}{2}\begin{pmatrix} 0 & \sin^2\gamma \\ \cos^2\gamma & 1 \end{pmatrix}$, which renders the payoffs $p_A^{\hat{M}, \hat{D}}(0, 0) = \frac{1}{2}\left[(\mathfrak{T} - S)\sin^2\gamma + P + S\right]$, and $p_B^{\hat{M}, \hat{D}}(0, 0) = \frac{1}{2}\left[(S - \mathfrak{T})\sin^2\gamma + P + \mathfrak{T}\right]$.

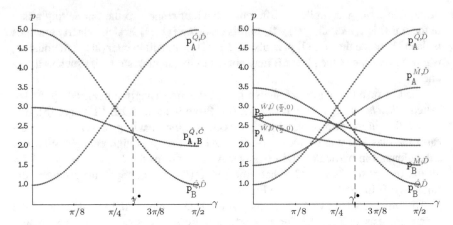

Fig. 9.4 Payoffs in a noiseless QPD(5, 3, 2, 1) game as a function of γ. Left: Payoffs of the $\{\hat{Q}, \hat{D}\}$ and $\{\hat{Q}, \hat{C}\}$ strategy pairs. Right: Payoffs of the $\{\hat{W}, \hat{U}(\pi/2, 0)\}$, $\{\hat{M}, \hat{D}\}$, and $\{\hat{Q}, \hat{D}\}$ strategy pairs

$\overline{p}_A > \overline{p}_B$ whereas $p_A^* < p_B^*$, in accordance with the results of the noiseless scenario (right panel of Fig. 9.4), reaching $p_A^{\hat{M},\hat{D}(u)}(0) = 3.5$, $p_B^{\hat{M},\hat{D}(u)}(0) = 1.5$ at $\gamma = \pi/2$. With noise, the actual payoffs $p_A^{\hat{M},\hat{D}}$ and $p_B^{\hat{M},\hat{D}}$ approach each other, becoming for maximum entangling fairly close to the values of $\overline{p}_A$ and $\overline{p}_B$ achieved in the CA simulations of Fig. 9.3.[6] These discrepancies between $\overline{p}$ and p^* are produced by the spatial effects in the CA simulations, much as were in Fig. 9.2, and will be in the 3P simulations treated in Sect. 9.3.2.

Figure 9.5 presents the results for an unfair game where the player B is allowed to resort only to the parameter α instead of θ as in the unfair simulations discussed so far. In this case, the $\{\hat{D}, \hat{Q}\}$ pair emerges in NE below the threshold γ^*, defined at the intersection of $p_A^{\hat{D},\hat{Q}}$ and $p_B^{\hat{D},\hat{Q}}$, and the $\{\hat{Q}, \hat{Q}\}$ pair beyond γ^*. These NE show up even in the noiseless scenario, where $\gamma_{0.0,0.0}^* = \pi/4$. In the $\mu = 0.5$ scenarios of Fig. 9.5, they are equal to $\gamma_{0.5,0.0}^* = 1.0472$ [from the intersection of Eqs. (9.2c) and (9.2d)], $\gamma_{0.5,0.5}^* = 0.9921$, $\gamma_{0.5,1.0}^* = 0.9449$ [from the intersection of Eqs. (9.11c) and (9.11d)], with equalized payoffs $(p_{A,B}^{\hat{D},\hat{Q}} = p_{A,B}^{\hat{Q},\hat{Q}})$ at $\gamma = \gamma^*$: $p(0.5, 0.0) = 2.9362$, $p(0.5, 0.5) = 2.9732$, $p(0.5, 1.0) = 2.9982$. For $\gamma < \gamma^*$, as the full quantum player A fully defects $(\overline{\theta}_A = \pi)$, the $\overline{\alpha}_A$ values shown in the bottom part of Fig. 9.5 become irrelevant, whereas the player B sets his free parameter to its maximum $\overline{\alpha}_B = \pi/2$. As a result, the $\{\hat{D}, \hat{Q}\}$ pair emerges, and player A beats player B, who cannot resort to $\theta_B > 0$ strategies such as $\hat{D}$.

---

[6]The values of (p_A, p_B) at $\gamma = \pi/2$ in the $(\hat{M}, \hat{D})$ two-player game, and those of $(\overline{p}_A, \overline{p}_B)$ in the CA simulations of Fig. 9.3 are (from top to bottom): (3.125, 2.125), (3.042, 2.181); (3.203, 1.996), (3.110, 2.071); (3.280, 1.866), (3.166, 1.944).

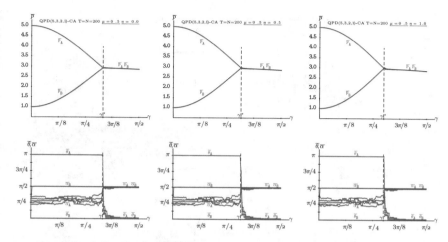

Fig. 9.5 Mean payoffs (upper) and parameter values (lower) for five simulations at T = 200 of an unfair quantum PD(5, 3, 2, 1)CA, with a (θ, α)-player A (red) and a (α)-player B (blue) with variable entanglement γ. Noise strength $\mu = 0.5$ and memory factor $\eta = 0.0$ (left), $\eta = 0.5$ (center), and $\eta = 1.0$ (right)

9.3.2 Three-Parameter Strategy Simulation

This section is devoted to the full space of SU(2) strategies, i.e., with three parameters according to the operators given in Eq. (2.3).

Figure 9.6 shows the results for 3P QPD(5, 3, 2, 1)-CA simulations with $\mu = 0.5$ in the three characteristic memory scenarios: $\eta = 0.0$ (top), $\eta = 0.5$ (center), and $\eta = 1.0$ (bottom). Two main features characterize the parameter snapshots (right) in Fig. 9.6: (*i*) the $\overline{\theta}$ parameters of both players drift to π (particularly with low γ) which makes irrelevant the values of the $\overline{\alpha}$ parameters, and (*ii*) the $\overline{\beta}$ parameters oscillate around $\pi/4$. Thus, in a mean-field approach, both players would adopt the strategy $\hat{V} = \hat{U}(\pi, \alpha, \pi/4) = \frac{1}{\sqrt{2}} \begin{pmatrix} 0 & 1+i \\ 1-i & 0 \end{pmatrix}$. For γ over the $\gamma^\star$ landmark featured afterwards, the $\overline{\theta}$ average values become erratic, which, in turn, is reflected in violent fluctuations of the mean-field payoffs presented on the left panels. Nevertheless, heavy spatial effects compensate these $\overline{\theta}$-turbulences, so that the actual mean payoffs increase all over the γ-range, even for $\gamma > \gamma^\star$, roughly according to the coincident payoff for both players that the $\{\hat{V}, \hat{V}\}$ pair renders.

It is proved that the pair $\{\hat{V}, \hat{V}\}$ is in NE before the $\gamma^\star_{\mu,\eta}$ landmark given at the intersection of the payoff induced by $\{\hat{V}, \hat{V}\}$ and that of $p_A^{\hat{Q}\hat{D}}$, which in the noiseless scenario becomes $\gamma^\star_{0,0} = \arcsin\sqrt{(P - S)/(\mathfrak{T} + P - R - S)}$ [6].

In the 3P-QPD-CA simulations obtained for a memoryless ($\eta = 0.0$) noise of strength $\mu = 0.5$ which are shown in the top panels of Fig. 9.6, the pair $\{\hat{V}, \hat{V}\}$ induces a Π matrix equal to that given by Eq. (7.10) for the $\{\hat{Q}, \hat{D}\}$ contest but with its rows interchanged. Then, the payoff for both players is identical

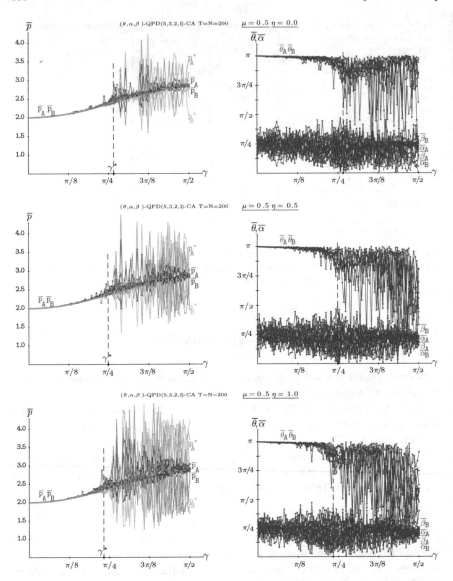

Fig. 9.6 A three parameter QPD(5, 3, 2, 1)-CA with noise strength $\mu = 0.5$. Variable entanglement factor γ. Five simulations at T = 200. Left: Mean actual and mean-field payoffs. Right: mean quantum parameter values. Memory factor $\eta = 0.0$ (top), $\eta = 0.5$ (middle), $\eta = 1.0$ (bottom)

and equal to $p_{A,B}^{\nabla,\nabla}(0.5, 0.0) = P + \frac{1}{4}\sin^2\frac{\gamma}{2}(\mathfrak{T} + R + S - P) + 2(R - P)\sin^4\frac{\gamma}{2} = 2 + \frac{3}{4}\sin^2\frac{\gamma}{2} + \frac{1}{2}\sin^4\frac{\gamma}{2}$, which reaches $p_{A,B}^{\nabla,\nabla}(0.5, 0.0) = 2.875$ at $\gamma = \pi/2$. These payoffs are smaller than those for the noiseless setting, where $p_{A,B}^{\nabla,\nabla}(0.0, 0.0) = P + (R - P)\sin^2\gamma = 2 + \sin^2\gamma$, which equals $p_{A,B}^{\nabla,\nabla}(0.0, 0.0) = 3$ for $\gamma = \pi/2$ (see Ref. [7] for further details).

The bottom panels of Fig. 9.6 show the results for 3P-QPD-CA simulations with a maximally correlated noise ($\eta = 1.0$) of strength $\mu = 0.5$. Here, the pair $\{\hat{\nabla}, \hat{\nabla}\}$ induces a diagonal Π matrix obtained by interchanging the rows of the matrix given in Eq. (9.9a) for the $\{\hat{Q}, \hat{D}\}$ contest. In this case, both players have an equal payoff of $p_{A,B}^{\nabla,\nabla}(0.5, 1.0) = R\pi_{c(2,1)}^{\hat{Q},\hat{D}} + P(1 - \pi_{c(2,1)}^{\hat{Q},\hat{D}}) = P + (R - P)\pi_{c(2,1)}^{\hat{Q},\hat{D}} = 2 + (\frac{3}{2} + \sqrt{2})\sin^2\frac{\gamma}{2} - (1 + \sqrt{2})\sin^4\frac{\gamma}{2}$, which reaches $p_{A,B}^{\nabla,\nabla}(0.5, 1.0) = 2.854$ at $\gamma = \pi/2$. The payoffs induced by the pair $\{\hat{\nabla}, \hat{\nabla}\}$ with $\eta = 1.0$ and $\eta = 0.0$ are dramatically close, being their maximal difference equal to solely 0.021 for $\gamma = \pi/2$.

The middle panels of Fig. 9.6 show the results of our 3P-QPD-CA simulations for intermediate values of the noise strength ($\mu = 0.5$) and the memory factor ($\eta = 0.5$). Because of Eq. (9.8), the results for the average memory factor ($\eta = 0.5$) correspond to the average values for uncorrelated ($\eta = 0.0$) and maximally correlated noise ($\eta = 1.0$), as in the 2P fair games previously discussed in Sect. 9.3.1. In this case, we have $p_{A,B}^{\nabla,\nabla}(0.5, 0.5) = 2.864$ at $\gamma = \pi/2$. Incidentally, $\{\hat{\nabla}, \hat{\nabla}\}$ and $\{\hat{Q}, \hat{Q}\}$ induce the same payoff at $\gamma = \pi/2$.

Figure 9.7 shows the results for 3P strategies QPD(5, 3, 2, 1)-CA in unfair simulations, which can be regarded as a generalization of the 2P unfair simulations already presented in Fig. 9.3. The structure of the graphs of the actual mean payoffs (left) of Fig. 9.7 notably resembles that of Fig. 9.3: equal payoffs for both players for moderate values of γ, a γ transition interval inducing $\{\hat{Q}, \hat{D}\}$ (recall that $\theta_A = 0$ makes irrelevant β_A), and prevalence of the actual mean payoff of the quantum player A over that of the classical player B for high γ.

In Fig. 9.7, the player B adopts the $\hat{D} = \hat{U}(\pi, 0, 0)$ strategy for $\gamma < \gamma^*$ while the player A follows the $\hat{\Delta} = \hat{U}(\pi, \alpha, \pi/2) = \begin{pmatrix} 0 & i \\ i & 0 \end{pmatrix} = i\hat{D}$ strategy (recall that $\theta_A = \pi$ makes irrelevant α_A). Mutual $\hat{\Delta}$ produces the same output as mutual $\hat{D}$. But as both $(\hat{\Delta} \otimes \hat{D})$ and $(\hat{\nabla} \otimes \hat{\nabla})$ share an antidiagonal structure that leads to the same Π density matrix, the pair $\{\hat{\Delta}, \hat{D}\}$ produces the same payoff as the pair $\{\hat{\nabla}, \hat{\nabla}\}$. Consequently, the equations that define the equal payoffs for both players before γ^* in Fig. 9.7 are the same as in Fig. 9.6. Similarly, the right thresholds $\gamma^\bullet$ in Fig. 9.7 are equal to those in Fig. 9.6.

The values of the γ^* threshold in the unfair 3P-QPD(5, 3, 2, 1)-CA simulations in Fig. 9.7 turn out to be: (i) $\gamma_{0.5,0.0}^* = \arcsin\sqrt{\frac{1}{6}} = 0.841$, arising from the intersection of $p_A^{\Delta,\hat{D}}(0.5, 0.0) = 2 + \frac{3}{4}\sin^2\frac{\gamma}{2} + \frac{1}{2}\sin^2\gamma$ and $p_A^{\hat{Q},\hat{D}(u)}(0.0)$ given in Eq. (9.2c). It is $\gamma_{0.5,0.0}^* < \gamma_{0.0,0.0}^* = \sqrt{\frac{1}{3}} = 0.616$. (ii) $\gamma_{0.5,1.0}^* = 0.732$, arising from the

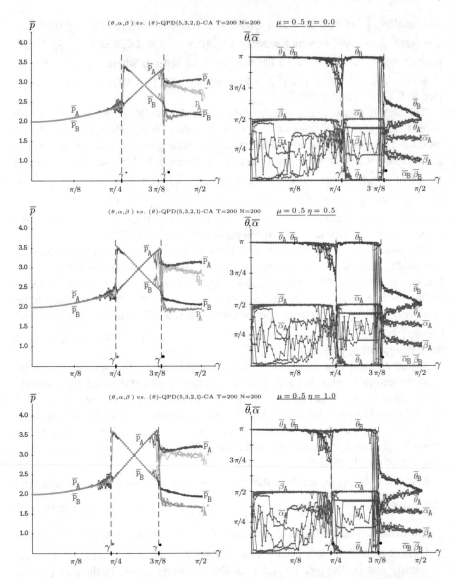

Fig. 9.7 An unfair three parameter QPD(5, 3, 2, 1)-CA with noise strength $\mu = 0.5$. Left: Mean actual and mean-field payoffs. Right: Mean quantum parameter values. Variable entanglement factor γ. Five simulations at T = 200. Memory factor $\eta = 0.0$ (top), $\eta = 0.5$ (middle), $\eta = 1.0$ (bottom)

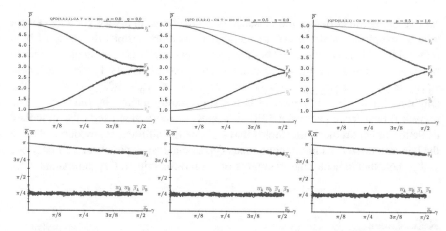

Fig. 9.8 Mean actual and mean-field payoffs (top), and average parameter values (bottom) for five simulations at T = 200 of an unfair QPD(5, 3, 2, 1)-CA with a (θ, α, β)-player A (red) and a $(0, \alpha, \beta)$-player B (blue) with variable entanglement γ. Noise strength μ and memory η. Left: $\mu = \eta = 0.0$. Center: $\mu = 0.5, \eta = 0.0$. Right: $\mu = 0.5, \eta = 1.0$

intersection of $p_A^{\Delta,\hat{D}}(0.5, 1.0) = 2 + \frac{1}{2} \sin^2 \frac{\gamma}{2} + \frac{1+2\sqrt{2}}{4} \sin^4 \frac{\gamma}{2}$ and $p_A^{\hat{Q},\hat{D}(c)}(0.5)$ given in Eq. (9.11c). (iii) $\gamma_{0.5,0.5}^* = 0.780$, arising from the intersection of $p_A^{\Delta,\hat{D}}(0.5, 0.5) = (p_A^{\Delta,\hat{D}}(0.5, 0.0) + p_A^{\Delta,\hat{D}}(0.5, 1.0))/2$ and $(p_A^{\hat{Q},\hat{D}(u)}(0.5) + p_A^{\hat{Q},\hat{D}(c)}(0.5))/2$.

The general considerations on the structure of the mean-field payoffs estimations p^* and the actual payoffs $\overline{p}$ in the CA simulations in the 2P scenario of Fig. 9.3 beyond $\gamma^\bullet$ also apply in the 3P scenario of Fig. 9.7: (i) both $\overline{\theta}_A$ and $\overline{\theta}_B$ drift to $\pi/2$, while $\overline{\alpha}_A$ remains close to $3\pi/8$, and $\overline{\beta}_A$ drifts to a low value of $\pi/8$, (ii) the p^* estimations notably differ from the actual $\overline{p}$, with $\overline{p}_A > \overline{p}_B$ whereas $p_A^* < p_B^*$.

Figure 9.8 shows the results for an unfair 3P-game where the player B is allowed to resort to the (α, β)-parameters but not θ, i.e., it is a 3P extension of the 2P results shown in Fig. 9.5. The full quantum 3P-player A shows a high $\overline{\theta}$ mean value for all γ, whereas the $\overline{\alpha}$ and $\overline{\beta}$ of both players drift to $\pi/4$. In a conventional (non-CA) noiseless game, the $\hat{U}(\pi, \pi/4, \pi/4)$ vs. $\hat{U}(0, \pi/4, \pi/4)$ contest generates $\Pi_u^{U(\pi,\pi/4,\pi/4),U(0,\pi/4,\pi/4)}(0) = \begin{pmatrix} 0 & 0 \\ 1 & 0 \end{pmatrix}$, which favours the full quantum player A with the temptation $\mathfrak{T}$, whereas the *imaginary* player B must resign to only the *sucker's* payoff S. The left panel of Fig. 9.8 reflects this fact in terms of the mean-field approach, i.e., $p_A^* \simeq 5$ and $p_B^* \simeq 1$ for all γ, prohibiting any payoff equalization, as it was feasible in Fig. 9.5. The simulations with correlated noise in Fig. 9.8 (central and right panels) show a moderation in the privileged status of player A. Very strong spatial effects support the notable difference of the mean-field and actual mean payoffs as soon as γ takes off, so that the increase of γ induces the approximation of both actual mean payoffs, even in the noiseless game.

Let us summarize this chapter concluding that the two thresholds, γ^* and $\gamma^\bullet$, that separate the different behaviour of the QPD-CA in the uncorrelated noise scenario [7] survive but get distorted in the presence of correlations, in a very good agreement with a two-player analysis. As a general rule, memory shrinks the value of both thresholds. This is so when players follow two-parameter (2P) strategies, either in fair games and in games facing quantum versus classical players. In these scenarios: $\gamma^\star_{0.0,0.0} < \gamma^\star_{0.5,1.0} < \gamma^\star_{0.5,0.5} < \gamma^\star_{0.5,0.0}$, and $\gamma^\bullet_{0.0,0.0} < \gamma^\bullet_{0.5,1.0} < \gamma^\bullet_{0.5,0.5} < \gamma^\bullet_{0.5,0.0}$. The amplitude of the $(\gamma^\star, \gamma^\bullet)$ interval is not significantly altered by memory. When players follow three-parameter (3P) strategies, memory also exerts the described depletion effect on the γ-thresholds that separate the different behaviours, albeit in the fair case there is a kind of continuous dependence on γ, punctuated by only one γ-threshold that indicates the emergence of spatial effects in the CA game simulation.

References

1. Nielsen, M.A., Chuang, I.L.: Quantum Computation and Quantum Information. Cambridge University Press, Cambridge (2000)
2. Ramzan, M., Nawaz, A., Toor, A.H., Khan, M.K.: The effect of quantum memory on quantum games. J. Phys. A: Math. Theor. **41**, 055307 (2008)
3. Eisert, J., Wilkens, M., Lewenstein, M.: Quantum games and quantum strategies. Phys. Rev. Lett. **83**(15), 3077–3080 (1999)
4. Alonso-Sanz, R., Revuelta, F.: On the effect of memory in a quantum prisoner's dilemma cellular automaton. Quantum Inf. Process. **17**(3), 60 (2018)
5. Du, J.F., Xu, X.D., Li, H., Zhou, X., Han, R.: Entanglement playing a dominating role in quantum games. Phys. Lett. A **89**(1–2), 9–15 (2001)
6. Du, J.F., Li, H., Xu, X.D., Zhou, X., Han, R.: Phase-transition-like behaviour of quantum games. J. Phys. A: Math. Gen. **36**(23), 6551–6562 (2003)
7. Alonso-Sanz, R.: On the effect of quantum noise in a quantum prisoner's dilemma cellular automaton. Quantum Inf. Process. **16**(6), 161 (2017)
8. Alonso-Sanz, R.: Variable entangling in a quantum prisoner's dilemma cellular automaton. Quantum Inf. Process. **14**, 147–164 (2015)
9. Alonso-Sanz, R.: A quantum prisoner's dilemma cellular automaton. Proc. R. Soc. A **470**, 20130793 (2014)
10. Flitney, A.P., Abbott, D.: Advantage of a quantum player over a classical one in 2 × 2 quantum games. Proc. R. Soc. Lond. A **459**(2038), 2463–2474 (2003)
11. Flitney, A.P., Abbott, D.: Quantum games with decoherence. J. Phys. A: Math. Gen. **38**(2), 449 (2004)

Chapter 10
Games with Werner-Like States

The density matrix of a Werner quantum state is invariant under all unitary operators of the form $U \otimes U$. That is, its ρ matrix satisfies $\rho = (U \otimes U)\rho(U^\dagger \otimes U^\dagger)$ for all unitary operators U acting on the Hilbert space [1]. Werner-like states as implemented in quantum games in this chapter are linear combinations of a maximally entangled and an uncorrelated mixed state that will act as alternative initial states in the EWL model.

10.1 Werner-Like States in Quantum Games

According to [2, 3], we will consider in this chapter Werner-like states by means of,

$$\rho = (1 - \delta)\mathbf{I}/4 + \delta\rho_f, \quad 0 \le \delta \le 1 \tag{10.1}$$

Where δ ponders the *fidelity* to the density matrix ρ_f. If $\delta = 0$ it is $\rho = \mathbf{I}/4$, so that $\Pi = \frac{1}{4}\begin{pmatrix} 1 & 1 \\ 1 & 1 \end{pmatrix}$, so that in the PD game it is $p_A(\gamma, 0) = p_B(\gamma, 0) = K$, with $K = (\mathfrak{T} + P + R + S)/4$.

From Eq. (10.1), it is,

$$\Pi = \frac{1 - \delta}{4}\begin{pmatrix} 1 & 1 \\ 1 & 1 \end{pmatrix} + \delta\Pi_f, \quad 0 \le \delta \le 1 \tag{10.2}$$

Therefore, in the PD game the payoffs achieved with ρ_f are to be multiplied by δ and increased by $(1 - \delta)K$. It is, $p_{A,B}^{\hat{D},\hat{D}} = (1 - \delta)K + \delta P$, and $p_{A,B}^{\hat{Q},\hat{Q}} = (1 - \delta)K + \delta R$, leading to $p_{A,B}^{\hat{D},\hat{D}}(\gamma, 1) = P$ and $p_{A,B}^{\hat{Q},\hat{Q}}(\gamma, 1) = R$. In the QPD(5, 3, 1, 0) context of Fig. 10.1, with $K = (5 + 3 + 1 + 0)/4 = 2.25$, $p_{A,B}^{\hat{D},\hat{D}}(\delta = 0.5) = (2.25 +$

© Springer Nature Switzerland AG 2019
R. Alonso-Sanz, *Quantum Game Simulation*, Emergence, Complexity and Computation 36, https://doi.org/10.1007/978-3-030-19634-9_10

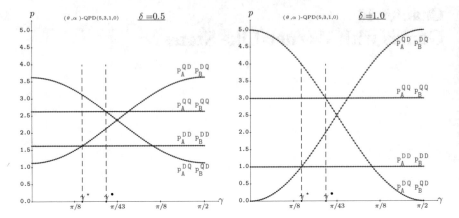

Fig. 10.1 Payoffs of the interactions of the D and Q strategies in the QPD(5, 3, 1, 0) game. Left: $\delta = 0.5$, Right: $\delta = 1.0$

$0) = 1.625$, $p_{A,B}^{\hat{Q},\hat{Q}}(\delta = 0.5) = (2.25 + 3)/2 = 2.265$ (left frame), and $p_{A,B}^{\hat{D},\hat{D}}(\delta = 1.0) = P = 1.0$, $p_{A,B}^{\hat{Q},\hat{Q}}(\delta = 1.0) = R = 3.0$ (right frame). It is, $p_A^{\hat{Q},\hat{D}} = (1-\delta)K + \delta(S \cos^2 \gamma + \mathfrak{T} \sin^2 \gamma)$. As a result, the intersection of $p_A^{\hat{D},\hat{D}}$ and $p_A^{\hat{Q},\hat{D}}$ is given with $P = S \cos^2 \gamma + \mathfrak{T} \sin^2 \gamma$, i.e., at $\gamma^{\star} = \arcsin\left(\sqrt{\frac{P-S}{\mathfrak{T}-S}}\right)$ which does not depend on δ. Similarly, $p_A^{\hat{D},\hat{Q}} = (1-\delta)K + \delta(S \sin^2 \gamma + \mathfrak{T} \cos^2 \gamma)$, so that the intersection of $p_A^{\hat{D},\hat{Q}}$ and $p_A^{\hat{Q},\hat{Q}}$ is given with $S \sin^2 \gamma + \mathfrak{T} \cos^2 \gamma = R$, i.e., at $\gamma^{\bullet} = \arcsin\left(\sqrt{\frac{\mathfrak{T}-R}{\mathfrak{T}-S}}\right)$ which neither depends on δ. In the QPD(5, 3, 1, 0) context of Fig. 10.1 it is: $\gamma^{\star} = \arcsin\sqrt{1/5} = 0.464$, closely over $\pi/8 = 0.393$, and $\gamma^{\bullet} = \arcsin\sqrt{2/5} = 0.685$, closely under $\pi/4 = 0.785$.

10.2 Spatial Games

As used in this book, in the spatial simulations we deal with in this section, the initial quantum parameter values will be assigned at random [4]. Thus, initially: $\bar{\theta} \simeq \pi/2$ and $\bar{\alpha} = \pi/4$. The mean payoffs $(\bar{p})$ and the underlying mean value of the quantum parameters $(\bar{\theta}, \bar{\alpha})$ are shown at $T = 200$ in simulations with variable entanglement factor γ. Next Sect. 10.2.1 deals only with the symmetric PD and HD games by fixing one of the parameters γ and δ, whereas Sect. 10.2.2 scrutinizes also the SD and the BOS games with both γ and δ variable.

In this chapter the PD parameters are to be (5, 3, 1, 0), that differ from those in Chap. 3 where the PD parameters are (5, 3, 2, 1). In the same vein, the HD parameters in this chapter are to be (5, 3, 0, 1) instead of (3, 2, 0, −1) as in Chap. 3.

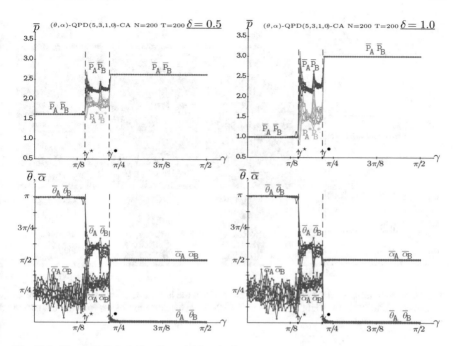

Fig. 10.2 The QPD(5, 3, 1, 0) with variable γ in five spatial simulations at T = 200. Left: $\delta = 0.5$. Right: $\delta = 1.0$. Upper: Mean payoffs. Lower: Mean quantum parameters

10.2.1 Either γ or δ Fixed

In the QPD(5, 3, 1, 0) spatial simulations of Fig. 10.2, mutual defection arises below the lower $\gamma^{\bullet} = 0.464$ threshold and mutual $\hat{Q}$ beyond the higher $\gamma^{\star} = 0.685$ threshold. In the $(\gamma^{\star}, \gamma^{\bullet})$ transition interval, where both $(\hat{Q}, \hat{D})$ and $(\hat{D}, \hat{Q})$ are in NE, the dynamics is highly conditioned by spatial effects as explained below when commenting Fig. 10.2. Remarkably, not only the values of the $\gamma^{\star}$ and $\gamma^{\bullet}$ thresholds are unaffected by δ, neither the evolution of the $\overline{\theta}$ and $\overline{\alpha}$ mean quantum parameters are affected by δ, as it is shown in the lower frames of Fig. 10.2. In short, implementing ρ instead of ρ_f only affects the payoffs. It is noticeable that for $\gamma < \gamma^{\bullet}$, as both $\overline{\theta}$'s are selected to be π, both $\overline{\alpha}$'s are irrelevant, as $\cos(\pi/2) = 0$ annihilates the influence of α in Eq. (2.3), whereas for $\gamma > \gamma^{\star}$, both $\overline{\theta}$'s are set to zero and both $\overline{\alpha}$'s to $\pi/2$, i.e., the $(\hat{Q}, \hat{Q})$ pair is stated.

In Fig. 10.2, the mean-field payoffs coincide with the actual mean payoffs with γ below $\gamma^{\star}$ and over $\gamma^{\bullet}$. But in the $(\gamma^{\star}, \gamma^{\bullet})$ interval the mean-field payoff approaches underestimate the actual mean payoffs. The lack of coincidence of both the mean-field and actual mean payoffs reflects the emergence of quantum parameter patterns that impede an approach based on the mean values of said parameters. An example of this is given in the next Fig. 10.3.

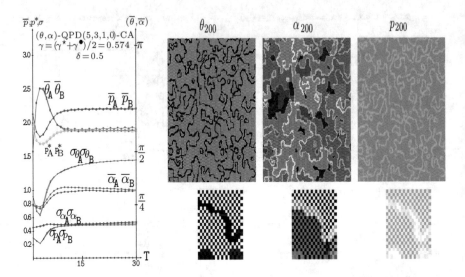

Fig. 10.3 A simulation in the QPD(5, 3, 1, 0) scenario of the left frame of Fig. 10.2. Far left frame: Dynamics up to $T = 30$. Right: Patterns at $T = 200$. Increasing grey levels indicate increasing values in the patterns

Figure 10.3 takes care of one simulation in the $\delta = 0.5$ left frame scenario of Fig. 10.2 at γ in the center of the $(\gamma^\star, \gamma^\bullet)$ transition interval. Its far left frame shows the dynamics up to $T = 30$ of mean payoffs and parameters, the standard deviations (σ) of these magnitudes, and the mean-field payoff approaches. It demonstrates that the dynamics induced by the imitation of the best paid mate implemented across this book actuates in a straightforward manner, so that the permanent regime is achieved very soon. This in fact applies not only in the context considered in this figure, but in a general manner, regardless of the game and conditions under scrutiny. Thus, iterating up to $T = 200$ may be excessive, less iterations would suffice to reach stable configurations. Thus, the simulations from now on will be run up to $T = 100$. In the patterns at $T = 200$ shown in Fig. 10.3, increasing grey levels indicate increasing values of parameter and payoffs, both in the whole 200×200 patterns (upper) and in their zooms of the 20×20 central part shown below. The whole patterns exhibit a kind of *patchwork* aspect, where irregular *borders* separate $(\hat{Q}, \hat{D})$ and $(\hat{D}, \hat{Q})$ clusters. This is particularly enhanced in the zoom of the θ parameter, where two $\hat{D}(\theta = \pi)$-$\hat{Q}(\theta = 0)$ regions are separated by a black spot (*border*) formed by defectors $(\theta = \pi)$. As the players in the *borders* are defectors mostly surrounded by defectors, they get a low payoff, which is reflected in clear border cells in the payoff pattern (far right). The spatial structure of the patterns in the γ transition interval, far from the initial random configuration and of the fixed point reached with low or high γ, explains why the mean-field estimations of the payoffs (p^*) differ from the actual ones $(\overline{p})$ in the simulations with γ in the $(\gamma^\star, \gamma^\bullet)$ transition interval.

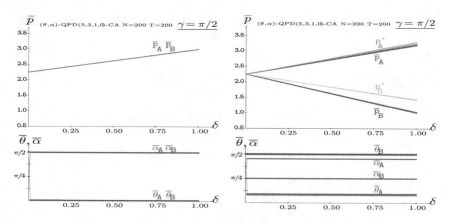

Fig. 10.4 The QPD(5, 3, 1, 0) with $\gamma = \pi/2$ and variable δ in five spatial simulations at T = 200. Left: The two players update strategies. Right: Only player A updates strategies

In the QPD(5, 3, 1, 0) spatial simulations of Fig. 10.4, the entanglement factor is fixed to its maximum, i.e., $\gamma = \pi/2$, whereas δ is the free parameter. In the left frame of Fig. 10.4, both players update their strategies. The updating leads to the $\{\hat{Q}, \hat{Q}\}$ pair, and consequently $p_{A,B}^{\hat{Q},\hat{Q}} = (1 - \delta)K + \delta R$, that in PD(5, 3, 1, 0) game, varies from $p_{A,B}^{\hat{Q},\hat{Q}}(\delta = 0.0) = K = 2.25$ up to $p_{A,B}^{\hat{Q},\hat{Q}}(\delta = 1.0) = 3.0$. In the right frame of Fig. 10.4, only player A updates its strategies, whereas player B remains *passive*, so that in a mean-field analysis, he is to be assigned the Middle-level strategy $\hat{M} = U(\pi/2, \pi/4)$. In turn, player A resorts to a strategy not far from $\hat{Q}$ as $\overline{\alpha}_A$ appears closely under $\pi/2$, and $\overline{\theta}_A$ not far from zero. It is $\Pi_f^{\hat{Q},\hat{M}}(\gamma) = \frac{1}{2}\begin{pmatrix} 1 - \frac{1}{2}\sin^2\gamma & 1 - \sin^2\gamma \\ \sin^2\gamma & \frac{1}{2}\sin^2\gamma \end{pmatrix}$, so that $\Pi_f^{\hat{Q},\hat{M}}(\gamma = \pi/2) = \frac{1}{2}\begin{pmatrix} \frac{1}{2} & 0 \\ 1 & \frac{1}{2} \end{pmatrix}$. Consequently, $p_A^{\hat{Q},\hat{M}}(\pi/2, \delta) = (1 - \delta)K + \delta\frac{1}{2}(\frac{1}{2}(R + P) + \mathfrak{T})$, $p_B^{\hat{Q},\hat{M}}(\pi/2, \delta) = (1 - \delta)K + \delta\frac{1}{2}(\frac{1}{2}(R + P) + S)$. In the QPD(5, 3, 1, 0) context of Fig. 10.4 it is: $p_A^{\hat{Q},\hat{M}}(\pi/2, \delta) = \frac{9}{4} + \frac{5}{4}\delta$, $p_B^{\hat{Q},\hat{M}}(\pi/2, \delta) = \frac{9}{4} - \frac{5}{4}\delta$, with $p_{A,B}^{\hat{Q},\hat{M}}(\pi/2, 0) = K = 2.25$, $p_A^{\hat{Q},\hat{M}}(\pi/2, 1) = 14/4 = 3.50$, $p_B^{\hat{Q},\hat{M}}(\pi/2, 1) = 4/4 = 1.00$. The latter values are not far from the actual mean payoffs in the simulations in the right panel of Fig. 10.4, where $\overline{p}_A(\delta = 1) = 3.2$, $\overline{p}_B(\delta = 1) = 1.01$.

In the QHD(5, 3, 0, 1)-CA simulations of Fig. 10.5, mutual $\hat{Q}$ arises, as it does in the QPD, at the intersection of $p_A^{\hat{D},\hat{Q}}$ and $p_A^{\hat{Q},\hat{Q}}$, thus at $\gamma^\bullet = \arcsin\left(\sqrt{\frac{\mathfrak{T} - R = 5 - 3}{\mathfrak{T} - S = 5 - 1}} = \frac{1}{2}\right) = \pi/4$. But at variance with what happens in the PD, mutual defection is never in NE, not even for low entanglement. As a result, in the QHD, below $\gamma^\bullet$ the mean payoff and parameter pattern graphs reflect what happens in the transition interval of the QPD: They exhibit a noisy aspect, and high spatial effects induce mean-field

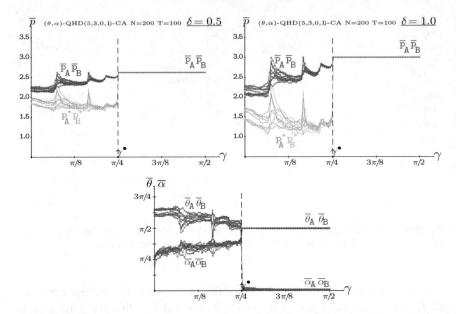

Fig. 10.5 The QHD(5, 3, 0, 1) with variable γ in five spatial simulations at T = 100. Above: Mean payoffs and their standard deviations at $\delta = 0.5$ (left) and $\delta = 1.0$ (right). Below: Mean quantum parameters

approaches that underestimate the actual mean payoffs. The mean payoffs trend to the payoff achieved with mutual $\hat{Q}$ as γ increases from $K = 2.25$ at $\gamma = 0.0$. The standard deviations (σ) of the actual payoffs decrease in a fairly smooth way as γ increases in the two scenarios of Fig. 10.5, albeit they plummet to zero at $\gamma = \gamma^{\bullet}$. As pointed out before, the graphs of the mean quantum parameters are unaffected by δ, therefore only one quantum parameters frame is shown below in Fig. 10.5.

10.2.2   Both γ and δ Variable

This subsection deals with a spatial simulation run up to $T = 100$ with both γ and δ variable in the PD, HD, SD and BOS games. Both γ and δ have been sampled in one hundred equidistant points.

Figure 10.6 shows the mean payoffs of the player A in a spatial simulation of the QPD(5, 3, 1, 0) (left frame) and the QHD(5, 3, 0, 1) (right frame) games. In the QPD (left) it stands out how the ($\gamma^{\star}$, $\gamma^{\bullet}$) transition interval is unaffected by δ and the mean-field approaches underestimate the actual mean payoffs in said transition interval. In the QHD (right) the transition to mutual $\hat{Q}$ at $\gamma^{\bullet}$ is fairly smooth, as explained when commenting Fig. 10.5. The mean-field approaches underestimate the actual payoffs before $\gamma^{\bullet} = \pi/4$ in the QHD game as do in the transition interval in the QPD. With

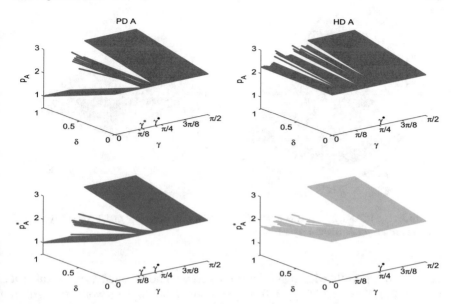

Fig. 10.6 Mean payoffs of the player A in a spatial simulation at T = 100 of the QPD(5, 3, 1, 0) (left) and the QHD(5, 3, 0, 1) (right). Upper: Actual mean payoffs. Lower: Mean-field payoffs

no entanglement and full δ in the QHD(5, 3, 0, 1) it is $\overline{p}_A(0.0, 1.0) = 2.28$, close to the average of the payoffs achieved in the three NE in the HD classical game: $(5+1+5/3)/3 = 2.56$.

The mentioned greater values of the actual mean payoffs compared to the mean-field approaches seems to indicate that spatialization boosts the payoffs of both players in the PD and the HD. This is so largely as in the way reported in the seminal paper [5] in the classical PD scenario. This article considers only pure Cooperators and Defectors arranged in a 2D lattice that evolve following the imitation of the best neighbour rule. As a result of the local interaction in the lattice, it is found that Cooperation is not fully discarded, but survives in a no negligible proportion of the cells. Provided that the temptation is not high. Otherwise, Defection will fully occupy the lattice.

Figure 10.7 shows the mean payoffs of a spatial simulation of the QSD(3, 2, 1, −1) game. For $\delta = 0.0$, the players get the average of the elements of their payoff matrices, i.e., $\overline{p}_A = 0.25$, $\overline{p}_B = 1.5$ regardless of γ. In the QSD game, $\{\hat{Q}, \hat{Q}\}$ is the only pair in NE for $\gamma > \gamma^{\bullet} = \pi/4$. This is so because, (i) $p_B^{\hat{Q}\hat{Q}} = 2$ and $p_B^{\hat{Q}\hat{D}} = 3\cos^2\gamma + \sin^2\gamma = 3 - 2\sin^2\gamma$, so that $p_B^{\hat{Q}\hat{Q}} > p_B^{\hat{Q}\hat{D}}$ for $\gamma > \gamma^{\bullet} = \pi/4$, and (ii) $p_A^{\hat{Q}\hat{Q}} = 3$ and $p_A^{\hat{D}\hat{Q}} = -1\sin^2\gamma - \cos^2\gamma = -1$, so that $p_A^{\hat{Q}\hat{Q}} > p_A^{\hat{D}\hat{Q}}$ $\forall\gamma$. Incidentally, the QHD(5, 3, 0, 1) and the QSD(3, 2, 1, −1) games share the same critical $\gamma^{\bullet} = \pi/4$. Consequently, the Samaritan player A overrates the beneficiary player B if $\gamma > \pi/4$, with $\overline{p}_A(\gamma > \pi/2, 1.0) = 3.0$, $\overline{p}_B(\gamma > \pi/2, 1.0) = 2.0$. Opposite to this, if $\gamma > \pi/4$ the beneficiary player B overrates the Samaritan player A, which

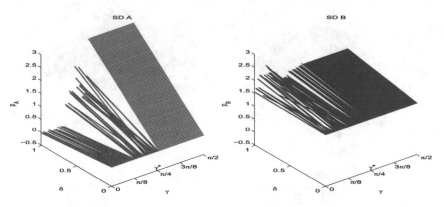

Fig. 10.7 The QSD(3, 2, 1, −1) in a spatial simulation with variable γ and δ. Mean payoffs at T = 100. Left: Player A, Right: Player B

gets negative payoffs with high δ. In particular, with no entanglement and full δ it is $\overline{p}_A(0.0, 1.0) = -0.028$, $\overline{p}_B(0.0, 1.0) = 1.549$, close to the payoffs achieved in the NE in the classical SD(3, 2, 1, −1) game given in the Sect. 1.2, i.e., $p_A = -0.2$, $p_B = 1.50$. Unexpectedly after studying the QHD game, no particular spatial effects arise in the QSD(3, 2, 1, −1) game before $\gamma^{\bullet} = \pi/4$, so that the mean-field payoffs approach very well the actual mean payoffs, and consequently they have been not included in Fig. 10.7. As an example, in the particular case of no entanglement and full δ just mentioned, it is $p_A^*(0.0, 1.0) = -0.013$, $p_B^*(0.0, 1.0) = 1.564$.

Figure 10.8 shows the mean payoffs of a spatial simulation of the QBOS(5, 1) game, where, as due, it is $\overline{p}_A(\gamma, 0.0) = \overline{p}_B(\delta, 0) = (5 + 1)/4 = 1.5$. With no entanglement and $\delta = 1$ it is $\overline{p}_A(0.0, 1.0) = 2.27$ and $\overline{p}_B(0.0, 1.0) = 2.64$, very close to the average of the payoffs achieved in the three NE in the classical BOS(5,1) game: $(5+1+5/6)/3 = 2.28$. It is remarkable that the original formulation of the EWL model is somehow biased toward the player B (*female*) in the BOS game. This is so as with middle-level election of the quantum parameters ($\theta = \pi/2, \alpha = \pi/4$) it is $\pi_{11} = \frac{1}{4}\cos^2 \gamma$, $\pi_{22} = \frac{1}{4}(1 + \sin \gamma)^2$. Thus, in the QBOS game with middle-level election of the parameters it is: $p \begin{Bmatrix} A \\ B \end{Bmatrix} = \begin{Bmatrix} R \\ r \end{Bmatrix} \frac{1}{4}\cos^2 \gamma + \begin{Bmatrix} r \\ R \end{Bmatrix} \frac{1}{4}(1 + \sin \gamma)^2$.

Finally, as $(1 + \sin \gamma) > \cos \gamma$ in $\gamma \in [0, \pi/2]$, it is $p_B > p_A$. As a reflect of this bias, in Fig. 10.8 it turns out that $\overline{p}_B > \overline{p}_A$ in simulations with very high entanglement. So for example with maximum entanglement, it is $\overline{p}_A(\pi/2, 1.0) = 2.47$, $\overline{p}_B(\pi/2, 1.0) = 3.01$. But opposite to this, with low entanglement it turns out that $\overline{p}_B < \overline{p}_A$, with maximal advantage of player B nearly before $\gamma = \pi/8$ with $\delta = 1.0$, where $\overline{p}_B = 4.25, \overline{p}_A = 1.25$. Both payoffs equalize at $\gamma = 3\pi/8$ in the $\delta = 1.0$ scenario [6]. Spatial effects are particularly important in the BOS game. Thus, with maximum entanglement it is $p_A^*(\pi/2, 1.0) = 0.97 < 2.45$, $p_B^*(\pi/2, 1) = 4.83 > 2.47$. The structures that emerge in the spatial simulations of the BOS game show a

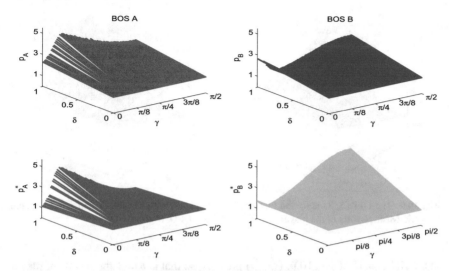

Fig. 10.8 The QBOS(5, 1) in a spatial simulation with variable γ and δ. Mean payoffs at T = 100. Left: Player A, Right: Player B. Upper: Actual Mean payoffs. Lower: Mean-field approaches

maze-like aspect described in [7], that are not comparable to the *patchworks* found in the QPD spatial simulations as that shown in Fig. 10.3.

10.3 Games on Random Networks

In the games on networks we deal with in this section, all the premises stated for the spatial games studied in the preceding Sect. 10.2 are preserved, except that the players are connected at random, without any spatial structure.

The payoffs of player A in a simulation of the QPD and the QHD games on a random network in Fig. 10.9 are to be compared to those in a spatial simulation shown in Fig. 10.6. Regarding the QPD(5, 3, 1, 0) game it is to be remarked that the mutual defection regime persists beyond the $\gamma^{\bullet} = 0.685$ landmark featuring the spatial simulation, and that the behaviour of the payoffs in the transition interval is rather erratic: some simulations render low payoff, other ones render high payoffs, in particular those just after the mutual defection regime that rocket up to around 4.0. In the spatial simulations of the QPD, the payoffs become stabilized in the transition interval (e.g., in Fig. 10.2) because both $(\hat{Q}, \hat{D})$ and $(\hat{D}, \hat{Q})$ coexist (as explained when commenting Fig. 10.3). In contrast to this, in the network simulations of the QPD, in the transition interval either $(\hat{Q}, \hat{D})$ or $(\hat{D}, \hat{Q})$ prevails. If $(\hat{Q}, \hat{D})$ prevails, player A gets a high payoff (and player B gets a low payoff), and the opposite happens if $(\hat{D}, \hat{Q})$ prevails. The just described considerations about the CA-NW contrast in the behaviour of the payoffs in the transition interval of the QPD, apply for the QHD before the emergence of the $(\hat{Q}, \hat{Q})$ pair in NE. Thus, in the network QHD simulation

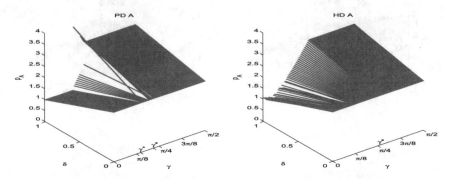

Fig. 10.9 The QPD(5, 3, 1, 0) and QHD(5, 3, 0, 1) games with variable γ and δ in a random network. Mean payoffs of the player A at T = 100. Left: QPD, Right: QHD

in the right panel of Fig. 10.9, $(\hat{Q}, \hat{D})$ prevails, so that at $\delta = 1$ the payoff of player A grows as $p_A = S + (\mathfrak{T} - S) \sin^2 \gamma = 1 + 4 \sin^2 \gamma$, from $p_A = 1$ at $\gamma = 0$ up to $p_A = 3$ at $\gamma = \pi/2$, whereas the (not shown in Fig. 10.9) payoff of player B decays at $\delta = 1$ as $p_B = \mathfrak{T} - (\mathfrak{T} - S) \sin^2 \gamma = 5 - 2 \sin^2 \gamma$, from $p_A = 5$ at $\gamma = 0$ up to $p_A = 3$ at $\gamma = \pi/2$. Anyhow, the behaviour of the network simulations in the QHD before the emergence of mutual $\hat{Q}$ depends on the initial random assignments of the (θ, α) parameters, so that $(\hat{D}, \hat{Q})$ instead of $(\hat{Q}, \hat{D})$ may emerge before $\gamma^{\bullet}$. Let us conclude the comments regarding the QPD and QHD stressing that simulations on networks are influenced by the initial conditions, whereas the spatialization (immediate previous section) induces the stabilization in the behaviour of the asymptotic strategies and consequently in the payoffs, so that the asymptotic results are free of the initial conditions stated in the simulation. This is so largely as in the seminal paper [5] dealing with the spatial simulation of the classical PD game.

The general form of both surface payoffs for the QSD(3, 2, 1, −1) in Fig. 10.10 is very similar to those in Fig. 10.7, albeit two relevant differences are to be remarked. First, the critical value of the entanglement $\gamma^{\bullet}$ that indicates the emergence of $(\hat{Q}, \hat{Q})$ in NE emerges in Fig. 10.10 close after the $\pi/4$, but not exactly at this middle level of gamma as happens in the spatial simulations of Fig. 10.7. Second, before $\gamma^{\bullet}$ both players get lower payoffs in the simulations in networks, particularly the Samaritan player A that gets payoffs around -0.4 in simulations with full δ and $\gamma < \gamma^{\bullet}$.

The general form of the surface payoffs of both QBOS(5, 1)-players in Fig. 10.11 notably differs from to that in Fig. 10.8. In the network simulations in Fig. 10.11 the discontinuities in the values of the payoffs that affect in Fig. 10.8 only to player A with low entanglement, appear in Fig. 10.11 affecting to both players and almost in the whole range of variation of γ. As a rule, with low entanglement, player A gets higher payoffs than player B which in turn tends to receive higher payoffs when the entanglement increases (this largely as in the spatial simulations of Fig. 10.8). The general payoff features observed in the simulation shown in Fig. 10.11 are preserved with different initial random assignments of the (θ, α) parameters, though the details

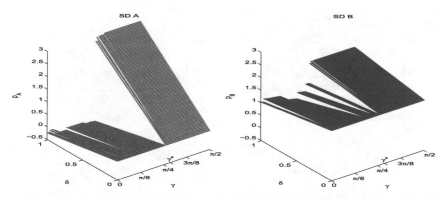

Fig. 10.10 The QSD(3, 2, 1, −1) with variable γ and δ in a random network. Mean payoffs at T = 100. Left: Player A, Right: Player B

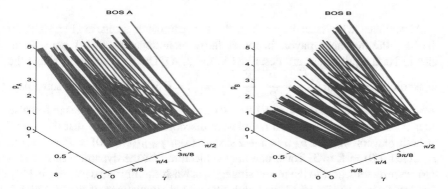

Fig. 10.11 The QBOS(5, 1) with variable γ and δ in a random network. Mean payoffs at T = 100. Left: Player A, Right: Player B

are altered in a detectable way when varying said initial conditions, as shown in [8] at $\delta = 1.0$ with five different random assignments of the (θ, α) parameters.

10.4 Three Quantum Parameter Strategies

In this section, both players follow general SU(2) strategies, i.e., three-parameter (3P) strategies with the β parameter active in the U structure given in Eq. (2.3).

Figures 10.12 and 10.13 deal with spatial simulations of the QPD(5, 3, 1, 0) game in their left frames, and the QHD(5, 3, 0, 1) game in the right frames when the players are allowed to follow 3P strategies. Figure 10.12 concerns only player A in a simulation with both γ and δ variable, whereas Fig. 10.13 concerns both players in five simulations with fixed $\delta = 1.0$.

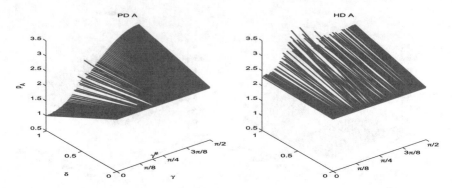

Fig. 10.12 The spatial QPD(5, 3, 1, 0) and QHD(5, 3, 0, 1) games with three quantum parameter strategies. Mean payoffs at T = 100 with variable γ and δ. Left: QPD, Right: QHD

At variance with what happens in the two parameter scenario (Fig. 10.2), in the 3P-QPD the mean payoff increases fairly monotonically. In [9] pure strategies in Nash equilibrium are described in the 3P-QPD($\delta = 1$) scenario below the threshold $\gamma^{\#} = \arcsin\left(\sqrt{\dfrac{P-S}{\mathfrak{T}+P-R-S}} = \dfrac{1}{3}\right) = 0.612$, providing the same payoffs for both players: $p_{A,B} = P + (R-P)\sin^2\gamma$. This equation applies in fact also after $\gamma^{\#}$ in the simulation, with no relevant discontinuity at $\gamma^{\#}$, so that the payoff of both players well fit $p_{A,B} = 1 + 2\sin^2\gamma$, from $p_{A,B}(0.0, 1.0) = P = 1$ up to $p_{A,B}(1.0, 1.0) = R = 3$. Thus, one may conjecture that the dynamics in this scenario resorts somehow to the mixed strategies in Nash equilibrium described in [10]. The left frames of Fig. 10.13 deal with five spatial simulations of the 3P-QPD(5, 3, 1, 0)($\delta = 1.0$) game with variable γ at $T = 100$. Its upper-left frame shows the actual mean and mean-field payoffs, and its lower-left frame the mean values of the quantum parameters. In this scenario, both $\overline{\theta}_A$ and $\overline{\theta}_B$ are set to $\pi/2$ (which makes α irrelevant), except in the proximity of $\gamma^{\#}$ where some turbulence is observed. With low entanglement, the β parameter of both players is not far from its middle value, and set to exactly $\pi/4$ with high entanglement. In short, the $\mathfrak{Q} = U(\pi, \alpha, \pi/4) = \dfrac{1}{\sqrt{2}}\begin{pmatrix} 0 & 1+i \\ 1-i & 0 \end{pmatrix}$ strategy dominates the scene in the 3P-QPD. The pair $(\mathfrak{Q}, \mathfrak{Q})$ generates the joint probability distribution $\Pi^{\mathfrak{Q},\mathfrak{Q}} = \begin{pmatrix} \sin^2\gamma & 0 \\ 0 & \cos^2\gamma \end{pmatrix}$, and consequently in the 3P-QPD(δ=1) scenarios of Figs. 10.12 and 10.13, the aforementioned $p_{A,B} = P + (R-P)\sin^2\gamma = 1 + 2\sin^2\gamma$ equalitarian payoff emerges in the simulation.

In contrast with what happens in the 3P-QPD, the right frames of Figs. 10.12 and 10.13 show that the $(\mathfrak{Q}, \mathfrak{Q})$ pair only emerges with high entanglement in the 3P-QHD. Thus, the lower-right frame of Fig. 10.13 indicates that, although the $\overline{\alpha}$ and $\overline{\beta}$ parameters oscillate close the middle value $\pi/4$ (as in the 3P-QPD), the $\overline{\theta}$ parameter values only approach π with very high γ. The upper-right frame of Fig. 10.13 in turn

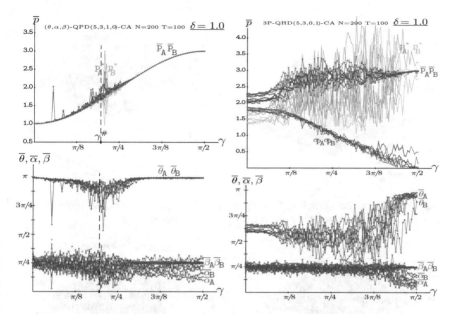

Fig. 10.13 The $\delta = 1.0$ QPD and QHD with three quantum parameter strategies and variable γ. Five spatial simulations at T = 100. Left: QPD(5, 3, 1, 0). Right: QHD(5, 3, 0, 1). Upper: Mean payoffs. Lower: Mean quantum parameters

indicates that spatial effects (absent in the 3P-QPD) induce mean-field payoffs that vary erratically, mainly underestimating the actual mean payoffs in the 3P-QHD($\delta =$ 1), while the standard deviations (σ) of the actual payoffs decrease in a fairly smooth way down to zero as γ increases. Also noticeable is that in agreement with what happens in the 2P-QHD simulation in Fig. 10.6, the actual mean payoffs of both players in the 3P-QHD depart at $\gamma = 0.0$ from $\overline{p}(0.0, 1.0) \simeq 2.25$, close to 2.56, i.e., to the average of the payoffs achieved in the three NE in the HD classical game, instead of from $\overline{p}(0.0, 1.0) = 1.0$ as happens in the 3P-QPD.

Figure 10.14 deals with a spatial simulation of the QSD(3, 2, 1, −1) game with three quantum parameter strategies, instead of with two parameters as in Fig. 10.10. At variance with what happens in the latter, the actual mean payoffs of both players monotonically increase their values as the entanglement increases Fig. 10.14, with no emergence of the ($\mathfrak{Q}$, $\mathfrak{Q}$) pair in NE. In the $\delta = 1.0$ scenario, the payoff of player A increases from approximately zero up to approximately 1.5, and that of player B from approximately 1.5 up to approximately 2.25. Thus, player B overrates player A all along the γ variation, albeit in a lower degree as γ grows. It is shown in [11] that in the $\delta = 1.0$ QSD(3, 2, 1, −1) scenario, heavy spatial effects arise, given rise to a rather erratic behaviour of the mean-field estimations.

The surface payoffs of both players in the spatial 3P-QBOS(5, 1) simulation shown in Fig. 10.15 appear to be much smoother than those in the 2P-QBOS(5, 1) simulations in Fig. 10.8. Also relevant is that both players achieve similar payoffs in

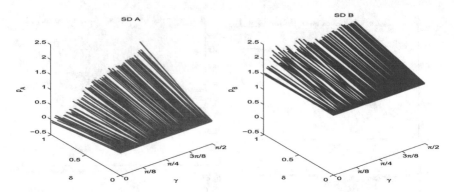

Fig. 10.14 The spatial QSD(3, 1, 1, −1) game with three quantum parameter strategies. Mean payoffs at T = 100 with variable γ and δ. Left: Player A, Right: Player B

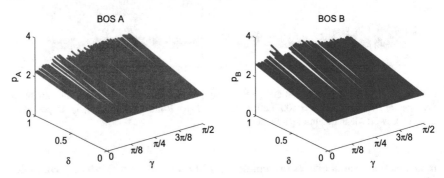

Fig. 10.15 The spatial QBOS(5, 1) game with three quantum parameter strategies. Mean payoffs at T = 100 with variable γ and δ. Left: Player A, Right: Player B

Fig. 10.15, at variance with the trends favouring to one of the players depending on the entanglement level that emerge in Fig. 10.8. The strong spatial effects that arise in the simulations in the $\delta = 1.0$ scenario of Fig. 10.15 are described in [12].

References

1. Werner, R.F.: Quantum states with Einstein-Podolsky-Rossen correlation admitting a hidden variable model. Phys. Rev. A **40**(8), 4277–4281 (1989)
2. Melo-Luna, C., Susa, C.E., Ducuara, A.F., Barreiro, A., Reina, J.H., D.: Quantum locality in game strategy. Scientific Reports **7**, 44730 (2017)
3. Nawaz, A.: Werner-like States and Strategic Form of Quantum Games (2013). arXiv preprint arXiv:1307.5508
4. Alonso-Sanz, R.: Collective quantum games with Werner-like states. Physica A **510**, 812–827 (2018)
5. Nowak, M.A., May, R.A.: Evolutionary games and spatial chaos. Nature **359**, 826–829 (1992)

6. Alonso-Sanz, R.: Variable entangling in a quantum battle of the sexes cellular automaton. ACRI-2014, LNCS, 8751, 125–135 (2014)
7. Alonso-Sanz, R.: A quantum battle of the sexes cellular automaton. Proc. R. Soc. A **468**, 3370–3383 (2012)
8. Alonso-Sanz, R.: On collective quantum games. Quantum Inf. Process. **18**, 64 (2019)
9. Du, J.F., Li, H., Xu, X.D., Zhou, X., Han, R.: Phase-transition-like behaviour of quantum games. J. Phys. A: Math. Gen. **36**(23), 6551–6562 (2003)
10. Eisert, J., Wilkens, M.: Quantum games. J. Modern Opt. **47**(14–15), 2543–2556 (2000)
11. Alonso-Sanz, R., Situ, H.: A quantum Samaritan's dilemma cellular automaton. Royal Soc. Open Sci. **4**(6), 863–160669 (2017)
12. Alonso-Sanz, R.: On a three-parameter quantum battle of the sexes cellular automaton. Quantum Inf. Process. **12**(5), 1835–1850 (2013)

Chapter 11
Imperfect Information and Imprecise Payoffs

The dynamics of a spatial quantum formulation of the iterated Battle of the Sexes game with imperfect information is studied in the first part of this chapter. The second part of this chapter deals with games with imprecise payoffs. In both scenarios, the games are played with variable entangling in a cellular automata manner. The effect of spatial structure is assessed in fair and unfair scenarios.

11.1 The Battle of the Sexes with Imperfect Information

This section deals with the Battle of the Sexes (BOS), with players unsecure about the true desire of the opponent. This kind of games are referred to as games with incomplete information or Bayesian games.

The dynamics of the BOS with imperfect information will be analyzed in the context of quantum approach implemented with the EWL protocol, in which one the expected payoffs of both players are,

$$
P \begin{Bmatrix} \sigma \\ \varphi \end{Bmatrix} = \begin{Bmatrix} R \\ r \end{Bmatrix} |\psi_1|^2 + \begin{Bmatrix} r \\ R \end{Bmatrix} |\psi_4|^2 \tag{11.1}
$$

The particular features of the QBOS described in Sect. 2.1 will be relevant along this section, in particular the bias towards the female player under the 2P model commented regarding Eq. (2.6).

Recall that the three NE in the conventional (or classical) BOS are achieved with: $x = y = 0$, $x = y = 1$, and $(x = R/R + r, y = 1 - x)$. If $y = 1 - x$ it is

© Springer Nature Switzerland AG 2019
R. Alonso-Sanz, *Quantum Game Simulation*, Emergence, Complexity
and Computation 36, https://doi.org/10.1007/978-3-030-19634-9_11

$P_{\sigma} = P_{Q}$, with maximum $p^+ = (R+r)/4$ for $x=y=1/2$. Also relevant is that if $\pi_{11} = \pi_{22} = \pi$, both players get the same payoff $p^= = \pi(R+r)$. So that if $\pi > 1/4$ the players get equalitarian payoffs that are not accessible in the uncorrelated strategies scenario, with maximum $p^+ = (R+r)/2$ if $\pi = 1/2$.

11.1.1   Imperfect Information in the BOS

Let us suppose that the male player is unsecure whether the female player prefers to join him or prefers to avoid him. He believes that the two $\mathcal{Q}$-types deserve the probabilities $(\lambda_{\sigma}, 1 - \lambda_{\sigma})$ [1, 2]. In such as scenario, the payoff matrices of the two $\mathcal{Q}$-types are given in Table 11.1.

If neither player knows whether the other wants to meet on not, with the male player assigning $(\lambda_{\sigma}, 1 - \lambda_{\sigma})$, and the female assigning the probabilities $(\lambda_{Q}, 1 - \lambda_{Q})$, the payoff matrices are given in Table 11.2.

Table 11.1 Payoff matrices for an unsecure male player

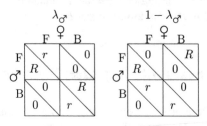

Table 11.2 Payoff matrices in an imperfect information BOS game

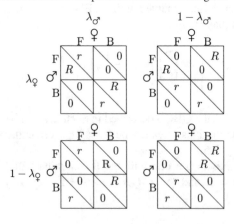

Table 11.3 A general assignment of probabilities in the scenario of Table 11.2

		$\female_1$		$\female_2$	
$\male_1$	ϵ_1	ϵ_2	ϵ_5	ϵ_6
	ϵ_3	ϵ_4	ϵ_7	ϵ_8
$\male_2$	ϵ_9	ϵ_{10}	ϵ_{13}	ϵ_{14}
	ϵ_{11}	ϵ_{12}	ϵ_{15}	ϵ_{16}

A general assignment of probabilities in the scenario of Table 11.2 is shown in Table 11.3.

The layout probabilities given in Table 11.3 together with the belief probabilities generate the expected payoffs in the BOS:

$$
\begin{aligned}
P_{\male_1} &= \lambda_{\male}\left(R\epsilon_1 + 0\epsilon_2 + 0\epsilon_3 + r\epsilon_4\right) + (1 - \lambda_{\male})\left(R\epsilon_5 + 0\epsilon_6 + 0\epsilon_7 + r\epsilon_8\right)\\
P_{\male_2} &= \lambda_{\male}\left(0\epsilon_9 + R\epsilon_{10} + r\epsilon_{11} + 0\epsilon_{12}\right) + (1 - \lambda_{\male})\left(0\epsilon_{13} + R\epsilon_{14} + r\epsilon_{15} + 0\epsilon_{16}\right)\\
P_{\female_1} &= \lambda_{\female}\left(r\epsilon_1 + 0\epsilon_2 + 0\epsilon_3 + R\epsilon_4\right) + (1 - \lambda_{\female})\left(r\epsilon_9 + 0\epsilon_{10} + 0\epsilon_{11} + R\epsilon_{12}\right)\\
P_{\female_2} &= \lambda_{\female}\left(0\epsilon_5 + R\epsilon_6 + r\epsilon_7 + 0\epsilon_8\right) + (1 - \lambda_{\female})\left(0\epsilon_{13} + R\epsilon_{14} + r\epsilon_{15} + 0\epsilon_{16}\right)
\end{aligned}
\tag{11.2}
$$

The probabilities in Table 11.3 must obey to the normalization constraints:

$$
\sum_{i=1}^{4} \epsilon_i = \sum_{i=5}^{8} \epsilon_i = \sum_{i=9}^{12} \epsilon_i = \sum_{i=13}^{16} \epsilon_i = 1, \text{ and in the EPR experiment inspired approach}
$$

followed in [1] also to the locality constraints: $\epsilon_1 + \epsilon_2 = \epsilon_5 + \epsilon_6, \epsilon_3 + \epsilon_4 = \epsilon_7 + \epsilon_8; \epsilon_9 + \epsilon_{10} = \epsilon_{13} + \epsilon_{14}; \epsilon_{11} + \epsilon_{12} = \epsilon_{15} + \epsilon_{16}; \epsilon_1 + \epsilon_3 = \epsilon_9 + \epsilon_{11}; \epsilon_2 + \epsilon_4 = \epsilon_{10} + \epsilon_{12}; \epsilon_5 + \epsilon_7 = \epsilon_{13} + \epsilon_{15}; \epsilon_6 + \epsilon_8 = \epsilon_{14} + \epsilon_{16}$.

At variance with the general procedure of constructing the layout of ϵ-probabilities adopted in [1], in this work every submatrix in Table 11.3 will be equalized to Π, thus:

$$
\begin{pmatrix} \epsilon_1 & \epsilon_2 \\ \epsilon_3 & \epsilon_4 \end{pmatrix} = \begin{pmatrix} \epsilon_5 & \epsilon_6 \\ \epsilon_7 & \epsilon_8 \end{pmatrix} = \begin{pmatrix} \epsilon_9 & \epsilon_{10} \\ \epsilon_{11} & \epsilon_{12} \end{pmatrix} = \begin{pmatrix} \epsilon_{13} & \epsilon_{14} \\ \epsilon_{15} & \epsilon_{16} \end{pmatrix} = \Pi = \begin{pmatrix} \pi_{11} & \pi_{12} \\ \pi_{21} & \pi_{22} \end{pmatrix}
\tag{11.3}
$$

Consequently,

$$
\begin{aligned}
P_{\male_1} &= \lambda_{\male}\left(R\pi_{11} + r\pi_{22}\right) + (1 - \lambda_{\male})\left(R\pi_{11} + r\pi_{22}\right) = R\pi_{11} + r\pi_{22}\\
P_{\male_2} &= \lambda_{\male}\left(R\pi_{12} + r\pi_{21}\right) + (1 - \lambda_{\male})\left(R\pi_{12} + r\pi_{21}\right) = R\pi_{12} + r\pi_{21}\\
P_{\female_1} &= \lambda_{\female}\left(r\pi_{11} + R\pi_{22}\right) + (1 - \lambda_{\female})\left(r\pi_{11} + R\pi_{22}\right) = r\pi_{11} + R\pi_{22}\\
P_{\female_2} &= \lambda_{\female}\left(R\pi_{12} + r\pi_{21}\right) + (1 - \lambda_{\female})\left(R\pi_{12} + r\pi_{21}\right) = R\pi_{12} + r\pi_{21}
\end{aligned}
\tag{11.4}
$$

Please note in Eq. (11.4) that $p_{\male_2} = p_{\female_2}$.

In this study, every male/female player will be featured by a sole payoff constructed as:

$$p_{\sigma} = \lambda_{\female} p_{\sigma_1} + (1 - \lambda_{\female}) p_{\sigma_2}$$
$$p_{\female} = \lambda_{\sigma} p_{\female_2} + (1 - \lambda_{\sigma}) p_{\female_2} \tag{11.5}$$

11.1.2   The Spatialized Imperfect QBOS

Again in the simulations in this section, the ♂ players A and ♀ players B are distributed in a two-dimensional lattice, and interact in the simple cellular automaton manner introduced in Sect. 3.1. Therefore, every player (i, j) will adopt the quantum parameters $(\theta_{k,l}^{(T)}, \alpha_{k,l}^{(T)}, \beta_{k,l}^{(T)})$) and the belief probability $\lambda_{k,l}^{(T)}$ of his nearest-neighbor mate (including himself) that received the highest payoff [3]. As customary across the simulations of this book, the quantum parameters and belief probabilities are initially assignment at random. Thus, initially: $\bar{\lambda} \simeq 1/2$, and $\bar{\theta} \simeq \pi/2 = 1.57$, $\bar{\alpha} \simeq \bar{\beta} \simeq \pi/4 = 0.78$.

Figure 11.1 deals with the two-parameter QBOS(5,1)-CA with imperfect information in both the male (red), and the female (blue) players. Five initial configurations evolve up to $T = 200$. It is remarkable that for any given value of γ, the variation of the initial quantum parameter configurations only alters minor details of the evolving dynamics, so that the main features of the long-term patterns are preserved.

The bias towards the female player in the model with maximal entangling (pointed out in Sect. 2.1) becomes apparent in Fig. 11.1 for high γ values, specifically for $\gamma > 3\pi/8$. In contrast to this, with $\gamma < 3\pi/8$ the male player overrates the female player, in a dramatic way for low values of γ, with the extreme scenario for the classical $\gamma = 0$ in which case $\overline{p}_{\sigma} = 5, \overline{p}_{\female} = 1$. The behaviour for low values of γ notably differs from this one in the perfect information scenario of Fig. 3.11, where: (i) both mean payoffs are similar close to $\gamma = 0$, (ii) the initial increase of γ leads

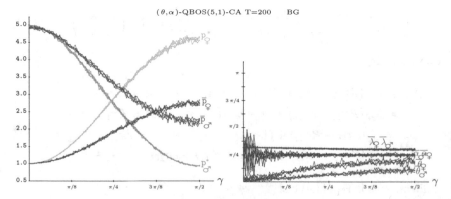

Fig. 11.1   The two-parameter QBOS(5,1)-CA with imperfect information in both the male (red), and the female (blue) players. Variable entangling factor γ. Five simulations at $T = 200$. Left: Mean payoffs $(\overline{p})$ and mean-field payoffs (p^*). Right: Mean belief and quantum parameters

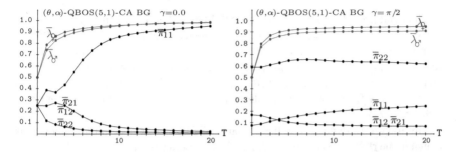

Fig. 11.2 Dynamics up to $T = 20$ of the mean probabilities in Π and the mean belief probabilities in a simulation of Fig. 11.1. Left: $\gamma = 0.0$. Right: $\gamma = \pi/2$

to a dramatic bifurcation in favor of the male player reaching a peak close to $(5, 1)$ when γ approaches $\pi/8$, (*iii*) before, but close, this value of γ, both mean payoffs approach as γ increases, reaching fairly equal values by $\gamma \simeq 3\pi/8$, much in as similar way as in the imperfect information scenario studied here.

Figure 11.2 shows the dynamics of the mean probabilities in Π and the mean belief probabilities in a simulation of Fig. 11.1 up to $T = 20$. In the left panel, as $\gamma = 0.0$ it is initially $\overline{\pi}_{11} \simeq \overline{\pi}_{12} \simeq \overline{\pi}_{21} \simeq \overline{\pi}_{22} \simeq 1/4$. In the right panel, with maximum entangling, the highest initial $\overline{\pi}$ is $\overline{\pi}_{22}$ and the lowest is $\overline{\pi}_{11}$, accordingly to the bias to the female player in the 2P-model. In the $\gamma = 0.0$ scenario (left panel), $\overline{\pi}_{11}$ notably increases its value in the evolving dynamics, whereas the remaining $\overline{\pi}$ values plummet to zero, what supports the distant $\overline{p}_{\sigma^{\!}} = 5, \overline{p}_{\varphi} = 1$ payoffs for $\gamma = 0$ in Fig. 11.1. In the $\gamma = \pi/2$ scenario (right panel), $\overline{\pi}_{22}$ remains over $\overline{\pi}_{11}$ in the evolving dynamics, though approaching their values (at $T = 200$ it is $\overline{\pi}_{11} = 0.381$, $\overline{\pi}_{22} = 0.555$), which explains why $\overline{p}_{\varphi}$ overrates $\overline{p}_{\sigma^{\!}}$ for $\gamma = \pi/2$ in Fig. 11.1, albeit not in a great extent. Further evolution of the simulation treated in the right panel of Fig. 11.2 is shown in Fig. 11.3.

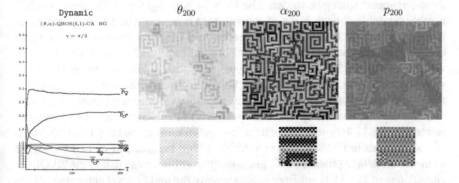

Fig. 11.3 Simulation of a 2P QBOS(5,1)-CA in the scenario of Fig. 11.1 with $\gamma = \pi/2$. Far Left: Evolving mean parameters and payoffs. Center: Parameter patterns at $T = 200$. Far Right: Payoff pattern at $T = 200$. Increasing grey levels indicate increasing parameter pattern values

The evolution of the mean belief probabilities in Fig. 11.2 is highly remarkable. In the unentangled scenario (left) both $\overline{\lambda}$ parameter values rocket towards 1.0 from its initial 0.5 random assignment. In the fully entangled (right) the drift to 1.0 is also very noticeable, albeit not so dramatic as in the unentangled model and showing a small lag in both $\overline{\lambda}$ values. The drift to 1.0 of the beliefs of both types of players becomes a key feature in the simulations in this study (see also Figs. 11.4 and 11.9 at this respect).

The long-term dynamics in the scenario of Fig. 11.1 whose initial dynamics is shown in the right panel of Fig. 11.2, i.e., dealing with maximum $\gamma = \pi/2$, is shown in Fig. 11.3. The far left panel of the figure shows the evolution up to $T = 200$ of the mean values across the lattice of the quantum parameters and actual payoffs. The $\overline{\theta}$ values evolve initially in opposite direction from their mean values, close to $\pi/2 = 1.57$, a $\overline{\theta}_{\male}$ decreases whereas $\overline{\theta}_{\female}$ grows, but the ulterior dynamics depletes the $\overline{\theta}_{\female}$ value, so that both $\overline{\theta}$ parameter values decrease slowly in a fairly parallel extent, and at $T = 200$ it is $\overline{\theta}_{\female} = 0.64$; $\overline{\theta}_{\male} = 0.36$. The dynamics of both $\overline{\alpha}$ parameter values is smoother compared to that of $\overline{\theta}$, so from the initial $\overline{\alpha} \simeq \pi/4 = 0.78$, at $T = 200$ it is: $\overline{\alpha}_{\male} = 0.81 \simeq \overline{\alpha}_{\female} = 0.82$. The parameter dynamics in Fig. 11.3 quickly drives the $\overline{p}$ mean payoffs to distant values, so that from approximately $T = 100$ they become fairly stabilized, reaching at $T = 200$ the values: $\overline{p}_{\female} = 2.80, \overline{p}_{\male} = 2.14$.

Figure 11.3 also shows the snapshots of the parameter and payoff patterns at $T = 200$, both for the full lattice and its zoomed 23×23 central part. Rich maze-like structures may be appreciated in the parameter patterns, particularly in the α parameter pattern, whereas in the θ and payoff patterns, the maze-like structure appears fairly fuzzyfied.

The spatial heterogeneity in the parameter values make it difficult the capacity of the mean-field payoff estimations computed as proposed in Sect. 11.1.2 to approach the actual mean payoffs. Maze-like structures in the parameter patterns emerge as soon as γ takes off, and consequently the mean-field p^* values tend to poorly reflect the actual mean payoffs of both kinds of players as γ grows (see Fig. 11.1).

Figure 11.4 shows the asymptotic results in the scenario of Fig. 11.1, but in the three-parameter strategies model. The plots in both figures of the mean payoffs notably resemble, though in the 3P model, there is not any bias favoring the female player, not even for high values of the entangling factor (as shown in Fig. 11.1), so that for $\gamma > 3\pi/8$ both mean payoffs are not significantly different. At variance with what happens in the 3P-QBOS with perfect information in Fig. 3.20, in the 3P-QBOS with imperfect information in Fig. 11.4, the mean payoffs are notably influenced by the variation of γ, in such a way that the mean payoffs converge from (5, 1) with no entanglement towards approximately 2.5 with full entanglement. The parameter patterns in Fig. 11.4 show rich structures (not shown here) as soon as γ takes off. Much like as happens in Fig. 3.20, but in a different way, because in Fig. 11.4 the aspect of the mean-field approaches with low entanglement resemble that in the 2P-QBOS simulations of Fig. 11.1, but from approximately the middle level entanglement the mean-field approach of the female turns out rather erratic. Unexpectedly, the form of the payoff graphs (both actual mean and mean-field) in the left frame of Fig. 11.4 very much resembles that in the 3P-QRBOS quantum-relativistic scenario of Fig. 8.19.

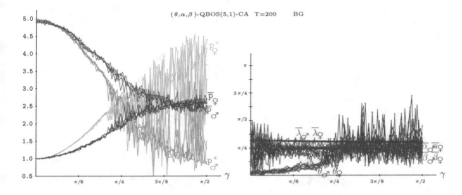

Fig. 11.4 The three-parameter QBOS(5,1)-CA with imperfect information in both the male (red), and the female (blue) players. Variable entangling factor γ. Five simulations at $T = 200$. Left: Mean payoffs $(\overline{p})$ and mean-field payoffs (p^*). Right: Mean belief and quantum parameters

   Please, recall that the maximum equalitarian payoff in the uncorrelated context is $p^+ = (R+r)/4 = 1.5$, whereas the mean payoffs in the simulations of a three-parameter QBOS(5,1)-CA with high entanglement factor in Fig. 11.4 reach values over 2.5, not far from the maximum feasible equalitarian payoff $p^= = (R + r)/2 = 3.0$.

   The main features of the plots in the Figs. 11.1 and 11.4 have been checked to be preserved with the BOS parameters (2, 1), (4, 1) and (6, 1). In these three cases, the general form of the $\overline{p}$ versus γ plots is that shown here for (5,1), reaching long-term $\overline{p}$ values not far from $(R + r)/2$. Thus, with $R = 2$ around 1.5, with $R = 4$ around 2.25, and with $R = 6$ around 3.25.

Unfair Contests

Let us assume the unfair situation: A type of players is restricted to classical strategies $\tilde{U}(\theta, 0)$, whereas the other type of players may use quantum $\hat{U}(\theta, \alpha)$ ones [4].

   Figure 11.5 shows the asymptotic payoffs in five simulations of an unfair two-parameter quantum versus classical players in a BOS(5,1)-CA with variable entanglement factor γ. The case of the quantum male (red) versus classical female (blue), shown in the left panel, behaves as expected a priori: the quantum player overrates the classical one. That is so in an extreme way, $\overline{p}_{\sigma} = 5, \overline{p}_{\varphi} = 1$, regardless of the γ parameter value. Incidentally, in the BOS(5,1)-CA unfair scenario with perfect information of Fig. 4.6 the full advantage of the quantum male player is achieved only from $\gamma \simeq \pi/8$. At variance with what happens in the quantum male versus classical female contest, the case of the classical male (red) versus quantum female (blue) shown in the right panel of Fig. 11.5 shows a dramatic dependence of the γ: The classical male overrates the quantum female for $\gamma < \pi/2$, where the opposite ordering of payoffs (that expected a priori) appears for $\gamma > \pi/2$,

   The parameter patterns in the scenarios of Fig. 11.5 are free of spatial effects, so that the mean-field approximation operates fairly well. Thus, in the left panel the

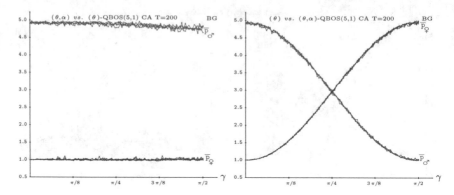

Fig. 11.5 Long-term payoffs in five simulations of a 2P-quantum versus classical BOS(5,1)-CA with imperfect information. Variable entanglement factor γ. Left: Quantum male (red) versus classical female (blue). Right: Classical male (red) versus quantum female (blue)

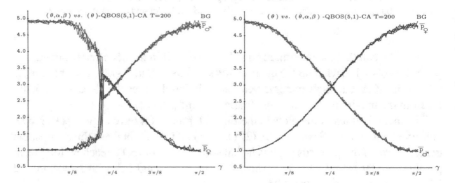

Fig. 11.6 Long-term payoffs in five simulations of a three-parameter quantum versus classical BOS(5,1)-CA with imperfect information. Variable entanglement factor γ. Left: Quantum male versus classical female. Right: Classical male versus quantum female

(θ, α) parameters approaches to zero in the long-term, so that π_{11}^* approach to 1.0, and consequently $p_{\male}^* \simeq R = 5$, $p_{\female}^* \simeq r = 1$, in agreement with the graphs of the actual mean payoffs. The *scissors*-like form of the graphs emerging in the right panel of Fig. 11.5 will be commented when dealing with Fig. 11.9.

Figure 11.6 shows the asymptotic payoffs in five simulations of an unfair three-parameter quantum versus classical players in a BOS(5,1)-CA with variable entanglement factor γ. The graphs in the right panel of Fig. 11.6 (classical male vs. quantum female) are similar to those in the same scenario with two-parameters shown in the right panel of Fig. 11.5. At variance with this, the graphs in the left panel of Fig. 11.6 (quantum male vs. classical female) show that the advantage of the quantum male in the 3P scenario is not as evident as in the 2P case (left panel of Fig. 11.6). This advantage appears notably weakened circa the middle $\gamma = \pi/4$, with the semiquantum female overrating the full quantum male in the proximity of this value.

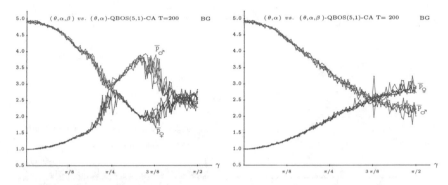

Fig. 11.7 Long-term payoffs in five simulations of (θ, α, β) vs. (θ, α) players in a BOS(5,1)-CA with imperfect information. Variable entanglement factor γ. Left: Quantum male versus semi-quantum female. Right: Semi-quantum male versus quantum female

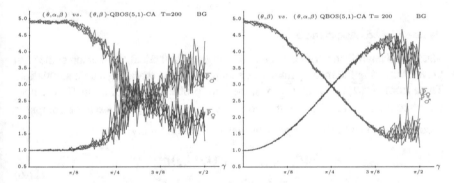

Fig. 11.8 Long-term payoffs in five simulations of (θ, α, β) vs. (θ, β) players in a BOS(5,1)-CA with imperfect information. Variable entanglement factor γ. Left: Quantum male versus semi-quantum female. Right: Semi-quantum male versus quantum female

Figures 11.7 and 11.8 consider the case of three-parameter quantum versus semi-quantum contests, semi-quantum referring to players that may not implement one of the *quantum* parameters, either α or β, but have access to the other one, β or α respectively (in both cases with the θ parameter operative). Let us first remark that these figures reveal that the roles of the α and β parameters in the three-parameter model are fairly different.

The aspect of the graphs in Figs. 11.7 and 11.8 is roughly comparable to those in Fig. 11.6. Thus, in the left panels the advantage of the full quantum male decreases as γ grows up to approximately $\gamma = \pi/4$, then it is recovered after this value of the entanglement factor, though in a different way Figs. 11.7 and 11.8 compared to the robust way shown in Fig. 11.6. The *scissors*-like form of the graphs in the right panel of Fig. 11.6 is much altered in Figs. 11.7 and 11.8. It is also remarkable that the graphs in Fig. 11.8 show a noisy aspect for high γ values which is not found in the other plots in this study.

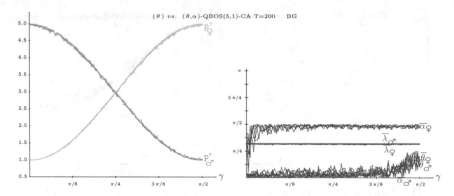

Fig. 11.9 Mean-field payoffs (left) and mean quantum parameters (right) in the 2P-QBOS(5,1)-CA unfair scenario of the right panel of Fig. 11.5 Five simulations at $T = 200$

Mean-Field Approach.

Mean-field payoffs may be computed in a single hypothetical two-person game with players adopting the mean parameters appearing in the spatial dynamic simulation. Thus, from $U_{\male}^*\left(\overline{\theta}_{\male}, \overline{\alpha}_{\male}\right)$, $U_{\female}^*\left(\overline{\theta}_{\female}, \overline{\alpha}_{\female}\right)$, the joint probability distribution Π^* is calculated according to (11.4), then the payoffs $p_{\male_1}^*$, $p_{\male_2}^*$, $p_{\female_1}^*$, $p_{\female_2}^*$ are to be computed according to (11.4) using the probabilities in Π^*, and finally:

$$
\begin{aligned}
p_{\male}^* &= \overline{\lambda}_{\female} p_{\male_1}^* + (1 - \overline{\lambda}_{\female}) p_{\male_2}^* \\
p_{\female}^* &= \overline{\lambda}_{\male} p_{\female_2}^* + (1 - \overline{\lambda}_{\male}) p_{\female_2}^*
\end{aligned}
\tag{11.6}
$$

Figure 11.1 shows in its left panel the mean-field (p^*) payoffs (left panel) computed according to (11.6) in the 2P-QBOS(5,1)-CA scenario of Fig. 11.1. Spatial effects are in the origin of the divergence of the actual mean payoffs and their mean-field approximation. The divergence increases as γ increases, as the values of both types of actual payoffs are approached compared to their corresponding mean-field approximations. The same effect operates in the 3P scenario of Fig. 11.4.

Figure 11.9 shows the mean-field payoffs computed according to (11.6) (left panel) and the long-term mean parameters (right panel) in the unfair 2P scenario of the right panel of Fig. 11.5. The absence of spatial effects in this scenario explains why the mean-field approximation fits almost perfectly to the actual mean payoffs.

According to Eqs. (2.8a) and (2.8d), in the conventional (non-CA) 2P-EWL quantum model with perfect information, if $\alpha_{\male} + \alpha_{\female} = \pi/2$ and $\theta_{\male} = \theta_{\female} = 0$, the diagonal elements of Π reduce to,

$$
\pi_{11} = \cos^2 \gamma, \quad \pi_{44} = \sin^2 \gamma
\tag{11.7}
$$

And consequently, in the scenario of Eq. (11.7) in the QBOS game it is :

$$P\left\{\begin{matrix}\sigma\\\varphi\end{matrix}\right\} = \left\{\begin{matrix}R\\r\end{matrix}\right\} \cos^2 \gamma + \left\{\begin{matrix}r\\R\end{matrix}\right\} \sin^2 \gamma \qquad (11.8)$$

The drift to 1.0 of the beliefs of both types of players found in this study, revealed in particular in the right panels of Figs. 11.1, 11.4 and 11.9, allows to consider the just above formulas obtained in the conventional (non-CA) 2P-EWL perfect information context to explain the mean-field approximations.

The Eq. (11.8) generate the kind of *scissors* shape found in some graphs dealing with the BOS game. Thus, for example: (*i*) in the scenario of Fig. 11.9, where (left panel) $\bar{\alpha}_{\sigma} = 0$, $\bar{\alpha}_{\varphi} \simeq \pi/2$, and both values of $\bar{\theta}$ are low, and (*ii*) in the scenario of Fig. 11.1, where (left panel) $\bar{\alpha}_{\sigma} \simeq \bar{\alpha}_{\varphi} \simeq \pi/4$, and both values of $\bar{\theta}$ are not high. It is remarkable that in the full entangling 2P-EWL model, strategies with $\alpha_{\sigma} + \alpha_{\varphi} = \pi/2$ and $\theta_{\sigma} = \theta_{\varphi} = 0$ are in Nash equilibrium [5, 6].

Let us conclude this section by pointing out that the main finding reported on the QBOS-CA dynamics with imperfect information is that of recovering the perfect information scenario, i.e., the original formulation of the BOS game, in which coordination is desirable. But the transition from a random assignment of beliefs into the belief in what both players want to meet the other induces quantum parameter and payoff patterns that notably differ from those achieved with the perfect information. Moreover, the intrinsic asymmetry of the BOS game induces a very interesting modification in the player's payoffs when varying the entangling factor γ. Games with imperfect information embedded in cellular automata are to be studied when assigning joint probabilities inspired in the EPR experiment [1, 7].

11.2 Games with Imprecise Payoffs

In real world problems, the exact values of the payoffs are usually unknown. Hence the need to study the games with uncertain payoffs arises. A way to deal with uncertainties associated with payoffs is to use the concept of fuzzy numbers [8]. Many studies on fuzzy game theory have been reported in the literature [9–11].

In this study we will consider only the simple type of symmetric triangular fuzzy (STF) numbers $\tilde{M} = T(m, d)$, where m denotes the center and d the deviation, i.e., with boundaries $m \pm d$. Multiplication and addition of symmetric triangular fuzzy numbers using the extension principle are considerable simple. Thus, let $\tilde{M} = T(m, d)$ and $\tilde{M}' = T(m', d')$ be two symmetric triangular fuzzy numbers, and c be a positive real number. Then,

$$cT(m, d) = T(cm, cd) \qquad (11.9a)$$
$$T(m, d) + T(m', d') = T(m + m', d + d') \qquad (11.9b)$$

Hence, the expected payoffs of a game are also symmetric triangular fuzzy numbers. Let us consider the case of the PD game with fuzzy payoffs $(\mathfrak{T}, d_{\mathfrak{T}})$, (R, d_R), (P, d_P), (S, d_S). In this FPD scenario, the expected fuzzy payoffs induced by $\boldsymbol{\varPi}$ will be,

$$\tilde{p}_A = T(\pi_{11}R + \pi_{12}S + \pi_{21}\mathfrak{T} + \pi_{22}P, \pi_{11}d_R + \pi_{12}d_S + \pi_{21}d_{\mathfrak{T}} + \pi_{22}d_P) \tag{11.10a}$$

$$\tilde{p}_B = T(\pi_{11}R + \pi_{12}\mathfrak{T} + \pi_{21}S + \pi_{22}P, \pi_{11}d_R + \pi_{12}d_{\mathfrak{T}} + \pi_{21}d_S + \pi_{22}d_P) \tag{11.10b}$$

Ranking fuzzy numbers becomes necessary in decision making when alternatives are fuzzy numbers. Various methods for ranking fuzzy numbers have been proposed [12–14]. Here we will adopt that proposed in [15], so that the average index featuring a symmetric triangular fuzzy number $\tilde{M} = T(m, d)$, will be,

$$\hat{M} = m + (\lambda - 0.5)d, \quad 0 \le \lambda \le 1 \tag{11.11}$$

Where λ ponders the pessimism-optimism degree. The index (11.11) is symmetric respect to λ, with $M(\lambda = 0.0) = m - 0.5d$, $M(\lambda = 0.5) = m$, and $M(\lambda = 1.0) = m + 0.5d$.

Please, note that index (11.11) is very close to the simple index constructed as the convex linear combination of the two extremes of a STF number, i.e., $\hat{M} = (1 - \lambda)(m - d) + \lambda(m + d) = m + (2\lambda - 1)d$.

Quantum games with fuzzy payoffs will be here referred to as QF games. The general expressions of the index payoffs according to (11.11) of the D and Q strategy interactions in a QFPD game with $(\mathfrak{T}, d_{\mathfrak{T}})$, (R, d_R), (P, d_P), (S, d_S) STF payoffs are,

$$\hat{p}_A^{DD} = \hat{p}_B^{DD} = P + (\lambda - 0.5)d_P \tag{11.12a}$$

$$\hat{p}_A^{QD} = \hat{p}_B^{QD} = (1 - \sin^2)(S + (\lambda - 0.5)d_S) + \sin^2(\mathfrak{T} + (\lambda - 0.5)d_{\mathfrak{T}}) = \tag{11.12b}$$

$$= S + (\lambda - 0.5)d_S + (\mathfrak{T} - S + (\lambda - 0.5)(d_{\mathfrak{T}} - d_S))\sin^2$$

$$\hat{p}_A^{DQ} = \hat{p}_B^{DQ} = \sin^2(S + \lambda d_S - 0.5) + (1 - \sin^2)(\mathfrak{T} + (\lambda - 0.5)d_{\mathfrak{T}}) = \tag{11.12c}$$

$$= \mathfrak{T} + (\lambda - 0.5)d_{\mathfrak{T}} - (\mathfrak{T} - S + (\lambda - 0.5)(d_{\mathfrak{T}} - d_S))\sin^2$$

$$\hat{p}_A^{QQ} = \hat{p}_B^{QQ} = R + (\lambda - 0.5)\lambda d_R \tag{11.12d}$$

In the QFPD$((5, 3), (3, 1.5), (1, 0.5), (0, 0.0))$ game with $\lambda = 0.8$, Eq. (11.12) become,

$$\hat{p}_A^{DD} = \hat{p}_B^{DD} = 1 + (0.8 - 0.5)0.5 = 1.15 \tag{11.13a}$$

$$\hat{p}_A^{QD} = \hat{p}_B^{QD} = 0 + (0.8 - 0.5)0 + (5 - 0 + (0.8 - 0.5)(3 - 0))\sin^2 = 5.90\sin^2 \tag{11.13b}$$

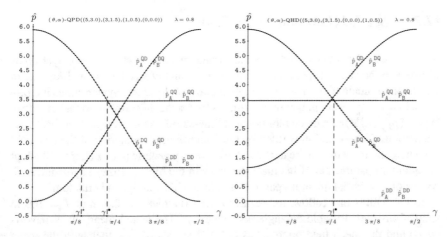

Fig. 11.10 The $\hat{p}$ index payoffs of the D and Q strategy interactions in the QFPD((5, 3), (3, 1.5), (1, 0.5), (0, 0.0)) and QFPD((5, 3), (3, 1.5), (0, 0.0), (1, 0.0.5)) games with $\lambda = 0.8$. Left: QFPD, Right: QFHD

$$\hat{p}_A^{DQ} = \hat{p}_B^{DQ} = 5 + (0.8 - 0.5)3 - (5 - 0 + (0.8 - 0.5)(3 - 0))\sin^2 = 5.90 - 5.90\sin^2 \tag{11.13c}$$

$$\hat{p}_A^{QQ} = \hat{p}_B^{QQ} = 3 + (0.8 - 0.5)1.5 = 3.45 \tag{11.13d}$$

In the QFHD((5, 3), (3, 1.5), (0, 0.0), (1, 0.5)) game with $\lambda = 0.8$ Eq. (11.12),

$$\hat{p}_A^{DD} = \hat{p}_B^{DD} = 0 + (0.8 - 0.5)0.0 = 0.0 \tag{11.14a}$$

$$\hat{p}_A^{QD} = \hat{p}_B^{QD} = 1 + (0.8 - 0.5)0.5 + (5 - 1 + (0.8 - 0.5)(3 - 0.5))\sin^2 = 1.15 + 4.75\sin^2 \tag{11.14b}$$

$$\hat{p}_A^{DQ} = \hat{p}_B^{DQ} = 5 + (0.8 - 0.5)3 - (5 - 1 + (0.8 - 0.5)(3 - 0.5))\sin^2 = 5.90 - 4.75\sin^2 \tag{11.14c}$$

$$\hat{p}_A^{QQ} = \hat{p}_B^{QQ} = 3 + (0.8 - 0.5)1.5 = 3.45 \tag{11.14d}$$

Figure 11.10 shows the $\hat{p}$ index payoffs given by the above Eqs. (11.13)–(11.14). The $\gamma^{\bullet}$ landmark in the two frames of Fig. 11.10 is computed at the intersection of $\hat{p}_B^{DQ}$ and $\hat{p}^{QQ}$ whereas $\gamma^{\star}$ in the QFPD scenario (left frame) is computed at the intersection of $\hat{p}^{DD}$ and $\hat{p}_A^{QD}$. The general formulas of these γ landmarks are given in Eqs. 11.15 below, that generalize those given in Eq. 2.9 in the crisp payoffs context.

$$\gamma^{\star} = \arcsin\left(\sqrt{\frac{P - S + (\lambda - 0.5)(d_P - d_S)}{\mathfrak{T} - S + (\lambda - 0.5)(d_{\mathfrak{T}} - d_S)}}\right), \quad \gamma^{\bullet} = \arcsin\left(\sqrt{\frac{\mathfrak{T} - R + (\lambda - 0.5)(d_{\mathfrak{T}} - d_R)}{\mathfrak{T} - S + (\lambda - 0.5)(d_{\mathfrak{T}} - d_S)}}\right) \tag{11.15}$$

11.2.1   Spatial Quantum Fuzzy Games

As in the rest of this book, in the spatial simulations of this section the player-types alternate in the site occupation of a two-dimensional $N \times N$ lattice, and interact in a cellular automata (CA) manner, i.e., with uniform, local and synchronous interactions. In this way, every player plays with his four adjacent partners, so that the fuzzy payoff $(p_{i,j}^{(T)}, d_{i,j}^{(T)})$ of a given individual at time-step T is the average over these four interactions. The evolution is ruled by the (deterministic) imitation of the best paid mate neighbour, so that in the next generation, every generic player (i, j) will adopt the quantum parameters of his mate player with the highest fuzzy payoff index $\hat{p}$. As customary, initial quantum parameter values will be assigned at random.

In the figures that follow, the result of the dynamics are shown at $T = 100$ in simulations with variable entanglement factor γ. The actual mean fuzzy payoffs $(\overline{p}, \overline{d})$ and the mean-field approaches $(p^\star, d^\star)$ are shown together with the mean value of the quantum parameters $(\overline{\theta}, \overline{\alpha}, \overline{\beta})$

Two-Parameter Simulations.

In the 2P-QFPD((5, 3), (3, 1.5), (1, 0.5), (0, 0.0))-CA at $\lambda = 0.8$ simulations of Fig. 11.11, mutual defection arises below the lower $\gamma^\bullet = 0.475$ threshold and mutual $\hat{Q}$ beyond the higher $\gamma^\star = 0.700$ threshold. In the $(\gamma^\star, \gamma^\bullet)$ transition interval, where both $(\hat{Q}, \hat{D})$ and $(\hat{D}, \hat{Q})$ are in NE, the dynamics is highly conditioned by spatial effects, as explained below when commenting Fig. 11.13. The above reported thresholds in the fuzzy scenario are very close to those found in a scenario with crisp payoffs ($d_{\mathfrak{T}} = d_R = d_P = d_S = 0.0$), where $\gamma^\star = 0.464$, and $\gamma^\bullet = 0.685$. In the studied QFDD it is $\gamma^\star(\lambda = 0.0) = 0.481$, $\gamma^\bullet(\lambda = 0.0) = 0.641$, and $\gamma^\star(\lambda = 1.0) = 0.454$, $\gamma^\bullet(\lambda = 1.0) = 0.708$. It is noticeable that for $\gamma < \gamma^\star$, as both $\overline{\theta}$'s are selected to be π, both $\overline{\alpha}$'s are irrelevant, as $\cos(\pi/2) = 0$ annihilates the influence of α, whereas for $\gamma > \gamma^\star$, both $\overline{\theta}$'s are set to zero and both $\overline{\alpha}$'s to $\pi/2$, i.e., the $(\hat{Q}, \hat{Q})$ pair is stated.

In Fig. 11.11, the mean-field payoffs coincide with the actual mean payoffs with γ below $\gamma^\star$ and over $\gamma^\bullet$. But in the $(\gamma^\star, \gamma^\bullet)$ interval the mean-field payoff approaches underestimate the actual mean payoffs. The lack of coincidence of both the mean-field and actual mean payoffs reflects the emergence of quantum parameter patterns that impede an approach based on the mean values of said parameters. An example of this is given in Fig. 11.13 (and in Fig. 11.16 regarding the BOS game).

Figure 11.12 takes care of the dynamics of two simulations in the scenario of Fig. 11.11. The left frames concern with γ in the center of the $(\gamma^\star, \gamma^\bullet)$ transition interval, whereas the right frames concern with $\gamma = \pi/2$. With $\gamma = \pi/2$ (right frames), the dynamics induced by the imitation of the best paid mate implemented in this book actuates in a straightforward manner, so that the $\{\hat{Q}, \hat{Q}\}$ permanent regime is achieved very soon. As a result, the fuzzy payoff of mutual cooperation, i.e., (3,1.5), quickly emerges. The dynamics with $\gamma = (\gamma^\star + \gamma^\bullet)/2 = 0.579$ (left frames), the quantum parameters also stabilize very soon, but at variance with what happens at $\gamma = \pi/2$, the mean-field approaches underestimate the actual mean values of both p and d.

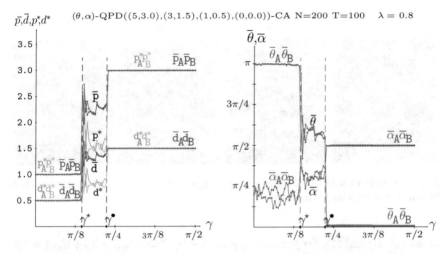

Fig. 11.11 The 2P-QFPD((5, 3), (3, 1.5), (1, 0.5), (0, 0.0)), $\lambda = 0.8$ game with variable γ in a spatial simulation at T = 100. Left: Mean fuzzy payoffs. Right: Mean quantum parameters

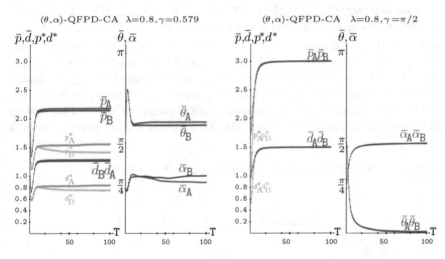

Fig. 11.12 Dynamics up to $T = 100$ in the QFPD scenario of Fig. 11.11. Left frames: $\gamma = (\gamma^* + \gamma^{\bullet})/2$. Right frames: $\gamma = \pi/2$

The patterns at $\gamma = 0.579$ in Fig. 11.13 show a kind of *patchwork* aspect, where irregular *borders* separate $(\hat{Q}, \hat{D})$ and $(\hat{D}, \hat{Q})$ clusters. The borders separating both types of clusters are formed by mutual defectors, i.e., with $\theta = \pi$ (black pixels in the far left patterns), thus providing low payoffs which is reflected in clear border cells in the payoff pattern. These spatial structures have been reported and analyzed in deep in previous works dealing with games with crisp payoffs [16]. Here we will note just that in the $\gamma = 0.579$ far left frame of Fig. 11.12 it is $p_A = p_B = 2.16$, $d_A = d_B = 1.27$ in the long term. These values are not far from the average of the payoff achieved

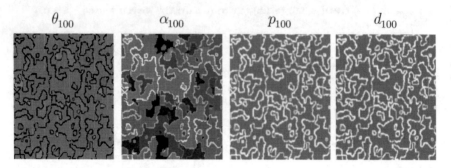

Fig. 11.13 Patterns in the QFPD scenario of Fig. 11.11 at $\gamma = (\gamma^* + \gamma^\bullet)/2$. Increasing grey levels indicate increasing values in the patterns

in the $(\hat{Q}, \hat{D})$ and $(\hat{D}, \hat{Q})$ games, i.e., $p = (5+0)/2 = 2.5$, $d = (3+0)/2 = 1.5$. The borders of mutual defection are responsible for the lower actual mean values of p and d found in the simulation.

In the two-parameter 2P-QFHD(5, 3), (3, 1.5), (0, 0.0), (1, 0.5))-CA at $\lambda = 0.8$ simulations of Fig. 11.14, mutual $\hat{Q}$ arises, as it does in the QFPD, at the intersection of $p_A^{\hat{D},\hat{Q}}$ and $p_A^{\hat{Q},\hat{Q}}$, thus at $\gamma^\bullet = 0.801$, very close to the threshold in a scenario with crisp payoffs $\gamma^\bullet = \pi/4 = 0.785$. In the studied QFHD it is $\gamma^\bullet(\lambda = 0.0) = 0.704$, and $\gamma^\bullet(\lambda = 1.0) = 0.804$. But at variance with what happens in the QFPD, mutual defection is never in NE, not even for low entanglement. Therefore, in the QFHD, below $\gamma^\bullet$ the mean payoff and parameter pattern graphs reflect what happens in the transition interval of the crisp QPD: They exhibit a noisy aspect, and high

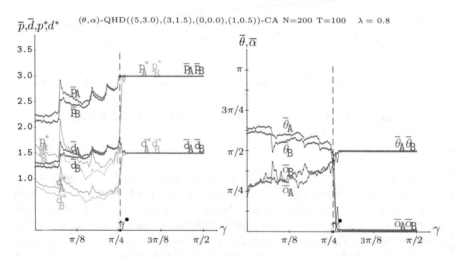

Fig. 11.14 The 2P-QFHD((5, 3), (3, 1.5), (0, 0.0), (1, 0.5)), $\lambda = 0.8$ game with variable γ in a spatial simulation at T = 100. Left: Mean fuzzy payoffs. Right: Mean quantum parameters

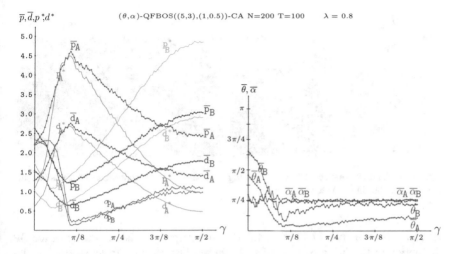

Fig. 11.15 The 2P-QFBOS((5, 3), (1, 0.5)), $\lambda = 0.8$ game with variable γ in a spatial simulation at T = 100. Left: Mean fuzzy payoffs. Right: Mean quantum parameters

spatial effects induce mean-field approaches that underestimate the actual mean and deviation payoffs.

In the 2P-QFBOS(5, 3), (1, 0.5))-CA at $\lambda = 0.8$ simulations of Fig. 11.15, the initial increase of the entangling factor from the classical $\gamma = 0$ context leads to a notable bifurcation of the mean payoffs in favor to the male player A, reaching a peak not far from ((5, 3), (1, 0.5)) when γ approaches $\pi/8$. Before, but close, this value of γ, both mean payoffs commence a smooth approach as γ increases, reaching fairly equal values by $\gamma \simeq 3\pi/8$. For $\gamma > 3\pi/8$, the ordering of payoffs reverses, and layer B overrates player A. The right frame of Fig. 11.15 shows that, (i) both $\overline{\alpha}_A$ and $\overline{\alpha}_B$ oscillate nearly $\pi/4$ after a noisy regime for low values of γ, (ii) both $\overline{\theta}_A$ and $\overline{\theta}_B$ initially decrease as γ increases, but close to $\pi/8$ both $\overline{\theta}'s$ commence to increase: $\overline{\theta}_B$ grows faster, reaching $\overline{\theta}_B \simeq \pi/4$ at $\gamma = \pi/2$, whereas $\overline{\theta}_A \simeq \pi/16$ with full entangling.

The patterns at $T = 100$ shown in Fig. 11.16, very much differ from that in Fig. 11.13. They show a kind of *maze* aspect that has been reported and analyzed in deep in previous works dealing with games with crisp payoffs [17–19]. Additionally, a quantification of the variability of the center payoffs is given in the left frame of Fig. 11.16 by means of the variation with γ of their standard deviations (σ). The σ of both payoffs initially plummets from circa 2.0 with no entanglement down to close to zero before $\gamma = \pi/8$, then they recover up to circa 0.95 with full entanglement. This scheme of variation of $\sigma(\gamma)$ applies also regarding payoff deviations, albeit this has not been shown in the figure in order not to overload it even to a larger extent.

Three Quantum Parameter Strategies.

In the 3P-QFPD((5, 3), (3, 1.5), (1, 0.5), (0, 0.0))-CA at $\lambda = 0.8$ simulations of Fig. 11.17, the structure of the fuzzy payoffs notably differs from that shown in its

θ_{100} α_{100} p_{100} d_{100}

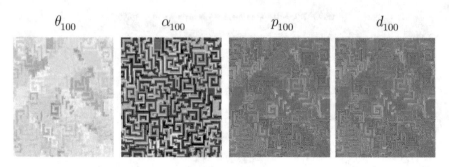

Fig. 11.16 Patterns in the QFBOS scenario of Fig. 11.11 at $\gamma = \pi/2$. Increasing grey levels indicate increasing values in the patterns

2P counterpart in Fig. 11.11. Particularly with respect to the similarity in the 3P simulation of the fuzzy payoffs of both players that increase fairly monotonically with γ without 'discontinuities'. Two main features characterize the parameter graphs in Fig. 11.17: (i) the $\overline{\theta}$ parameters drift to π which makes irrelevant the values of $\overline{\alpha}$, and (ii) the $\overline{\beta}$ parameters oscillate around $\pi/4$. Thus, in the mean-field approach, both players would adopt the strategy $\hat{\mathfrak{x}} = U(\pi, \alpha, \pi/4) = \frac{1}{\sqrt{2}} \begin{pmatrix} 0 & 1+i \\ 1-i & 0 \end{pmatrix}$. In the 3P scenario, it is proved that the pair $(\hat{\mathfrak{x}}, \hat{\mathfrak{x}})$ is in NE for γ below the threshold

$$\gamma^{\#} = \arcsin \left(\sqrt{\frac{P-S}{\mathfrak{T}+P-R-S}} = \frac{1}{3} \right) = 615 \, [5].$$ But in the 3P CA simulations of

Fig. 11.17 it seems that the pair $(\hat{\mathfrak{x}}, \hat{\mathfrak{x}})$ roughly dominates the scene regardless of γ. The pair $(\hat{\mathfrak{x}}, \hat{\mathfrak{x}})$ generates the joint distribution $\Pi^{\hat{\mathfrak{x}},\hat{\mathfrak{x}}} = \begin{pmatrix} \sin^2 \gamma & 0 \\ 0 & \cos^2 \gamma \end{pmatrix}$, and consequently both players get the same payoff $p^{\hat{\mathfrak{x}},\hat{\mathfrak{x}}} = P + (R-P)\sin^2 \gamma = 1 + 2\sin^2 \gamma$, and deviation $p^{\hat{\mathfrak{x}},\hat{\mathfrak{x}}} = d_P + (d_R - d_P)\sin^2 \gamma = 0.5 + \sin^2 \gamma$.

Figure 11.18 deals with 3P-QFHD((5, 3), (3, 1.5), (0, 0.0), (1, 0.5))-CA at $\lambda = 0.8$ simulations. In contrast with what happens in the 3P-QFPD simulations in Fig. 11.17, the right frame of Fig. 11.18 indicates that, although the $\overline{\alpha}$ and $\overline{\beta}$ parameters oscillate close the middle value $\pi/4$ (as in the 3P-QFPD), the $\overline{\theta}$ parameters do not significantly approach π, not even with very high γ. The left frame of Fig. 11.18 in turn indicates that spatial effects (absent in the 3P-QFPD) induce mean-field payoffs that vary fairly erratically. Also noticeable is that (much as it happens in the 2P-QFHD simulation in Fig. 11.11), at $\gamma = 0.0$ the fuzzy payoffs of both players are not far in the simulation from those computed as the average of the expected values in the three NE in the studied HD game. Namely, for the payoff centers: $\dfrac{1+5+15/9}{3} = 2.56$, and for the deviations: $\dfrac{0.5+3+8.5/9}{3} = 1.48$.

In the 3P-QFBOS(5, 3), (1, 0.5))-CA at $\lambda = 0.8$ simulations of Fig. 11.19, the structure of the fuzzy payoffs notably differs from that shown in its 2P counterpart in Fig. 11.15. Thus, at variance with what happens in the two parameter scenario

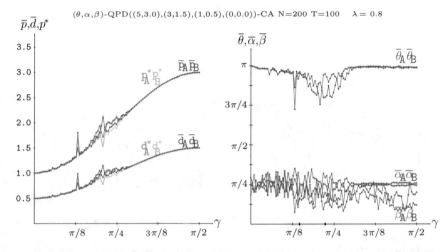

Fig. 11.17 The 3P-QFPD((5, 3), (3, 1.5), (1, 0.5), (0, 0.0)), $\lambda = 0.8$ game with variable γ in a spatial simulation at T = 100. Left: Mean fuzzy payoffs. Right: Mean quantum parameters

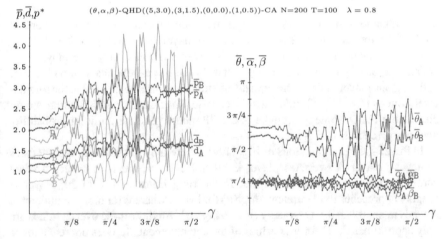

Fig. 11.18 The 3P-QFHD((5, 3), (3, 1.5), (0, 0.0), (1, 0.5)), $\lambda = 0.8$ game with variable γ in a spatial simulation at T = 100. Left: Mean fuzzy payoffs. Right: Mean quantum parameters

(Fig. 11.15), the center mean and deviation payoffs are not dramatically altered by the variation of γ. The overall effect of the increase of γ being a moderation in the variation of the $\overline{p}$ and $\overline{d}$ values that oscillate nearly 2.5 and 1.5 respectively. The spatial ordered structure induces a kind of self-organization effect, which allows to achieve fairly soon, approximately at $T = 20$, pairs of mean payoffs that are accessible only with correlated strategies in the two-person game. Recall that the maximum equalitarian payoff in the uncorrelated context is $p^+ = (R+r)/4 = 1.5$, whereas the center payoffs

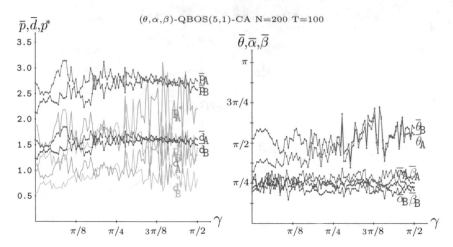

Fig. 11.19 The 3P-QFBOS$((5, 3), (1, 0.5))$, $\lambda = 0.8$ game with variable γ in a spatial simulation at T = 100. Left: Mean fuzzy payoffs. Right: Mean quantum parameters

in the simulations of Fig. 11.19 with high entanglement factor reach values over 2.5, not far from the maximum feasible equalitarian payoff $p^= = (R + r)/2 = 3.0$.

In the simulations that follow, the unfair scenario where only one player type updates his strategies is taken into account. In the mean-field analysis with unfair updating, the player that does not update his strategies (passive) is fixed to the middle-level strategy $\overline{U} = U(\pi/2, \pi/4)$. In this scenario, The Defection and Quantum strategies facing the $\overline{U}$ strategy produce in the 2P model generate the joint probability matrices given in Eqs. 5.1–5.2.

In the 2P-QFPD simulations of Fig. 11.20, where only the player A type updates strategies, player A overrates player B regardless of γ, much as expected when confronting active to passive players. In the right frame of Fig. 11.20, player A roughly approaches the $\hat{D}$ strategy (high θ) for low γ, whereas for high entanglement A tends to approach the $\hat{Q}$ strategy (low θ, high α). Anyhow these overall trends are only approximated: (i) for low values of the entanglement, $\overline{\theta}_A$ does not reach the π level featuring proper defection but remains close to approximately $7\pi/8$, and (ii) for high values of the entanglement, $\overline{\alpha}_A$ does not reach $\pi/2$ and $\overline{\theta}_A$ stays over zero.

According to the $\Pi^{D,\overline{U}}$ and $\Pi^{Q,\overline{U}}$ joint probability matrices given in Eqs. 5.1–5.2, it is: $\hat{p}_A^{D,\overline{U}} = \frac{1}{2}(\mathfrak{I} + P - \frac{1}{2}(\mathfrak{I} - S)\sin^2\gamma + \lambda(d_{\mathfrak{I}} + d_P - \frac{1}{2}(d_{\mathfrak{I}} - d_S)\sin^2\gamma))$,

$\hat{p}_A^{Q,\overline{U}} = \frac{1}{2}(R + S + (\mathfrak{I} - R\frac{1}{2} + P\frac{1}{2} - S)\sin^2\gamma + \lambda(d_R + d_S + (d_{\mathfrak{I}} - d_R\frac{1}{2} + d_P\frac{1}{2} - d_S)\sin^2\gamma))$. These index payoffs intersect at, The $p_A^{Q,\overline{U}}$ and $p_A^{D,\overline{U}}$ payoffs

equalize at $\gamma^\bullet = \arcsin\sqrt{\dfrac{2(\mathfrak{I} - R + P - S + \lambda(d_{\mathfrak{I}} - d_R + d_P - d_S))}{3\mathfrak{I} - R + P - 3S + \lambda(3d_{\mathfrak{I}} - d_R + d_P - 3d_S)}}$. In the

context of Fig. 11.20 it is $\gamma^\bullet = \arcsin\sqrt{0.474} = 0.760$, with $p_A^{D,\overline{U}}(\gamma^\bullet)$

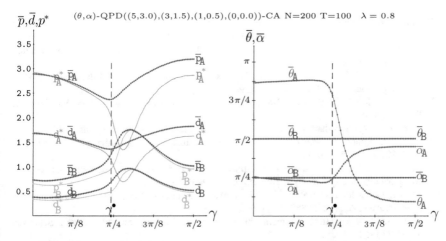

Fig. 11.20 The 2P-QFPD((5, 3), (3, 1.5), (1, 0.5), (0, 0.0)), $\lambda = 0.8$ game with variable γ in a spatial simulation at T = 100. Only player A updates strategy. Left: Mean fuzzy payoffs. Right: Mean quantum parameters

$= p_A^{Q,\overline{U}}(\gamma^\bullet) = 2.420$, and $d_A^{D,\overline{U}}(\gamma^\bullet) = d_A^{Q,\cdot}(\gamma^\bullet) = 1.402$. Said value of $\gamma^\bullet$ roughly marks the player A transition from nearly $\hat{D}$ to nearly $\hat{Q}$, so that the fuzzy payoffs behave at a first approach according to $\Pi^{D,M}$ before $\gamma^\bullet$ and to $\Pi^{Q,M}$ after $\gamma^\bullet$.

Quantum fuzzy game simulations with unfair strategy updating in the 3P-QFPD context as well as regarding the QFHD and QFBOS games are reported in [20].

Imprecise payoffs have been modelled in this section by means of symmetric triangular fuzzy numbers that are featured by an index that takes into account the degree of optimism of the players. It has been shown that this parameter does not notably alter the entanglement landmarks in the spatial evolving simulations. This is so particularly when dealing with the Prisoner's Dilemma and Hawk-Dove games; but also with the Battle of the Sexes (BOS) game, albeit the intricacies induced by having two social welfare solutions, make the BOS game more challenging to study with imprecise payoffs. Further work is planned regarding the features of the fuzzy payoffs. Thus, more general fuzzy numbers and ranking criteria are to be taken into consideration.

References

1. Iqbal, A., Chappell, J.M., Li, Q., Pearce, C.E.M., Abbott, D.: A probabilistic approach to quantum Bayesian games of incomplete information. Quantum Inf. Process. **13**(12), 2783–2800 (2014)
2. Situ, H.Z.: Quantum Bayesian game with symmetric and asymmetric information. Quantum Inf. Process. **14**, 1827–1840 (2015)
3. Alonso-Sanz, R.: A cellular automaton implementation of a quantum battle of the sexes game with imperfect information. Quantum Inf. Process. **14**(10), 3639–3659 (2015)

4. Flitney, A.P., Abbott, D.: Advantage of a quantum player over a classical one in 2×2 quantum games. Proc. R. Soc. Lond. A **459**(2038), 2463–2474 (2003)
5. Du, J.F., Li, H., Xu, X.D., Zhou, X., Han, R.: Phase-transition-like behaviour of quantum games. J. Phys. A: Math. Gen. **36**(23), 6551–6562 (2003)
6. Du, J.F., Xu, X.D., Li, H., Zhou, X., Han, R.: Entanglement playing a dominating role in quantum games. Phys. Lett. A **89**(1–2), 9–15 (2001)
7. Cheon, T., Iqbal, A.: Bayesian Nash Equilibria and Bell inequalities. J. Phys. Soc. Jpn. **77**, 024801 (2008)
8. Tanaka, K.: An Introduction to Fuzzy Logic for Practical Applications. Springer (1997)
9. Cunlin, L., Qiang, Z.: Nash equilibrium strategy for fuzzy non-cooperative games. Fuzzy Sets Syst. **176**(1), 46–55 (2011)
10. Maeda, T.: On characterization of equilibrium strategy of two-person zero-sum games with fuzzy payoffs. Fuzzy Sets Syst. **139**, 283–296 (2003)
11. Maeda, T.: On characterization of equilibrium strategy of bi-matrix games with fuzzy payoffs. J. Math. Anal. Appl. **251**, 885–896 (2000)
12. Chen, S.H.: Ranking fuzzy numbers with maximizing set and minimizing set. Fuzzy Sets Syst. **17**(2), 113–129 (1985)
13. Cheng, C.H.: A new approach for ranking fuzzy numbers by distance method. Fuzzy Sets and Syst. **95**(3), 307–317 (1998)
14. Liou, T.S., Wang, M.J.J.: Ranking fuzzy numbers with integral value. Fuzzy Sets Syst. **50**(3), 247–255 (1992)
15. de Campos Ibañez, L.M., Muñoz, A.G.: A subjective approach for ranking fuzzy numbers. Fuzzy Sets Syst. **29**(2), 145–153 (1989)
16. Alonso-Sanz, R.: On the effect of quantum noise in a quantum prisoner's dilemma cellular automaton. Quantum Inf. Process. **16**(6), 161 (2017)
17. Alonso-Sanz, R.: Variable entangling in a quantum battle of the sexes cellular automaton. ACRI-2014, LNCS, Vol. 8751, 125–135 (2014)
18. Alonso-Sanz, R.: On a three-parameter quantum battle of the sexes cellular automaton. Quantum Inf. Process., **12**(5), 1835–1850 (2013)
19. Alonso-Sanz, R.: A quantum battle of the sexes cellular automaton. Proc. R. Soc. A **468**, 3370–3383 (2012)
20. Alonso-Sanz, R.: Quantum fuzzy game simulation. Int. J. Quantum Inf. (submitted) (2019)

Chapter 12
Classical Correlated Games

This chapter studies correlated two-person games constructed from independent players in a purely classical context as proposed in the Ref. [1].

12.1 Classical Correlation from Independence

Non-factorizable joint probability distributions Π may be generated from independent classical strategies (x, y) as with the ad-hoc method based on an external parameter $k \in [0, 1]$ given in [1], and shown here in (12.1).

$$
\begin{aligned}
\pi_{11} &= (2k - 1)^2 xy & , \ \pi_{12} &= (1 - k)x(1 - y) + k(1 - x)y \\
\pi_{21} &= (1 - k)(1 - x)y + kx(1 - y) , \ \pi_{22} &= (1 - x)(1 - y) + 4k(1 - k)xy
\end{aligned}
\tag{12.1}
$$

Equations (12.2) give the values of the elements of Π from Eq. (12.1) for three relevant values of k. Please, note that $k = 1$ interchanges the $k = 0$ values of π_{12} and π_{21}, whereas those of π_{11} and π_{22} remain unaltered. Also relevant is that if $x = y = 1/2$, Π is uniform (all its elements equal to 1/4) for $k = 0$ and $k = 1$, but for $k = 1/2$ it is $\pi_{11} = 0$, $\pi_{12} = \pi_{21} = 1/4$, $\pi_{22} = 2/4$. As a result, in a balanced $x = y = k = 1/2$ scenario, the player B is privileged in the KBOS game. Thus, following with the male-female stereotypes, a male modeller would describe the BOS game assigning player B to the female, whereas the female modeller would reverse such role assignments.

$$
\begin{aligned}
k = 0 : &\ \pi_{11} = xy, \quad \pi_{12} = x(1 - y), \quad \pi_{21} = (1 - x)y, \quad \pi_{22} = (1 - x)(1 - y) & (12.2a) \\
k = 1/2 : &\ \pi_{11} = 0, \quad \pi_{12} = \pi_{21} = (x + y - 2xy)/2, \quad \pi_{22} = (1 - x)(1 - y) + xy & (12.2b) \\
k = 1 : &\ \pi_{11} = xy, \quad \pi_{12} = (1 - x)y, \quad \pi_{21} = x(1 - y), \quad \pi_{22} = (1 - x)(1 - y) & (12.2c)
\end{aligned}
$$

Equations (12.3) give the elements of Π from Eq. (12.1) for relevant values of x and y.

© Springer Nature Switzerland AG 2019
R. Alonso-Sanz, *Quantum Game Simulation*, Emergence, Complexity
and Computation 36, https://doi.org/10.1007/978-3-030-19634-9_12

$$\pi_{11}^{(1.0,1.0)} = (2k-1)^2, \quad \pi_{12}^{(1.0,1.0)} = \pi_{21}^{(1.0,1.0)} = 0, \quad \pi_{22}^{(1.0,1.0)} = 4k(1-k) \tag{12.3a}$$

$$\pi_{11}^{(1.0,0.0)} = 0, \quad \pi_{12}^{(1.0,0.0)} = 1-k, \quad \pi_{21}^{(1.0,0.0)} = k, \quad \pi_{22}^{(1.0,0.0)} = 0 \tag{12.3b}$$

$$\pi_{11}^{(0.0,1.0)} = 0, \quad \pi_{12}^{(0.0,1.0)} = k, \quad \pi_{21}^{(0.0,1.0)} = 1-k, \quad \pi_{22}^{(0.0,1.0)} = 0 \tag{12.3c}$$

$$\pi_{11}^{(0.0,0.0)} = \pi_{12}^{(0.0,0.0)} = \pi_{21}^{(0.0,0.0)} = 0, \quad \pi_{22}^{(0.0,0.0)} = 1 \tag{12.3d}$$

$$\pi_{11}^{(0.5,0.5)} = \frac{1}{4} - k(1-k), \quad \pi_{12}^{(0.5,0.5)} = \pi_{21}^{(0.5,0.5)} = \frac{1}{4}, \quad \pi_{22}^{(0.5,0.5)} = \frac{1}{4} + k(1-k) \tag{12.3e}$$

From the joint probabilities given in Eqs. (12.3), the payoffs in a KPD(5,3,2,1) with pure strategies and $x = y = 0.5$ are given in Eqs. (12.4) below, and plotted in the left frame of Fig. 12.4.

$$p_A^{(1.0,1.0)} = p_B^{(1.0,1.0)} = 4k^2 - 4k + 3 \tag{12.4a}$$

$$p_A^{(1.0,0.0)} = 1 + 4k, \quad p_B^{(1.0,0.0)} = 5 - 4k \tag{12.4b}$$

$$p_A^{(0.0,1.0)} = 5 - 4k, \quad p_B^{(0.0,1.0)} = 1 + 4k \tag{12.4c}$$

$$p_A^{(0.0,0.0)} = p_B^{(0.0,0.0)} = 2 \tag{12.4d}$$

$$p_A^{(0.5,0.5)} = p_B^{(0.5,0.5)} = \frac{1}{4}(11 - k(1-k)) \tag{12.4e}$$

Figure 12.1 shows the best responses to pure strategies in the KPD(5,3,2,1). The two upper and bottom frames prove, respectively, that the strategy pairs $(0,1)$ and $(1,0)$ are in Nash equilibrium in the $(k^\star, k^\bullet)$ interval. The $k^\star$-threshold is achieved in the intersection of $p_{A,B}^{0,0} = 2$ and $p_A^{1,0} = p_B^{0,1} = 1 + 4k$, thus $k^\star = 1/4$, whereas the $k^\bullet$-threshold is achieved in the intersection of $p_A^{0,1} = p_B^{1,0} = 5 - 4k$ and $p_{A,B}^{1,1} = 4k^2 - 4k + 3$, thus $k^\bullet = 1/\sqrt{2} = 0.707$.

Figure 12.2 deals also with the KPD(5,3,2,1). Its left frame shows the payoffs for the relevant strategies that determine the payoffs in NE shown in its right frame. In a general KPD game, $p^{(0,0)} = 2 = 1 - 4k = p_A^{(1,0)} \rightarrow k^\star = 1/4$, and $p_A^{(0,1)} = 5 - 4k = 4k^2 - 4k + 3 = p^{(1,1)} \rightarrow k^\bullet = 1\sqrt{2}$.

From the joint probabilities given in Eqs. (12.3), the payoffs in a KHD(3,2,−1,0) with pure strategies and with $x = y = 0.5$ are given in Eqs. (12.5) below, and plotted in the left frame of Fig. 12.5.

$$p_A^{(1.0,1.0)} = p_B^{(1.0,1.0)} = 12k^2 - 12k + 2 \tag{12.5a}$$

$$p_A^{(1.0,0.0)} = 3k, \quad p_B^{(1.0,0.0)} = 3(1-k) \tag{12.5b}$$

$$p_A^{(0.0,1.0)} = 3(1-k), \quad p_B^{(0.0,1.0)} = 3k \tag{12.5c}$$

$$p_A^{(0.0,0.0)} = p_B^{(0.0,0.0)} = -1 \tag{12.5d}$$

$$p_A^{(0.5,0.5)} = p_B^{(0.5,0.5)} = 1 - 3k(1-k) \tag{12.5e}$$

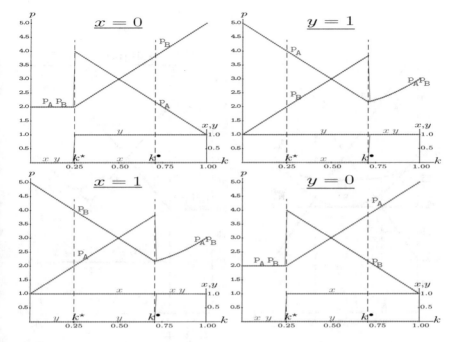

Fig. 12.1 Best responses to pure strategies in the KPD(5,3,2,1). Top left: x = 0.0, Top right: y = 1.0, Bottom left: x = 1.0, Bottom right: y = 0.0

From the joint probabilities given in Eqs. (12.3), the payoffs in a KSD(3,2,−1,0) with pure strategies and $x = y = 0.5$ are given in Eqs. (12.6) below, and plotted in the left frame of Fig. 12.6.

$$p_A^{(1.0,1.0)} = 12k^2 - 12k + 3, \quad p_B^{(1.0,1.0)} = 8k^2 - 8k + 2 \tag{12.6a}$$

$$p_A^{(1.0,0.0)} = -1, \quad p_B^{(1.0,0.0)} = 3 - 2k \tag{12.6b}$$

$$p_A^{(0.0,1.0)} = -1, \quad p_B^{(0.0,1.0)} = 1 + 2k \tag{12.6c}$$

$$p_A^{(0.0,0.0)} = p_B^{(0.0,0.0)} = 0 \tag{12.6d}$$

$$p_A^{(0.5,0.5)} = \frac{1}{4} - 3k(1 - k), \quad p_B^{(0.5,0.5)} = \frac{3}{2} - 2k(1 - k) \tag{12.6e}$$

In the KSD(3,2,−1,0) it is, $p_A = ((3(2k - 1)^2 + 2)y - 1)x - y$, $p_B = ((2(2k - 1)^2 - 4)x + 1 + 2k)y + (3 - 2k)x$. Consequently, the strategy pairs in NE in the KSD(3,2,−1,0) are given in (12.7), where the threshold $k^* = 0.89$ emerges from the $x \leq 1$ restrain applied to x. Before k^* it is, $p_A = -\dfrac{1}{2 + 3(2k - 1)^2}$, $p_B = \dfrac{(3 - 2k)(1 + 2k)}{4 - 2(2k - 1)^2}$.

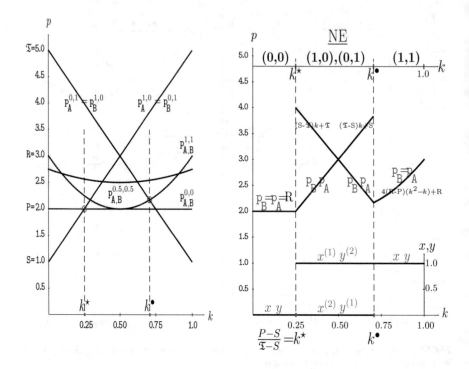

Fig. 12.2 The KPD(5,3,2,1). Left frame: Payoffs pure and (0.5,0.5) strategies. Right frame: NE

$$\text{NE}(k): \begin{pmatrix} x = \dfrac{1+2k}{4-2(2k-1)^2}, y = \dfrac{1}{2+3(2k-1)^2} \end{pmatrix} \quad k \le k^* = 0.89$$
$$\begin{pmatrix} x = 1, y = 1 \end{pmatrix} \qquad\qquad\qquad k \ge k^* = 0.89 \tag{12.7}$$

Figure 12.3 deals with the KSD(3,2,−1,0) game. It shows in its left panel the strategies and payoffs in NE for variable k and in its right panel, the payoffs for $(x = 0.5, y = 0.2)$. It is remarkable that the payoffs in the latter scenario do not differ very much from that in NE, particularly in the case of p_A.

From the joint probabilities given in Eqs. (12.3), the payoffs in a KBOS(5,1) with pure strategies and $x = y = 0.5$ are given in Eqs. (12.8) below, and plotted in the left frame of Fig. 12.7.

$$p_A^{(1.0,1.0)} = 16k^2 - 16k + 5, \; p_B^{(1.0,1.0)} = -16k^2 + 16k + 1 \tag{12.8a}$$

$$p_A^{(1.0,0.0)} = p_B^{(1.0,0.0)} = p_A^{(0.0,1.0)} = p_B^{(0.0,1.0)} = 0 \tag{12.8b}$$

$$p_A^{(0.0,0.0)} = 1, \; p_B^{(0.0,0.0)} = 5 \tag{12.8c}$$

$$p_A^{(0.5,0.5)} = \frac{3}{2} - 4k(1-k), \; p_B^{(0.5,0.5)} = \frac{3}{2} + 4k(1-k) \tag{12.8d}$$

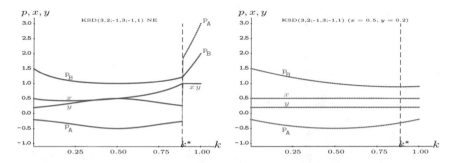

Fig. 12.3 The KSD(3,2,−1,0). Left frame: Strategies and payoffs in NE. Right frame: Payoffs with ($x = 0.5$, $y = 0.2$)

The Ref. [1] gives a second method of constructing non-factorizable Π from independent strategies (x, y). It departs from the fact that in factorizable Π it is $\pi_{11} = xy$, $\pi_{12} = x - \pi_{11}$, $\pi_{21} = y - \pi_{11}$, $\pi_{22} = 1 + \pi_{11} - (x + y)$. Then, it is proposed just to alter the form of $\pi_{11} = xy$ maintaining those of the other three elements of Π as functions of π_{11}. It is concluded in [1] that $\pi_{11}(x, y) < xy$ is the only restriction to be imposed to $\pi_{11}(x, y)$ in order to make sure that all the elements of Π are in the [0, 1] interval and sum 1.0. The authors propose $\pi_{11} = (xy)^2$ and $\pi_{11} = x^2 y^3$ as examples. But, $\pi_{11}(x, y) < xy$ does not suffice to make sure that $\pi_{22} = 1 + \pi_{11} - (x + y)$ is no-negative. In order to prove this, let us consider the particular case of $x = y$, i.e., $\pi_{22} = 1 + \pi_{11} - 2x$: With $\pi_{11} = x^4$, π_{22} is negative if $0.554 < x < 1.0$, and with $\pi_{11} = x^5$, π_{22} is negative if $0.519 < x < 1.0$.

12.2   Spatial Games

According to the spatial scenario depicted in Sect. 3.1, the deterministic imitation of the best paid neighbour induces every player (i, j) to adopt the probabilities of his mate player (k, l) with the highest payoff among their mate neighbors [2]. Initially, the (x,y) probability values are assigned at random (sampled from a uniform distribution in the [0, 1] interval), so that initially: $\bar{x} \simeq 0.5$ and $\bar{y} \simeq 0.5$.

Table 12.1 shows a classically correlated version of the simple example with the PD(5,3,2,1) game given in Table 3.2, where initially every player cooperates ($x = y = 1$), except the defector ($x = 0$) player A located in the (3,4) cell. Thus at $T = 1$, the defector player A gets the $p = 20$ payoff instead of the common $p = 12$ payoff. The imitation mechanism spreads the $x_A = 1$ defection across the player A cells, whereas player B cooperation remains unaltered as no player B defects.

Prisoner's Dilemma

Figure 12.4 deals with spatial simulations of the PD(5,3,2,1) with joint probabilities generated according to (12.1). Its right frame shows the mean payoffs ($\bar{p}$) and mean

Table 12.1 A simple example in the spatial KPD(5,3,2,1) scenario. Far left: The (A,B) chessboard. Centre: Initially every player cooperates, except the defector player A located in the (3,4) cell. Far right: Probabilities and payoffs at $T = 2$

The (A,B) chessboard:

A	B	A	B	A	B
B	A	B	A	B	A
A	B	A	B	A	B
B	A	B	A	B	A
A	B	A	B	A	B
B	A	B	A	B	A

$T = 1$

x, y

1	1	1	1	1	1
1	1	1	1	1	1
1	1	0	1	1	1
1	1	1	1	1	1
1	1	1	1	1	1
1	1	1	1	1	1

p

12	12	12	12	12	12
12	12	**10**	12	12	12
12	**10**	**20**	12	12	12
12	12	**10**	12	12	12
12	12	12	12	12	12
12	12	12	12	12	12

$T = 2$

x, y

1	1	1	1	1	1
1	0	1	0	1	1
1	1	0	1	1	1
1	0	1	0	1	1
1	1	1	1	1	1
1	1	1	1	1	1

p

12	**10**	12	**10**	12	12
10	**20**	**6**	**20**	**10**	12
12	**6**	**20**	**6**	12	12
10	**20**	**6**	**20**	**10**	12
12	**10**	12	**10**	12	12
12	12	12	12	12	12

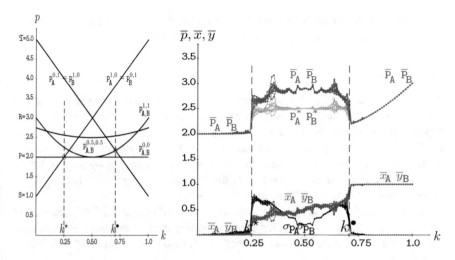

Fig. 12.4 The KPD(5,3,2,1) with variable k. Left frame: Payoffs in two person games. Right frame: Mean payoffs and mean values of x and y in five spatial simulations at $T = 200$

values of x and y at $T = 200$ starting from five different random assignments of x and y. Mutual defection ($x = y = 0$) arises below the lower $k^\star = 0.25$ threshold and mutual cooperation ($x = y = 1$) beyond the higher $k^\bullet = 0.707$ threshold. In the $(k^\star, k^\bullet)$ transition interval, where both (1,0) and (0,1) are in NE, $\bar{x}$ and $\bar{y}$ are fairly similar, increasing their values from 0.0 to 1.0 as k increases from $k^\star$ up to $k^\bullet$; the mean payoffs of both players in turn are fairly similar, reaching values not far from $R = 3$.

The right frame of Fig. 12.4 also shows the mean-field payoffs (p^*) achieved in a single hypothetical two-person game with players adopting the mean probabilities appearing in the spatial dynamic simulation. Namely, with join probability matrix:

$$\Pi^\star = \begin{pmatrix} (2k-1)^2 \bar{x}\,\bar{y} & (1-k)\bar{x}(1-\bar{y}) + k(1-\bar{x})\bar{y} \\ (1-k)(1-\bar{x})\bar{y} + k\bar{x}(1-\bar{y}) & (1-\bar{x})(1-\bar{y}) + 4k(1-k)\bar{x}\,\bar{y} \end{pmatrix} \quad (12.9)$$

The mean-field payoffs (colored brown for player A, green for player B) fully coincide with the actual mean payoffs out of the transition interval, but underestimate them in the transition interval. The lack of coincidence of both mean-field and actual mean payoffs is due to spatial effects that will be illustrated here when addressing the BOS game (Figs. 12.8 and 12.9). As expected, the standard deviations of the payoffs are zero out of the transition interval.

With the EWL method correlating independent quantum strategies, the transition interval from mutual defection up to mutual cooperation in the PD is shorter and a strategy pair in NE providing the payoff of mutual cooperation appear with lower degree of correlation (entanglement in the quantum approach implemented by the EWL method). In the QPD(5,3,2,1) studied here in Sect. 3.2, the thresholds of the

correlation parameter applying the EWL method (referred here to as k_q) in a 0.0 up
to 1.0 normalized scale are $k_q^\star = 0.333$ and $k_q^\bullet = 0.500$ [3].

Hawk–Dove

The right frame of Fig. 12.5 show the results in five spatial simulations of the
HD(3,2,0,−1) at $T = 200$. Spatial effects arise before $k^\star$ so that the mean-field
approaches underestimate the actual mean payoffs as in the spatial simulations of
the PD. After $k^\star$, the spatial simulations detect (1,1) as the unique NE, so that both
payoffs increase their values according to $p = 12k^2 - 12k + 2$ up to $p = 2.0$ at
$k = 1.0$. The $k^\star$ threshold appears from the intersection of $p^{(1.0,1.0)}$ and $p_B^{(1.0,0.0)}$,
given in Eqs. (12.5a) and (12.5b), thus $k^\star = 0.848$.

From the results on the spatial simulations of the HD using the EWL correlation
shown in Fig. 3.5, again, as stressed above regarding the PD, the outcome of mutual
cooperation (Dove in the HD) emerges before with the EWL method: $k_q^\star = 0.392 <
k_c^\star = 0.848$.

Samaritan's Dilemma

The central and right frames of Fig. 12.6 show the results in five spatial simulations
of a KSD(3,2,−1,0) at T = 200. As the SD has only one NE regardless of k, (i) the
results shown in the spatial simulation mimic those corresponding to NE in two-
person games shown in the left panel of Fig. 12.3, and (ii) no spatial effects arise so
that both mean-field and actual payoffs coincide for every k. In spatial simulations
of the SD using the EWL correlation method it is $k_q^\bullet = 0.500 > k_c^\bullet = 0.890$.

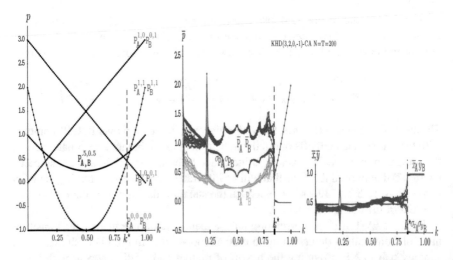

Fig. 12.5 The KHD(3,2,0,−1) with variable k. Left frame: Payoffs in two-person games. Right
frame: Mean payoffs and mean values of x and y in five spatial simulations at T = 200

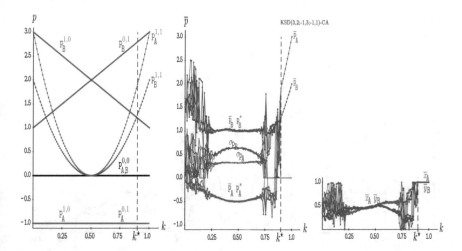

Fig. 12.6 The KSD(3,2,1,−1) with variable k. Left frame: Payoffs in two person games. Central and Right frames: Mean payoffs and mean values of x and y in five spatial simulations at $T = 200$

Battle of the Sexes

The central and right panels of Fig. 12.7 show the results in five spatial simulations of the KBOS(5,1) at $T = 200$. Due to the particular structure of the BOS game, where both π_{12} and π_{21} are irrelevant, the graphs in these panels are symmetric around $k = 0.5$. The general form of the payoffs (central panel) correspond to that of $\bar{x} = \bar{y} = 1$, diminished close to $k = 0.5$ (far right panel) where notable spatial effects arise, and particularly close to the extreme values of k, both 0.0 and 1.0. The output of the spatial simulations the BOS using the EWL correlation method notably differ from that show here using in Fig. 12.7 [4]. Let us say that the BOS game proves to be a highly challenging game.

Figures 12.8 and 12.9 deal with simulations of the KBOS(5,1). The former with $k = 0.0$, the latter with $k = 0.5$. Both show in its far left frame the dynamics up to $T = 200$, in their central panels the patterns of payoffs and probabilities at $T = 200$, and in their far right panels zooms of the 20×20 central region of the full patterns. In both scenarios, the dynamics induced by the imitation of the best paid implemented here actuates in a straightforward manner, so that the permanent regime is achieved very soon. This applies not only to the BOS game but in a general manner, regardless of the game under scrutiny.

The patterns of the payoffs and probabilities shown in the central snapshots of Fig. 12.8 are enhanced by the zooms of a small central region in its far right frames. The general patterns are featured by regions of black marked clusters where $x = y = 1.0$ and white marked clusters where $x = y = 0.0$. The emergence of these well defined spatial structures explain why the mean-field payoff fails to estimate actual mean payoff, as shown in the far left panel of Fig. 12.8. The pattern of probabilities at $T = 200$ shown in Fig. 12.9 for $k = 0.5$ turns out particularly surprising as two

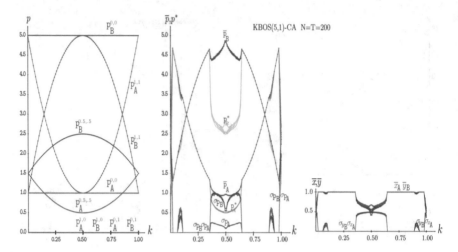

Fig. 12.7 The KBOS(5,1) with variable k. Left frame: Payoffs in two person games. Central and right frames: Five spatial simulations at T = 200. Central frame: Mean payoffs and mean-field approaches, Right frame: Mean values of x and y

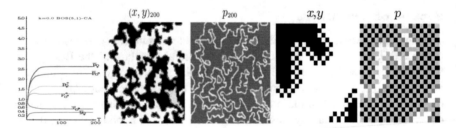

Fig. 12.8 The spatial KBOS(5,1) with $k = 0.0$. Far Left: Dynamics up to $T = 200$. Central panels: Patterns of the full 200×200 lattice at $T = 200$. Far right panels: Zooms of the 20×20 central area

horizontal compact bands with ($x = y = 0.0$) (upper and lower) and one with ($x = y = 1.0$) emerge. This dramatic spatial structure lays in the origin of the discrepancy between the mean-field and the actual mean payoffs shown in the far left panel of Fig. 12.9. In Figs. 12.8 and 12.9, the ($x = y = 0.0$) and ($x = y = 1.0$) clusters are separated by borders where either ($x = 0.0$, $y = 1.0$) or ($x = 1.0$, $y = 0.0$) and consequently the payoffs of both players are zero, which render white cells lines in the payoff patterns. These clear (almost white) border lines are clearly noticeable in Fig. 12.8, both in the full pattern and in the zoom of the central region, whereas in Fig. 12.9 they are only two not so apparent clear horizontal lines, one of them enhanced in the zoom which has been located in the upper transition from $x = y = 0.0$ into $x = y = 1.0$.

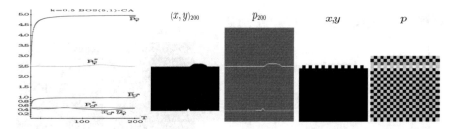

Fig. 12.9 The spatial KBOS(5,1) with $k = 0.5$. Far left: Dynamics up to $T = 200$. Central panels: Patterns of the full 200×200 lattice at $T = 200$. Far right panels: Zooms of the 20×20 central area

Matching Pennies

Figure 12.10 deals with the collective simulation of the MP game with the classically correlated mechanism given by Eq. (12.1). Its far left frame shows that the mean payoffs and strategies, as well as the standard deviations of these magnitudes are unaffected by k. Thus, (*i*) The actual mean payoffs of both players are zero, (*ii*) Both mean strategies are uniform, i.e., $\bar{x} = \bar{y} = 0.5$, and (*iii*) $\sigma_x = \sigma_y = 0.435$, $\sigma_{p_A} = \sigma_{p_B} = 0.614$. With $x^* = y^* = 1/2$, it is $\pi_{11} = \dfrac{1}{4} - k(1-k)$, $\pi_{12} = \pi_{21} = \dfrac{1}{4}$, $\pi_{22} = \dfrac{1}{4} + k(1-k)$, and consequently $p_A^* = p_B^* = 0$.[1] Therefore, the mean-field payoff approaches coincide with the zero actual mean payoffs. The central frame of Fig. 12.10 shows that the average strategies and actual mean payoffs are not significantly altered in the dynamics, whereas the standard deviations of these magnitudes quickly reach a plateau, so that the maze-like structures in the strategy pattern shown in Fig. 12.10 emerge fairly soon. In fact they are clearly appreciated at $T = 20$.

Figure 12.11 deals with the collective simulation of the MP game with the classically correlated mechanism given by Eq. (12.1) where only player A updates its strategy. As expected in an active A facing a passive player B, it is $\bar{p}_A = 0.201 > \bar{p}_B = -0.201$ regardless of k. The far left frame of Fig. 12.11 shows that, as in Fig. 12.10, the mean payoffs and strategies, as well as the standard deviations are unaffected by k. It is $\bar{x} = \bar{y} = 0.5$, which, again as in Fig. 12.10, leads to $p_A^* = p_B^* = 0.0$, different from the actual mean payoffs. It turns out relevant that the standard deviation of the actual payoffs of player B reaches the fairly large value $\sigma_{p_B} = 0.471$, whereas σ_{p_A} remains almost unaltered in the dynamics. The central frame of Fig. 12.10 shows that all the variable magnitudes quickly rocket to a plateau. Most importantly, the actual mean payoffs. But also the standard deviations, such as that of x, which starting from $\sigma_x = 0.355 \simeq \sigma[0, 1] = 1/2\sqrt{3} = 0.37$ reaches $\sigma_x = 0.451$ as soon as at $T = 10$, which indicates that the *percolated* structures in the strategy pattern shown in Fig. 12.10 emerge fairly soon.

---

[1] $p_A^* = \dfrac{1}{4} - k(1-k) - \dfrac{1}{4} - \dfrac{1}{4} + \dfrac{1}{4} + k(1-k) = p_B^* = -\dfrac{1}{4} + k(1-k) + \dfrac{1}{4} + \dfrac{1}{4} - \dfrac{1}{4} - k(1-k) = 0$.

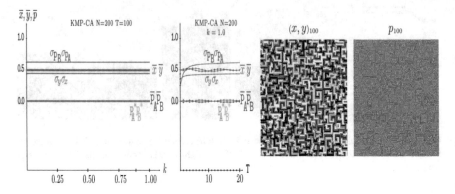

Fig. 12.10 The KMP-CA game. Far left frame: Mean payoffs and parameters with variable k at $T = 100$. Central frame: Dynamics up to $T = 20$ with $k = 1.0$. Far right frames: Strategies and payoff patterns at $T = 100$

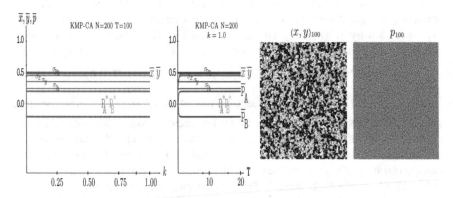

Fig. 12.11 The KMP-CA game where only player A updates strategy. Far left frame: Mean payoffs and parameters with variable k at $T = 100$. Central frame: Dynamics up to $T = 20$ with $k = 1.0$. Far right frame: Strategies and payoff patterns at $T = 100$

If the strategy of player B is fixed to $y = 1/2$, it is $\Pi = \dfrac{1}{2}\left(\begin{matrix}(2k-1)^2 x & k+(1-2k)x \\ 1-\pi_{12} & 1-\pi_{11}\end{matrix}\right)$, so that $p_A = 0$ regardless of x and k. Consequently in a collective simulation of the KMP game with every B player adopting the $y = 1/2$ strategy, no evolution is feasible, so that $p_A = 0 = p_B$ remains unaltered. The same holds if $x = 1/2$ and B is the active player. Please, note that the context of Fig. 12.11 is not that of $y = 1/2$ for every B player, but that of the average $\bar{y} = 1/2$, which enables the evolving dynamics of player A.

12.3 Kgames on Random Networks

In the simulations of this section, every player is connected at random with four partners and four mates, so that any spatial structure is absent in such random networks. In order to compare the simulations presented in this section to those based in spatially structured lattices in Sect. 12.2, also 200×200 players interact in the games on networks studied in this section, half of them of type A, half of them of type B.

Figure 12.12 deals with the KPD(5,3,2,1) game with variable k in network simulations. The left panel shows the mean payoffs of both players and their mean values of x and y at $T = 200$ in five simulations. The right panel shows the dynamics in one of such simulations up to $T = 20$ for $k = 0.0$ (far left), $k = 0.4$ (central), and $k = 1.0$ (far right).

The overall structure of the graphs in the left frame of Fig. 12.12 coincides with that in the right frame of Fig. 12.4. The $k^\star$ and $k^\bullet$ remain unaltered, with $x = y = 0$ before $k^\star$ and $x = y = 1$ after $k^\star$ in both scenarios. At variance with this, the behaviour of the system in the $(k^\star, k^\bullet)$ interval varies significantly in Fig. 12.12 compared to that in Fig. 12.4, as in the network simulation the $(1,0)$ and $(0,1)$ NE emerge with no spatial effects masking them. The right panel shows that also in network simulations the dynamics induced by the imitation of the best paid implemented here also actuates in a straightforward manner, so that the permanent regime is achieved almost immediately for $k = 0.0$ and $k = 1.0$, and as soon as just passed $T = 10$ for $k = 0.4$.

Figures 12.13, 12.14 and 12.15 show the results with the KHD(3,2,0,−1), the KSD(3,2,1,−1) and the KBOS(5,1) games with variable k in five network simulations at $T = 200$. The left frame of these figures shows the mean payoffs of both players, and the right frame, the mean values of x and y.

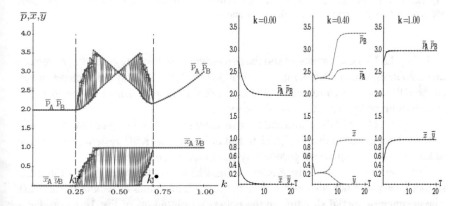

Fig. 12.12 The KPD(5,3,2,1) with variable k in network simulations. Left panel: Mean payoffs and mean values of x and y at $T = 200$ in five simulations. Right panel: Dynamics up to $T = 20$ of one of such simulations with $k = 0.0$ (far left), $k = 0.4$ (central) and $k = 1.0$ (far right)

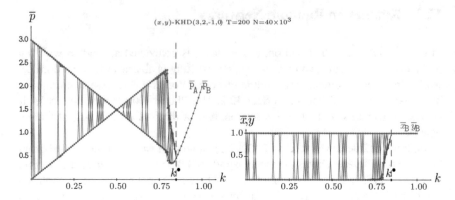

Fig. 12.13 The KHD(3,2,0−1) with variable k in five network simulations at T = 200. Left frame: Mean payoffs. Right frame: Mean values of x and y

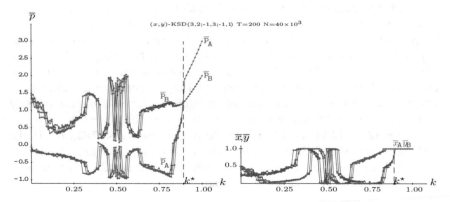

Fig. 12.14 The KSD(3,2,1,−1) with variable k in five network simulations at T = 200. Left frame: Mean payoffs. Right frame: Mean values of x and y

In Fig. 12.13 the $k^\bullet$ threshold and the permanent $x = y = 1$ regime after $k^\bullet$ remain unaltered compared to that in Fig. 12.5. But before $k^\bullet$, the KHD system behaves much as the KPD in its transition interval in network simulations: The (1,0) and (0,1) NE emerge with no spatial effects masking them.

In Fig. 12.14 the $k^\bullet$ threshold and the permanent $x = y = 1$ regime after $k^\bullet$ remain unaltered compared to that in Fig. 12.6. But before $k^\bullet$, the KSD system shows a kind of helter-skelter oscillations particularly pronounced around $k = 0.5$.

The overall structure of the graphs in Fig. 12.15 coincides with that in Fig. 12.7, so that $\bar{x} = \bar{y} = 1$ prevail, except close to the extreme values of k, both 0.0 and 1.0. The absence of spatial structure in the network simulations of Fig. 12.15 produces crisp payoffs (and probability) graphs, with no relevant alterations around $k = 0.5$, albeit in one of the simulations it is $\bar{x} = \bar{y} = 0$ rendering $p_A^{(0.0,0.0)} = 1$, $p_B^{(0.0,0.0)} = 5$ close to $k = 0.5$, coincident with $p_A^{(1.0,1.0)}(k = 0.5) = 1$, $p_B^{(1.0,1.0)}(k = 0.5) = 5$. In the graphs of payoffs in the left frame of Fig. 12.15, player B overrates player A in

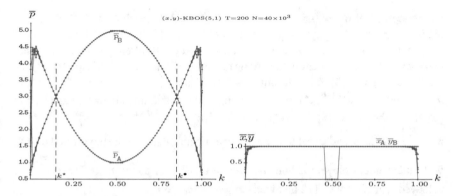

Fig. 12.15 The BOS(5,1) with variable k in five network simulations at $T = 200$. Left frame: Mean payoffs. Right frame: Mean values of x and y

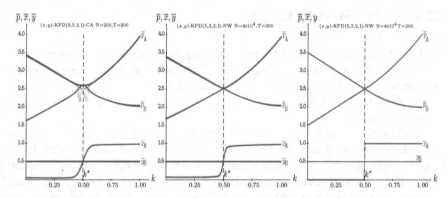

Fig. 12.16 The KPD(5,3,2,1) with variable k when only player A updates strategies. Five simulations at $T = 200$. Far left frame: Spatial simulations. Central frame: Games on networks. Far right frame: $y = 0.5$

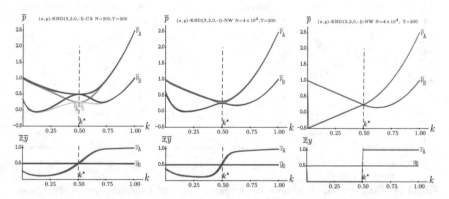

Fig. 12.17 The KHD(3,2,0,−1) with variable k when only player A updates strategies. Five simulations at $T = 200$. Far left frame: Spatial simulations. Central frame: Games on networks. Far right frame: $y = 0.5$

the wide interval $(k^\star = 0.5 - \sqrt{2}/4, k^\bullet = 0.5 + \sqrt{2}/4)$ (with $k^\star$ and $k^\bullet$ defined at the intersection of the payoffs given in Eqs. (12.8a)). This indicates a kind of bias of the proposed correlation mechanism that favors player B (already pointed out when commenting Eqs. (12.2)), a characteristic that is also found in the EWL model regarding the BOS game [4]. It is relevant to point out that $\pi_{11}^{(1.0,1.0)}(k^\star) = \pi_{11}^{(1.0,1.0)}(k^\bullet) = \pi_{22}^{(1.0,1.0)}(k^\star) = \pi_{22}^{(1.0,1.0)}(k^\bullet) = 1/2$, leading to the maximum attainable equalitarian payoff in the BOS: $p_A^{(1.0,1.0)}(k^\star) = p_B^{(1.0,1.0)}(k^\star) = p_A^{(1.0,1.0)}(k^\bullet) = p_B^{(1.0,1.0)}(k^\bullet) = 3 = \dfrac{R+r}{2}$. Please, note that the maximum attainable equalitarian payoff in the BOS with independent players is half the previous one, i.e., $p_A = p_B = 3/2 = \dfrac{R+r}{4}$, achieved with $x = y = 1/2$ that leads to $\pi_{11} = \pi_{12} = \pi_{21} = \pi_{22} = 1/4$.

12.4   Unfair Strategy Updating

In this section it is assumed that only one player type updates his strategies in the manner indicated in Sect. 12.2. Thus, in Figs. 12.16 and 12.17 only player A updates strategies in the symmetric games the PD and HD. The asymmetric games the SD and BOS are studied in Figs. 12.18, 12.19, 12.20 and 12.21, where both players are treated separately.

In all the figures of this section, the far left and central frames deal with spatial simulations and games on networks respectively, with the initial strategy probabilities assigned at random. In the far right frame, the probability of the player that does not update his probability strategies is fixed at 0.5, instead of being assigned at random as is done with the player that updates probability strategies. Thus, the far right frame

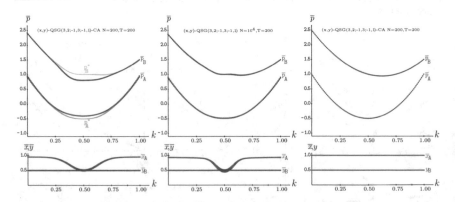

Fig. 12.18   The SD(3,2,1,−1) with variable k when only player A updates strategies. Five simulations at T = 200. Far left frame: Spatial simulations. Central frame: Games on networks. Far right frame: Games on networks

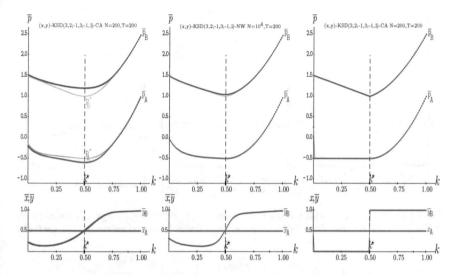

Fig. 12.19 The SD(3,2,1,−1) with variable k when only player B updates strategies. Five simulations at T = 200. Far left frame: Spatial simulations. Central frame: Games on networks. Far right frame: y = 0.5

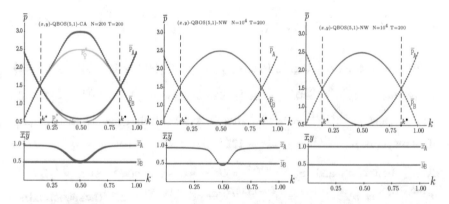

Fig. 12.20 The KBOS(5,1) with variable k when only player A updates strategies. Five simulations at T = 200. Far left frame: Spatial simulations. Central frame: Games on networks. Far right frame: y = 0.5

provides a kind of theoretical reference of what is to be expected in the collective behaviour, both in spatial simulations and in games on networks.

In the mean-field analysis with partial updating the player that does not update his probabilities will have his mean probability equal to the middle level 1/2. In this scenario, the joint probability matrices when the player B is fixed to $y = 1/2$ and the player A fixed to $x = 1/2$, become from Eqs. (12.2),

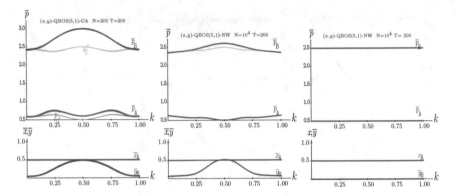

Fig. 12.21 The BOS(5,1) with variable k when only player B updates strategies. Five simulations at T = 200. Far left frame: Spatial simulations. Central frame: Games on networks. Far right frame: x = 0.5

$$\Pi(y = 1/2) = \frac{1}{2}\left(\begin{array}{cc} (2k-1)^2 x & k - (2k-1)x \\ (1-k) + (2k-1)x & 1 - (2k-1)^2 x \end{array}\right) \tag{12.10}$$

$$\Pi(x = 1/2) = \frac{1}{2}\left(\begin{array}{cc} (2k-1)^2 y & (1-k) + (2k-1)y \\ k - (2k-1)y & 1 - (2k-1)^2 y \end{array}\right) \tag{12.11}$$

In the KPD context of Fig. 12.16 it is $p_A^{(x,y=1/2)} = \frac{1}{2}(4k^2 + 4k - 3))x + 7 - 4k)$, where $(4k^2 + 4k - 3) = 0 \rightarrow k^\star = \frac{1}{2}$. Consequently, $p_A^{(x=0,y=1/2)} = \frac{1}{2}(7 - 4k)$ before $k^\star$, and $p_A^{(x=1,y=1/2)} = 2k^2 + 2$ after $k^\star$. As a result, the general form of the payoff of player B, $p_B^{(x,y=1/2)} = \frac{1}{2}(4k^2 - 12k + 5)x + 4k + 3)$ becomes $p_B^{(x=0,y=1/2)} = 2k + \frac{3}{2}$ before $k^\star$, and $p_B^{(x=1,y=1/2)} = 2k^2 - 4k + 4$ after $k^\star$. At $k = 1/2$ it is, $p_A^{(x=0,y=1/2)} = p_A^{(x=1,y=1/2)} = p_B^{(x=0,y=1/2)} = p_B^{(x=1,y=1/2)} = 2.5$. The spatial and network simulations in Fig. 12.16 agree fairly well with these theoretical results. The discrepancies rely in the smooth transition around $k^\star = 1/2$ and in the not exact convergence to $x = 0$ and $x = 1$ before and after $k^\star = 1/2$. As a result of the latter, at $k = 0$, p_A slightly exceeds its theoretical value 1.5 and p_B slightly undervalues its theoretical value 3.5. Only small spatial effects emerge in the spatial simulations of player A (far left frame) close to $k^\star$.

In the KHD context of Fig. 12.17 it is $p_A^{(x,y=1/2)} = 3(2k - 1)kx + 1 - \frac{3}{2}k$, where $2k - 1 = 0 \rightarrow k^\star = 1/2$. Consequently, $p_A^{(x=0,y=1/2)} = 1 - \frac{3}{2}k$, $p_A^{(x=1,y=1/2)} = 6k^2 - \frac{9}{2}k + 1$. As a result, the general form of the payoff of player B, $p_B^{(x,y=1/2)}$

$= (6k^2 k - 9k + 3)x + \frac{1}{2}(3k - 1)$ becomes $p_B^{(x=0,y=1/2)} = \frac{1}{2}(3k - 1)$, $p_B^{(x=1,y=1/2)}$

$= 6k^2 - \frac{15}{2}k + \frac{5}{2}$. At $k = 1/2$ it is, $p_A^{(x=0,y=1/2)} = p_A^{(x=1,y=1/2)} = p_B^{(x=0,y=1/2)}$

$= p_B^{(x=1,y=1/2)} = 0.25$. Much as reported in the PD just before, moderate spatial effects emerge in the spatial simulations of player A close to $k^\star$ in the far left frame of Fig. 12.17.

It is remarkable the strong effect that the absence of updating capacities from one of the players exerts on the collective dynamics here studied. Thus, the far left and central frames of Fig. 12.16 are to be compared to Figs. 12.4 and 12.12 regarding the PD, and the far left and central frames of Fig. 12.17 are to be compared to Figs. 12.5 and 12.13 regarding the HD. In any case, the intrinsic symmetry of both the PD and HD games ceases to be operative in this section, favoring the player A, i.e., the player allowed to find a best response to the fixed strategies of the other player, player B so far.

In the KSD context of Fig. 12.18 it is $p_A^{(x,y=1/2)} = (6k^2 - 6k + \frac{3}{2})x - \frac{1}{2}$, where $(6k^2 - 6k + \frac{3}{2}) \geq 0 \to x = 1$, so that $p_A^{(x=1,y=1/2)} = 6k^2 - 6k + 1$ and $p_B^{(x,y=1/2)} = (4k^2 - 6k + 2)x + k - \frac{1}{2}$ becomes $p_B^{(x=1,y=1/2)} = 4k^2 - 5k + \frac{5}{2}$. Please note, that the intrinsic unfairness of the SD game impedes the charity player A to overrate the beneficiary player B, even in the favorable to A scenario of Fig. 12.18. A common feature of all the simulations of this section is the no dependence of the permanent regime on the initial configuration: The five simulations run in each frame can not be distinguished. Thus, in the particular case of the KSD in Fig. 12.18 the outputs of the five CA and NW simulations are superimposed, so that it seems that only one has been implemented. This contrasts with the results shown in the Figs. 12.6 and 12.14 where the five simulations may be identified before $k^\star$, albeit their outputs are qualitatively similar.

In the KSD context of Fig. 12.19 it is $p_B(x = 1/2, y) = (4k^2 - 2k)y + \frac{3}{2} - k$, where $2k^2 - k = 0 \to k^\star = \frac{1}{2}$. Consequently, $p_B^{(x=1/2,y=0)} = \frac{3}{2} - k$, $p_B^{(x=1/2,y=1)}$

$= 4k^2 - 3k + \frac{3}{2}$; and the general form of the payoff of player A $p_A(x = 1/2, y)$

$= (6k^2 - 6k + \frac{3}{2})y - \frac{3}{2}$ turns out $p_A(x = 1/2, y = 0) = -\frac{3}{2}$, $p_A(x = 1/2, y)$

$= (6k^2 - 6k + 1$. In Fig. 12.19, the beneficiary player B overrates the charity player A. At a greater extent compared to Fig. 12.18, albeit not in a much bigger extent.

In the KBOS context of Fig. 12.20 it is $p_A^{(x,y=1/2)} = (8k^2 - 8k + 2)x + \frac{1}{2}$, where $16k^2 - 16k + 4 \geq 0$, and consequently the best response of player A is $x = 1$, which leads to $p_A^{(x=1,y=1/2)} = 8k^2 - 8k + \frac{5}{2}$. For player B it is $p_B^{(x,y=1/2)} = (-8k^2 - 8k - 2)x + \frac{5}{2}$, that for $x = 1$ renders $p_B^{(x=1,y=1/2)} = -8k^2 + 8k + \frac{1}{2}$. Fairly surprisingly, these payoffs are exactly half of those reported in the simulations of Fig. 12.15 with k not in its extreme values so that $x = y = 1$, i.e., those given in Eqs. (12.8a). In the far right frame of Fig. 12.20 it is $\bar{x} = 1$ for all k, also for $k = 1/2$, but in the CA and

NW simulations (far left and central frames) it is $\bar{x} = 1/2$ for $k = 1/2$. Remarkably, for $k = 1/2$ it is $8k^2 - 8k + 2 = -8k^2 - 8k - 20$, so that it is $p_A^{(x,y=1/2)} = \frac{5}{2}$ and $p_B^{(x,y=1/2)} = \frac{1}{2}$ regardless of x. As a result, at $k = 1/2$ it has no repercussion of being $x = 1/2$ instead $x = 1$ neither on the actual payoffs the NW simulations (central frame) nor in the mean-field payoff approaches in the CA simulations (far left frame). Nevertheless, in the CA simulations, spatial effects induce the increase of the actual mean payoff of player A up to nearly $\bar{p}_A = 3$. Anyhow, again in the context of this section player B overrates player A in the same wide interval of k of Fig. 12.15, i.e., the studied correlation mechanism favors player B in the KBOS, even if the latter is unable of updating his strategies.

In the KBOS context of Fig. 12.21 it is $p_B(x = 1/2, y) = (-8k^2 + 8k - 2)y + \frac{5}{2}$, where $-8k^2 + 8k - 2 \leq 0$ and consequently the best response of player B is $y = 0$, which leads to $p_B(x = 1/2, y = 0) = \frac{5}{2}$ and $p_A(x = 1/2, y = 0) = \frac{1}{2}$. As pointed just before when dealing with Fig. 12.20, for $k = 1/2$ it is $= -8k^2 - 8k - 2 = 0$, so that now it is $p_B^{(x=1/2,y)} = \frac{5}{2}$ and $p_A^{(x=1/2,y)} = \frac{1}{2}$ regardless of y so that there is no repercussion of being $y = 1/2$ at $k = 1/2$ on the actual payoffs in the NW simulations or in the mean-field payoff approaches in the CA simulations of Fig. 12.21. The spatial simulations show an odd aspect of the payoff graphs of not easy explanation. Player B notably overrates player A regardless of k in the KBOS simulations of Fig. 12.21. This is highly expected, when in addition to the structural bias favoring player B in the KBOS, only player B is allowed to search for best responses.

References

1. Iqbal, A., Chappell, J.M., Abbott, D.: On the equivalence between non-factorizable mixed-strategy classical games and quantum games. R. Soc. Open Sci. **3**, 150477 (2016)
2. Alonso-Sanz, R.: Spatial correlated games. R. Soc. Open Sci. **4**(6), 171361 (2017)
3. Alonso-Sanz, R.: On the effect of quantum noise in a quantum prisoner's dilemma cellular automaton. Quantum Inf. Process. **16**(6), 161 (2017)
4. Alonso-Sanz, R.: Variable entangling in a quantum battle of the sexes cellular automaton. In: ACRI-2014. LNCS, vol. 8751, pp. 125–135 (2014)

Index

Symbols
α, 12
β, 12
δ, 193
η, 177
γ, 11
λ, 210, 220
μ, 119
Π, 7
θ, 12
r, 141
2P, 13
3P, 13

A
Acceleration parameter (r), 141

B
Battle of the Sexes (BOS), 4
Belief (λ), 210

C
Cellular automata, 21
Classical correlation, 231
Correlated equlibrium (CE), 8
Correlated noise factor (η), 177

D
Density matrix, 117

E
Entanglement factor (γ), 11

EWL model, 11
Expected payoffs, 5

F
Fidelity (δ), 193
Fortran, 23
Fuzzy payoffs, 219

H
Hawk-Dove (HD), 2

I
Incomplete information, 209

J
Joint probability distribution (Π), 7

L
Lattice, 21

M
Marinatto-Weber, 45
Matching Pennies (MP), 4, 241
Mean-field, 23
Memory (classical), 70
Memory (quantum), 175
Middle-level strategy, 14, 16
Miracle strategy (M), 52

N
Nash Equilibrium (NE), 5, 16
Networks (NW), 73

© Springer Nature Switzerland AG 2019
R. Alonso-Sanz, *Quantum Game Simulation*, Emergence, Complexity
and Computation 36, https://doi.org/10.1007/978-3-030-19634-9

O
Optimism (λ), 220

P
Pareto-efficient, 2
Π, 7
Prisoner's Dilemma (PD), 2
Probabilistic strategies, 5
Probabilistic updating, 91

Q
Quantum strategies (U), 12
Quantum strategy (Q), 15

S
Samaritan's Dilemma (SD), 3
Social welfare (SW), 2
Standard deviation (σ), 23

T
Three-parameter strategies (3P), 13, 38, 59
Two-parameter strategies (2P), 13, 24

U
Uncorrelated noise factor (μ), 119
Unruh effect, 142

W
Werner-like states, 193

Printed in the United States
By Bookmasters